Environmental Contaminants and Terrestrial Vertebrates: Effects on Populations, Communities, and Ecosystems

Other titles from the Society of Environmental Toxicology and Chemistry (SETAC):

Ecotoxicology of Amphibians and Reptiles
Sparling, Linder, Bishop, editors
2000

*Development of Methods for Effects-Driven Cumulative Effects Assessment
Using Fish Populations: Moose River Project*
Munkittrick, McMaster, Van Der Kraak, Portt, Gibbons, Farwell, Gray
2000

*Reproductive and Developmental Effects of Contaminants
in Oviparous Vertebrates*
DiGiulio and Tillitt, editors
1999

Ecotoxicology and Risk Assessment for Wetlands
Lewis, Mayer, Powell, Nelson, Klaine, Henry, Dickson, editors
1999

Ecotoxicology Risk Assessment of the Chlorinated Organic Chemicals
Carey, Cook, Giesy, Hodson, Muir, Owens, Solomon, editors
1999

*Linkage of Effects to Tissue Residues: Development of a Comprehensive
Database for Aquatic Organisms Exposed to Inorganic and Organic Chemicals*
Jarvinen and Ankley, editors
1999

Radiotelemetry Applications for Wildlife Toxicology Field Studies
Brewer and Fagerstone, editors
1998

Advances in Earthworm Ecotoxicology
Sheppard, Bembridge, Holmstrup, Posthuma, editors
1998

*Whole Effluent Toxicity Testing: An Evaluation of Methods and
Prediction of Receiving System Impacts*
Grothe, Dickson, Reed-Judkins, editors
1996

For information about any SETAC publication, including SETAC's international journal, *Environmental Toxicology and Chemistry*, contact the SETAC Office nearest you.

1010 N. 12th Avenue	Avenue de la Toison d'Or 67
Pensacola, Florida, USA 32501-3367	B-1060 Brussels, Belgium
T 850 469 1500 F 850 469 9778	T 32 2 772 72 81 F 32 2 770 53 83
E setac@setac.org	E setac@setaceu.org

http://www.setac.org

Environmental Quality Through Science®

Environmental Contaminants and Terrestrial Vertebrates: Effects on Populations, Communities, and Ecosystems

Edited by

Peter H. Albers
USGS Patuxent Wildlife Research Center
Laurel, MD

Gary H. Heinz
USGS Patuxent Wildlife Research Center
Laurel, MD

Harry M. Ohlendorf
CH2M Hill
Sacramento, CA

SETAC Special Publications Series

Current Coordinating Editor of SETAC Books
C.G. Ingersoll
U.S. Geological Survey

Publication sponsored by the Society of Environmental Toxicology and Chemistry (SETAC)

Cover by Michael Kenney Graphic Design and Advertising
Indexing by IRIS

Library of Congress Cataloging-in-Publication Data

Environmental contaminants and terrestrial vertebrates: effects on populations, communities, and ecosystems/edited by Peter Albers, Gary Heinz, Harry Ohlendorf.
    p. cm. -- (SETAC special publications series)
    "Proceedings from the Symposium on Environmental Contaminants and Terrestrial Vertebrates: Effects on Populations, Communities, and Ecosystems, 19–21 October 1998, University of Maryland, College Park, Maryland."
    ISBN 1-880611-39-2 (alk. paper)
    1. Vertebrates—Effect of pollution on—Congresses. 2. Pollution—Environmental aspects—Congresses. I. Albers, Peter H. (Peter Heinz), 1943- II. Heinz, Gary H. III. Ohlendorf, Harry M. IV. Symposium on Environmental Contaminants and Terrestrial Vertebrates: Effects on Populations, Communities, and Ecosystems (1998 : College Park, Md.) V. Series

    QL605.A1 E58 2000
    596.17—dc21   00-021381

Information in this book was obtained from individual experts and highly regarded sources. It is the publisher's intent to print accurate and reliable information, and numerous references are cited; however, the authors, editors, and publisher cannot be responsible for the validity of all information presented here or for the consequences of its use. Information contained herein does not necessarily reflect the policy or views of the Society of Environmental Toxicology and Chemistry (SETAC).

No part of this publication may be reproduced, stored in a retrieval system, or transmitted in any form or by any means, electronic, electrostatic, magnetic tape, mechanical, photocopying, recording, or otherwise, without permission in writing from the copyright holder.

All rights reserved. Authorization to photocopy items for internal or personal use, or for the personal or internal use of specific clients, may be granted by the Society of Environmental Toxicology and Chemistry (SETAC), provided that the appropriate fee is paid directly to Copyright Clearance Center, Inc., 222 Rosewood Drive, Danvers, MA 01923 USA (telephone 978-750-8400). Before photocopying items for educational classroom use, please contact the Copyright Clearance Center (http://www.copyright.com).

SETAC's consent does not extend to copying for general distribution, for promotion, for creating new works, or for resale. Specific permission must be obtained in writing from SETAC for such copying. Direct inquiries to the Society of Environmental Toxicology and Chemistry (SETAC), 1010 North 12th Avenue, Pensacola, FL 32501-3367, USA.

© 2000 Society of Environmental Toxicology and Chemistry (SETAC)
SETAC Press is an imprint of the Society of Environmental Toxicology and Chemistry.
No claim is made to original U.S. Government works.

International Standard Book Number 1-880611-39-2
Printed in the United States of America
06 05 04 03 02 01 00 99   10 9 8 7 6 5 4 3 2 1

∞The paper used in this publication meets the minimum requirements of the American National Standard for Information Sciences—Permanence of Paper for Printed Library Materials, ANSI Z39.48-1984

Reference Listing: Albers, Peter H., Heinz, Gary H., Ohlendorf, Harry M., editors. 2000. Environmental contaminants and terrestrial vertebrates: Effects on populations, communities, and ecosystems. Published by the Society of Environmental Toxicology and Chemistry (SETAC), Pensacola, Florida, USA. 351p.

The SETAC Special Publications Series

The SETAC Special Publications Series was established by the Society of Environmental Toxicology and Chemistry (SETAC) to provide in-depth reviews and critical appraisals on scientific subjects relevant to understanding the impact of chemicals and technology on the environment. The series consists of books on topics reviewed and recommended by the Publications Advisory Council and approved by the SETAC Board of Directors for their importance, timeliness, and contribution to multidisciplinary approaches to solving environmental problems. The diversity and breadth of subjects covered in the series reflect the wide range of disciplines encompassed by environmental toxicology, environmental chemistry, and hazard and risk assessment. These volumes attempt to present the reader with authoritative coverage of the literature, as well as paradigms, methodologies, and controversies; research needs; and new developments specific to the featured topics. All books in the series are peer-reviewed for SETAC by acknowledged experts.

SETAC Special Publications are useful to environmental scientists in research, research management, chemical manufacturing, chemical regulation, risk assessment and education, as well as to students considering careers in these areas. The Series provides information for keeping abreast of recent developments in familiar subject areas and for rapid introduction to principles and approaches in new subject areas.

SETAC would like to recognize past SETAC Special Publications Series editors:

T.W. La Point, Institute of Applied Sciences
University of North Texas, Denton, TX

B.T. Walton, U.S. Environmental Protection Agency
Research Triangle Park, NC

C.H. Ward, Department of Environmental Sciences and Engineering
Rice University, Houston, TX

QL 605 .A1 E58 2000

Environmental contaminants
  and terrestrial vertebrates

Contents

List of Figures .... *viii*
List of Tables .... *x*
Foreword .... *xi*
Preface .... *xii*
Acknowledgments .... *xvi*
About the Editors .... *xvii*
Dedication .... *xviii*
Workshop Participants .... *xx*
Editors' Comments .... *xxii*

Chapter 1
The Role of Populations, Communities, and Ecosystems in Contemporary Management of Vertebrates 1
Stanley H. Anderson

Chapter 2
Perturbations in Terrestrial Vertebrate Populations: Contaminants as a Cause 19
Glen A. Fox

Chapter 3
Contaminant-Effect Endpoints in Terrestrial Vertebrates At and Above the Individual Level 61
Barnett A. Rattner, Jonathan B. Cohen, Nancy H. Golden

Chapter 4
Statistical Design of Wildlife Toxicology Studies 95
John R. Skalski

Chapter 5
Approaches for Assessment of Terrestrial Vertebrate Responses to Contaminants: Moving Beyond Individual Organisms 109
Peter H. Albers, Gary H. Heinz, Russell J. Hall

Chapter 6
Using Single-Species Measurements to Anticipate Community Consequences of Environmental Contaminants 149
Elizabeth E Holmes and Peter M. Kareiva

Chapter 7
Modeling Toxic Effects on Populations: Experience from Aquatic Studies 177
Glenn W. Suter II and Lawrence W. Barnthouse

Chapter 8
Group A Discussions of Endpoint Selection, Study Design, and Extrapolation 189
John B. French, Jr., Steven P. Bradbury, Hank Krueger, Elizabeth McGee, Bradley E. Sample

Chapter 9
Group B Discussions of Endpoint Selection, Study Design, and Extrapolation 203
James Chapman, Steve Sheffield, Rick P. Brown, Glenn W. Suter II

Chapter 10
Group C Discussions of Endpoint Selection, Study Design, and Extrapolation 217
Michalann Harthill, Donald W. Sparling, Joseph P. Sullivan, Harry M. Ohlendorf

Chapter 11
Estimation of Population-Level Effects on Wildlife Based on Individual-Level Exposures: Influence of Life-History Strategies 225
Bradley E. Sample, Kenneth A. Rose, Glenn W. Suter II

Chapter 12
Effects of Environmental Contaminants in Spatially Structured Environments 245
John F. McLaughlin and Wayne G. Landis

Chapter 13
Disruption of Rodent Assemblages in Disturbed Tallgrass Prairie Ecosystems Contaminated with Petroleum Wastes 277
Robert L. Lochmiller, Daniel P. Rafferty, Karen McBee, Nick T. Basta, James A. Wilson

Index 303

List of Figures

Figure 1-1 Populations, communities, and ecosystems all influence one another and all affect an animal's habitat 2
Figure 1-2 Habitats used by sage grouse vary by life stages and seasons during the year 8
Figure 1-3 Impacts can occur at the population, community, or ecosystem level 14

Figure 3-1 Relationship between responses at various levels of biological organization and their ecotoxicological relevance and response time-scales 64

Figure 4-1 Schematic emphasizing the interrelationships between study design and analysis elements 96
Figure 4-2 Schematics of time profiles for nonexposed and exposed populations under (a) null hypothesis and (b) alternative hypothesis of toxic effects on wild populations following a chemical accident at time 0 103

Figure 5-1 Changes in the number of breeding pairs of bald eagles in the Great Lakes and along rivers supporting runs of anadromous fishes from the Great Lakes and changes in the number of young eagles fledged per breeding pair 121

Figure 6-1 (a) Simple 3-species food web (b) 6-species food web 154
Figure 6-2 Food web of the Mono Lake ecosystem 155
Figure 6-3 Food web of a California salt marsh 155
Figure 6-4 Food web of the French Frigate Shoals 156
Figure 6-5 Impact of an increase in the death rate of 1 species on species abundances in the community as a whole 158
Figure 6-6 The relative abundance of a species in a community versus the community impact resulting from a 300% increase in that species' death rate 160
Figure 6-7 Abundance versus community impact 160
Figure 6-8 Trophic level of a species versus the community impact resulting from a 300% increase in that species' death rate 161
Figure 6-9 Number of links versus the community impact resulting from a 300% increase in death rate 161
Figure 6-10 Sensitivity of the 6-species food web to changes in death, fecundity, and interaction rates 163
Figure 6-11 Sensitivity of the 6-species food web to changes in death, fecundity, and interaction rates 164
Figure 6-12 Sensitivity of the Mono Lake food-web to changes in death, fecundity, and interaction rates 165
Figure 6-13 Sensitivity of the California salt marsh food web to changes in death, fecundity, and interaction rates 166
Figure 6-14 Sensitivity of the reef food web to changes in death, fecundity, and interaction rates 167
Figure 6-15 Summary of the parameters causing the greatest changes in community abundances 169

Figure 7-1 Example of a taxonomic regression 180
Figure 7-2 Regression model to estimate a chronic quartile effective concentration for fecundity from an acute median lethal concentration 181

Figure 9-1 Principles of study design 204

Figure 11-1 (a) Density-dependence multipliers for fecundity and (b) age-zero survival for hypothetical avian r- and K- strategist species 231
Figure 11-2 Baseline 100-y population trends for hypothetical avian (a) r- and (b) K-strategist species 236
Figure 11-3 Simulation results for reduced age-zero survival and fecundity for hypothetical r- and K-selected species 236
Figure 11-4 Simulation results for reduced adult survival for hypothetical r- and K-selected species .. 237
Figure 11-5 Simulation results for 10% reduction in age-zero survival and fecundity for hypothetical r- selected species with stochasticity increased or decreased relative to baseline condition 238

Figure 12-1 Kinds of population structure in patchy environments 247
Figure 12-2 Schematic diagram of patch structure in toxicant metapopulation model 263
Figure 12-3 Dynamics of the 3-patch linear model analyzed by Spromberg et al. 1998 263

Figure 13-1 Relative toxicity of soils in the habitat of rodent assemblages on 12 study sites that were comprised of a suspected petrochemical-contaminated area and a paired reference area in disturbed, tallgrass prairie in Oklahoma 284
Figure 13-2 Temporal fluctuations in abundance across 8 trapping occasions during 1 complete annual cycle for the common rodent populations inhabiting suspected petrochemical-contaminated areas and matched reference areas in disturbed, tallgrass prairie in Oklahoma 287
Figure 13-3 Dendrogram resulting from average-linkage cluster analysis of a matrix of similarity coefficients derived from the total abundances of species within rodent assemblages in summer on 12 study sites that were comprised of a suspected petrochemical-contaminated area and a paired reference area in disturbed, tallgrass prairie habitat in Oklahoma 288
Figure 13-4 Dendrogram resulting from average-linkage cluster analysis of a matrix of similarity coefficients derived from the total abundances of species within rodent assemblages in winter on 12 study sites that were comprised of a suspected petrochemical-contaminated area and a paired reference area in disturbed, tallgrass prairie habitat in Oklahoma 289
Figure 13-5 Mean measures of population stability across 4 trapping occasions in summer and winter for *S. hispidus*, *Peromyscus* spp., *Microtus* spp., and *M. musculus* 290
Figure 13-6 Scatter plot of observations depicting the relationship between seasonal abundance of *M. musculus* populations in summer and average concentrations of heavy metals, total PAHs, and TPHs, and a relative toxicity index, for 24 study areas in disturbed, tallgrass prairie habitat in Oklahoma 293

List of Tables

Table 2-1	Attributes associated with life-history strategies and their implications for population studies	26
Table 2-2	Factors contributing to the vulnerability of various species to the effects of contaminants	46
Table 3-1	USEPA guidelines for maximum concentrations in water of 4 chemicals which, if exceeded, could cause detrimental reproductive and population effects in terrestrial wildlife in the Great Lakes System	82
Table 3-2	Principal contaminants for which critical concentrations and diagnostic guidelines have been established for wild terrestrial vertebrates	84
Table 5-1	Suggested terms to describe natural groupings of animals	111
Table 5-2	Contaminants and the most common mammals in the 150 selected references	128
Table 5-3	Evaluation of attempts by authors to extrapolate measurements to unmeasured biological responses, taxa, or levels of organization and the incorporation of special conditions in the methods or discussion	129
Table 5-4	Contaminants, amphibians, and reptiles in the 95 selected references	136
Table 5-5	Evaluation of attempts by authors to extrapolate measurements to unmeasured biological responses, taxa, or levels of organization, and the incorporation of special conditions in the methods or discussion	137
Table 11-1	Life-history parameters for hypothetical r- and K-strategy avian species	229
Table 11-2	Distributions for inter-annual stochasticity multipliers for hypothetical avian r- and K-strategist species	230
Table 11-3	Contaminant exposure and effects scenarios and simulation results	234
Table 11-4	Distributions for reduced and increased inter-annual stochasticity multipliers for hypothetical avian r-strategist species	235
Table 12-1	Distribution of outcomes in 3-patch linear toxicant metapopulation model, with toxicant degradation	264
Table 13-1	Average concentration of heavy metals (Cr and Pb), total PAH, and TPH on 12 study sites that were comprised of a suspected petrochemical-contaminated area and a paired reference area in disturbed, tallgrass prairie habitats in Oklahoma	283
Table 13-2	Seasonal abundance for resident small mammal species occurring on 12 suspected petrochemical-contaminated and 12 matched reference sites in Oklahoma	285
Table 13-3	Total number of small mammals captured and total abundance in the small mammal assemblage following 1024 trap-nights of census monitoring on each of 12 study sites during summer and again in winter	286
Table 13-4	Logistic-regression models, analysis of maximum-likelihood estimates, and concordance outcomes for predicting contamination of habitats for rodent assemblages on 24 study areas that were comprised of suspected petrochemical-contaminated areas and reference areas in disturbed, tallgrass prairie habitats in Oklahoma	291
Table 13-5	Best-fit single-variable models relating contaminant concentrations in soil to intrinsic attributes of rodent assemblages inhabiting 24 study areas in disturbed tallgrass prairie habitats in Oklahoma using stepwise multiple-linear regression analysis	294
Table 13-6	Best-fit 3-variable models relating contaminant concentrations in soil to intrinsic attributes of rodent assemblages inhabiting 24 study areas in disturbed tallgrass prairie habitats in Oklahoma using stepwise multiple-linear regression analysis	295

Foreword

The symposium on "Environmental Contaminants and Terrestrial Vertebrates: Effects on Populations, Communities, and Ecosystems" was held from 19–21 October 1998, at The Inn and Conference Center, University of Maryland, College Park, MD. The 50 participants who attended this symposium represented federal and state agencies, private corporations, and several universities.

The idea to conduct such a symposium came from a discussion during the late winter of 1995 among several biologists who were conducting research on the effects of contaminants on wildlife. The focus of the symposium was later broadened to include application of research findings to ecological risk assessment at contaminated sites. Pete Albers developed a formal proposal that received approval from the Patuxent Wildlife Research Center. Subsequently, he recruited support from the Society of Environmental Toxicology and Chemistry (SETAC) and The Wildlife Society (TWS) in this joint effort. By late summer, both societies had agreed to participate and they gave their formal approval in 1996.

A Steering Committee of representatives from the three participating organizations was formed to develop the symposium and began to meet during the fall of 1995. The original Steering Committee consisted of Pete Albers and Gary Heinz, representing the Patuxent Wildlife Research Center; Kristin Brugger (Dupont Chemical Corp.) and Greg Foster (George Mason University), representing industry and academia for SETAC; and Jim Fleming (U.S. Geological Survey), representing TWS. This group produced the outline and began developing plans for the symposium. Kristin Brugger asked to be replaced because of her extensive work commitments, and Harry Ohlendorf replaced her in mid-1997 as the SETAC liaison.

Peter H. Albers, Coordinator
Kristin R. Brugger
Gregory D. Foster
W. James Fleming
Gary H. Heinz
Harry M. Ohlendorf, SETAC Liason

Preface

The desire of resource managers, risk assessors, and the general public to understand better the consequences of environmental contamination has produced a strong and growing need for information on the effects of contaminants on populations and groups of species, and over moderate to large areas of land or water. However, the problems associated with research involving populations and groups of species or large and complex geographic areas, especially in terrestrial environments, are well-known within the scientific community. Perhaps these research problems are so well-known that investigators have become discouraged about the prospects for the development of innovative approaches to such "higher-level" relations.

With the previous thoughts in mind, an interactive symposium devoted to the effects of contaminants on terrestrial vertebrates was developed. Invited background and technical presenters provided a common baseline of information for symposium participants. Discussion groups were then asked to evaluate critically the topics of two technical sessions. Several presentations of recent or ongoing research provided participants with examples of current approaches to assessments of the effects of contaminants on terrestrial vertebrates at the population or higher level of organization.

Background

The first 2 chapters provide background information on relations between environmental contaminants and terrestrial vertebrate populations, associated communities and ecosystems, and natural resource management.

It is easy for contaminants investigators to lose sight of the ultimate purpose of their research, namely, to help natural resource managers and environmental regulators make informed decisions regarding the environmental effects of specific contaminants. In Chapter 1, Stan Anderson examines the role of populations, communities, and ecosystems in contemporary management of natural resources with examples of metals contamination, exotic species, population management, and ecosystem management that involve birds, mammals, and fishes. The primary message is that all environmental disturbances, natural or manmade, should be evaluated for primary and secondary population effects within the immediate community.

Evidence of the effects of environmental contaminants on terrestrial vertebrate populations is reviewed by Glen Fox in Chapter 2. Glen presents background information on the concepts of "population" and "contaminants," factors affecting populations, the modifying effects of landscape composition and life history on contaminant exposure, and the use of modeling to estimate effects of contaminants. Documented effects on individuals and populations of classes of contaminants, genotoxic effects of contaminants, and effects of habitat acidification are discussed.

Because there are only a few good examples of well-documented population declines induced by contaminants, despite much evidence from many studies of potential harm to populations, Glen proposes that our inability to detect population effects may be "an artifact of our science," and concludes with a list of problems associated with population assessments and recommendations for improving our knowledge of effects of contaminants on terrestrial vertebrate populations.

Endpoint Selection and Study Design

Chapters 3 and 4 evaluate endpoints and study designs used to determine the effects of contaminants on terrestrial vertebrates. Rattner et al. (Chapter 3) begin with a discussion of the history and development of biotic endpoints. This discussion is followed by a detailed treatment of measures of contaminant exposure and measures associated with population effects. The latter category includes discussion of acetylcholinesterase inhibition, malformations and pathological lesions, eggshell thinning, reproductive success, and survival. Following these subjects, the authors address efforts to determine contaminant effects in communities and ecosystems. The regulatory use of these endpoints in support of federal statutes, water-quality criteria, and other established threshold concentrations also is discussed. The authors conclude that we are not yet able to predict the effects of contaminants on terrestrial vertebrate populations, and they make recommendations for improved use of current endpoints and future directions for endpoint development.

The design of wildlife toxicology studies is addressed by John Skalski in Chapter 4. Introductory comments highlight the importance of the connection between goals, study design, data analysis, and inferences, and the significance of Fisher's principles of replication, randomization, and error control. John then discusses response variables at the level of the population and makes recommendations for the design of multi-plot investigations of wildlife populations. Assessments of accidental contaminant releases are treated as a special type of investigation; recommended approaches include the use of parallelism in time profiles and demographic analysis for exposed and nonexposed sites. Preliminary surveys and associated calculations of sample size also are discussed. Because field investigations are difficult and expensive to perform, John strongly recommends that investigators employ sound study designs.

Extrapolation

Chapters 5, 6, and 7 address the extrapolation of information from experiments and field surveys. In Chapter 5, Albers et al. present the results of a review of published literature on the effects of contaminants on all 4 classes of terrestrial vertebrates. Definitions and associated perceptions of "population" and the role of populations in the regulatory arena are discussed. The portion on birds contains 3 examples of assessments involving extrapolation from the individual to the population and 1 example involving extrapolation to communities and ecosystems. Characteristics of each case are evaluated and recommendations are presented. The portions on mammals, reptiles, and amphibians are based on a review of publications from 1980 through 1997, with emphasis on extrapolation efforts. Results of the literature evaluation are discussed and recommendations are made. Two examples of mammalian assessments and 1 example of amphibian assessment are presented. The chapter ends by merging the findings into conclusions relevant to all 4 classes of vertebrates.

In Chapter 6, Elizabeth Holmes and Peter Kareiva evaluate the utility of single-species measurements of contaminant effects as predictors of community responses to contaminants. They employ generic 3- and 6-species food web models and food-web models that are based on three real communities. The death rate (categorized according to abundance, trophic level, and interspecies connectivity), fecundity rate, and interaction rate (predator–prey relations and intraspecific competition) of a species are examined for their ability to predict changes in species abundances within a community. Communities are assessed for their comparative sensitivity to changes in rates of death, fecundity, and interaction. The authors end with some cautionary guidance on the extrapolation potential of single-species measures and the usefulness of sublethal responses compared with death rates.

In Chapter 7, Glenn Suter and Lawrence Barnthouse examine population modeling of contaminant effects in aquatic studies with the intent of benefitting similar efforts in the terrestrial arena. The authors begin by presenting the advantages of estimating effects at the population level and describing the current role of populations in environmental regulation in the U.S. The subsequent section contains information on modeling procedures developed at the Oak Ridge National Laboratory for aquatic populations. Data and modeling problems are discussed along with applicable solutions. Results of the modeling efforts are evaluated. Finally, recommendations are made for improving toxicity-test design and moving forward with the concept of population-based management of contaminants.

Synthesis of Issues

Three discussion groups each produced a report on the topics addressed by the previous technical sessions on endpoint selection, study design, and extrapolation. In Chapters 8, 9, and 10, these 3 groups address
- definitions and concepts underlying contaminants investigations,
- strengths and weaknesses of endpoints,
- design and statistical analysis of contaminants studies, and
- the utility of different types of extrapolation of findings.

Although the results and recommendations from each group generally reinforce each other, the depth of topic assessment and specifics of recommendations differ.

Research Applications

Chapters 11, 12, and 13 are presentations of ongoing or recently completed research that is relevant to the effects of contaminants on terrestrial vertebrates in populations, communities, or ecosystems. In Chapter 11, Sample et al. examine the influence of life-history strategies on estimation of the population consequences of contaminant effects on individuals. The current risk assessment process of relating individual effects to populations is viewed as being based on overly simplistic assumptions about wildlife population dynamics. Models developed for avian populations are evaluated with regard to life-history strategies, interannual stochasticity, density dependence, and contaminant effects over a 100-year period. The authors believe that underestimates of population responses are possible with current risk assessment practices. A major obstacle to increased use of population-dynamics models is the inadequacy, for model use, of much of the currently generated toxicity data.

The effects of environmental contaminants in spatially structured environments are evaluated by John McLaughlin and Wayne Landis in Chapter 12. Animal populations and contaminants usually are distributed unevenly in the environment, but this distribution often is ignored in traditional population assessments. The authors describe 5 kinds of spatial structure for populations and present the modeling approaches used for each kind of structure. Results of modeling efforts, in terms of population persistence, are presented for metapopulations, metacommunities, source-sink systems, and patchy populations. Also discussed are the effects of spatial structure on population dynamics, effects of contaminants on patch dynamics, and landscape considerations. Modeling results indicate that spatial structure affects population processes but that more experimental and observational studies are needed. Guidance is presented for integrating population models with field studies, along with suggestions for future research direction. Lastly, McLaughlin and Landis offer guidance on the question "When can spatial structure be ignored?"

In Chapter 13, Lochmiller et al. assess the sensitivity of rodent communities in Oklahoma grassland habitats to the effects of soils contaminated with petrochemical wastes. Twelve contaminated sites and their matched reference sites were studied during summer and winter over a 3-year period (4 sites per year). Soil contaminants and vegetation were characterized and permanent trapping grids were established. The authors estimated species abundance and calculated temporal variability, an index of similarity, and species richness. The results reveal altered relative abundance of several species, seasonal population instability, and several direct toxic effects; indirect effects are thought likely but are difficult to verify. Despite the inherent variability of rodent assemblages through space and time, the authors believe that the consequences of petrochemical contamination caused the observed differences between contaminated and reference sites.

Acknowledgments

The U.S Environmental Protection Agency (USEPA) Office of Research and Development, National Center for Environmental Assessment made a grant to the SETAC Foundation for Environmental Education to provide base funding for the symposium. James Andreasen arranged for the grant within USEPA. The U.S. Geological Survey (USGS) Patuxent Wildlife Research Center provided travel and lodging support for invited speakers, and SETAC administered the grant funds and managed the publication of the symposium proceedings.

The Steering Committee is grateful for the support provided for the symposium by USEPA, SETAC, TWS, and USGS. We acknowledge the symposium speakers who provided information on a range of relevant topics for participants and then provided manuscripts for this book. We also thank the manuscript and book reviewers for their constructive comments. We would also like to thank SETAC, especially Kristine Tabb, Melissa Winter, Rod Parrish, Greg Schiefer, and Stacey Hagman, for their hard work and attention to the book's production A final expression of thanks is directed to all of the symposium attendees; their participation in the discussion groups helped to make the symposium a success.

The Society of Environmental Toxicology and Chemistry thanks Michael Hooper and Scott McMurray, Texas Tech University, for their careful and thorough peer reviews of this work.

About the Editors

Peter H. Albers received a B.S. in Wildlife Biology from the University of Montana, an M.S. in Wildlife Biology from the University of Guelph (Canada), and a Ph.D. in Wildlife Management from the University of Michigan. He was an instructor for 1 semester at the University of Maine. Dr. Albers has been a Research Wildlife Biologist at the USGS Patuxent Wildlife Research Center since 1976. He has conducted investigations into the effects of petroleum, metals, metalloids, pesticides, and acid deposition on a variety of terrestrial vertebrates and their associated terrestrial, wetland, and aquatic habitats. He is a member of SETAC, the Chesapeake and Potomac Regional Chapter of SETAC, The Wildlife Society, the Wildlife Toxicology Working Group of The Wildlife Society, and the Ecological Society of America.

Gary H. Heinz received his B.A. in Biology from Lehigh University and his M.S. and Ph.D. in Wildlife Biology from Michigan State University. He has been a Research Wildlife Biologist at the USGS Patuxent Wildlife Research Center since 1969. His research has focused on the effects of pesticides, PCBs, metals, and other contaminants on wildlife, especially birds. Dr. Heinz is a charter member of SETAC and also is a member of The Wildlife Society and the American Ornithologists' Union.

Harry M. Ohlendorf received a B.S., a M.S., and a Ph.D. in Wildlife Science from Texas A&M University. He began his career as a Wildlife Research Biologist with the U.S. Fish and Wildlife Service's Patuxent Wildlife Research Center in 1971. For 19 years, Dr. Ohlendorf's research for the USFWS focused on the occurrence and impacts of contaminants in aquatic and terrestrial ecosystems. These studies included the sampling of various wetland and terrestrial food chains and the assessment of the effects observed in higher trophic levels, especially birds. As a Senior Environmental Scientist for CH2M Hill, Dr. Ohlendorf's duties include planning, implementation, and reporting of site ecological characterizations and surveys as well as ecological risk assessments. He is a member of SETAC, the NorCal Regional Chapter of SETAC (and a member of its Board of Directors), Cooper Ornithological Society, The Wildlife Society, the Waterbird Society, and Wilson Ornithological Society.

Dedication

Dr. Robert Lochmiller
(1954-2000)

This book is dedicated to one of the Symposium's contributors, Dr. Robert "Bob" L. Lochmiller II, who unexpectedly passed away on 3 March 2000. Bob had been a Regents Professor in the Department of Zoology at Oklahoma State University (OSU) in Stillwater, Oklahoma, since 1985. He was a wildlife biologist with research interests in regulation, stress physiology, immunology, nutritional ecology, and toxicology of wild animal populations. Bob taught numerous wildlife-related courses at OSU, including Wildlife Management, Advanced Wildlife Management, Wildlife Nutrition, Wildlife Immunology, and Wildlife Techniques.

In the late 1980s, Bob and his colleague Dr. Charles Qualls (OSU School of Veterinary Medicine) began examining exposure and possible effects of contaminants on a model wild rodent species, the hispid cotton rat (*Sigmodon hispidus*), inhabiting a Superfund site (Criner) in Oklahoma. Their grant proposal, "Development of the cotton rat as a biomonitor of environmental contamination," was funded in 1989 by the Oklahoma Center for the Advancement of Science and Technology. Bob continued his interest in this field into the 1990s, and was co-Primary Investigator (co-PI) with Qualls on a $440,000 Air Force grant awarded in 1991 for a study entitled "Wild mammalian biomonitors for assessing impacts of environmental contaminants on population and community ecology" that funded continuation of this work at another Superfund site (Cyril) in Oklahoma. In 1995, Bob and co-PI Qualls were again awarded a grant from the Air Force ($660,000) for a study entitled "In situ dose-response relationships for a multi-parameter model of petrochemical-induced ecotoxicology." This grant funded the work reported at the symposium *Environmental Contaminants and Terrestrial Vertebrates: Effects on Populations, Communities, and Ecosystems*, held at College Park, Maryland, in October of 1998. His contribution to this symposium was an invited paper entitled "Disruption of rodent assemblages in disturbed tallgrass prairie ecosystems contaminated with petroleum wastes," which reported significant ecotoxicological effects on small rodents inhabiting petrochemical waste sites. By the time of the symposium, this work had developed into some of the most comprehensive analyses ever performed on the ecotoxicology of small mammal individuals, populations, and communities at hazardous waste sites. Bob also participated in Discussion Group C at the symposium and later reviewed 1 of the other book chapters.

Bob was active in many scientific societies, including the Society of Environmental Toxicology and Chemistry, The Wildlife Society, American Society of Mammalogists, Wildlife Diseases Association, and Oklahoma Academy of Sciences. Bob served as Associate Editor for the *Journal of Mammalogy*, Editor for the *Bulletin of Environmental Contamination and Toxicology*, and was Managing Editor for the *Journal of Wildlife Management* at the time of his death. Bob maintained a very active research program, as evidenced by his prolific record of publication and grant writing. Bob was the author, or co-author, of at least 164 journal articles in more than 40 different journals, 7 book chapters, 23 non-refereed technical publications, and 8 popular publications. His grantsmanship was equally impressive, as he was awarded approximately $4 million (50 grants) as a PI or co-PI investigator from sources such as the U.S. Environmental Protection Agency, U.S. Department of Defense, the National Science Foundation, and the U.S. Air Force.

Bob was known for having the largest and most active group of graduate students in the Department of Zoology. His work ethic and the rigorous schedule of work he kept were legendary at OSU. His purple pickup truck in the first parking spot next to the building at 5:30 a.m. every morning will not be soon forgotten. However, during deer season, you could count on the fact that he would be difficult to find until he got his buck.

Bob was quite selective of the scientific meetings he attended and had a tendency to not attend many meetings. This symposium was the last time that many of us, including me, saw Bob. Bob Lochmiller was a unique personality who possessed an amazing passion for his work as well as other aspects of his life, particularly for his family, for hunting, and for the outdoors. He left us much too soon and will be missed by his family, friends, and the scientific community.

Steve Sheffield, Ph.D.
U.S. Fish and Wildlife Service
Arlington, VA
and
Department of Biology
George Mason University
Fairfax, VA

Workshop Participants*

Peter H. Albers
USGS Patuxent Wildlife Research Ctr.

Stanley H. Anderson
University of Wyoming

James K. Andreasen
USEPA

Lawrence W. Barnthouse
LWB Environment Services, Inc.

Tim Bartish
USGS Midcontinent Ecological Science Center

Nick T. Basta
Oklahoma State University

W. Nelson Beyer
USGS Patuxent Wildlife Research Ctr.

Steven P. Bradbury
U.S. Environmental Protection Agency

Gary D. Brewer
U.S. Geological Survey

Rick P. Brown
Exxon Biomedical Sciences

James Chapman
USEPA

Jonathan B. Cohen
USGS Patuxent Wildlife Research Ctr.

Daniel D. Day
USGS Patuxent Wildlife Research Ctr.

Glen A. Fox
Canadian Wildlife Service

John B. French, Jr.
USGS Patuxent Wildlife Research Ctr.

Hector Galbraith
Stratus Consulting, Inc.

Sarah Gerould
U.S. Geological Survey

Nancy H. Golden
University of Maryland

Russell J. Hall
USGS Florida Caribbean Science Center

Michalann Harthill
U.S. Geological Survey

Melanie S. Hawkins
U.S. Army Aberdeen Proving Ground

Gary H. Heinz
USGS Patuxent Wildlife Research Ctr.

Gerry M. Hennigsen
USEPA

Mary G. Henry
U.S. Fish and Wildlife Service

David J. Hoffman
USGS Patuxent Wildlife Research Ctr.

Elizabeth E. Holmes
University of Washington

Almira Hoogestyn
Cornell University

Kathleen Jennings
University of San Francisco

Mark S. Johnson
U.S. Army Aberdeen Proving Ground

Peter M. Kareiva
University of Washington

Hank Krueger
Wildlife International, Ltd.

James A. Kushlan
USGS Patuxent Wildlife Research Ctr.

Wayne G. Landis
Western Washington University

Kelly Lippenholz
U.S. Army Aberdeen Proving Ground

Robert L. Lochmiller
Oklahoma State University

Michael J. Mac
U.S. Geological Survey

James MacMahon
Utah State University

David T. Mayack
NY State Department of Environmental Conservation

Karen McBee
Oklahoma State University

John McCarty
USEPA

Elizabeth McGee
U.S. Fish and Wildlife Service

John F. McLaughlin
Western Washington University

Harry M. Ohlendorf
CH2M Hill

Katharine C. Parsons
Manomet Center for Conservation Sciences

John T. Paul, Jr.
U.S. Army Aberdeen Proving Ground

Daniel P. Rafferty
Oklahoma State University

Barnett A. Rattner
USGS Patuxent Wildlife Research Ctr.

Susan Roddy
USEPA

Kenneth A. Rose
Louisiana State University

Bradley E. Sample
Oak Ridge National Laboratory

Anne L. Secord
U.S. Fish and Wildlife Service

Steve Sheffield
Clemson University

John R. Skalski
University of Washington

Joseph P. Skorupa
U.S. Fish and Wildlife Service

Donald W. Sparling
USGS Patuxent Wildlife Research Ctr.

Joseph P. Sullivan
Ardea Consulting

Glenn W. Suter II
USEPA

Mark L. Watson
New Mexico Department of Fish and Game

Janet Whaley
U.S. Army Aberdeen Proving Ground

James A. Wilson
Oklahoma State University

Mark Woythal
NY State Department of Environmental Conservation

*affiliations were current at the time of the workshop.

Editors' Comments

The intent of this book, as with the symposium, is to see more clearly a complex and sometimes frustrating aspect of environmental research, namely, the responses of groups of individuals, species assemblages, and combinations of communities to the challenges posed by anthropogenic chemical modification of habitat. Through the use of retrospective assessments, critical discussion, and examples of relevant contemporary research, participants in this effort have examined successes, failures, and promising opportunities. The editors hope that the material presented in this book will enable investigators to more productively utilize existing knowledge of terrestrial vertebrates and environmental contaminants and serve as a source of ideas for the development of future work.

CHAPTER 1

The Role of Populations, Communities, and Ecosystems in Contemporary Management of Vertebrates

Stanley H. Anderson

Most wildlife managers manipulate species to accomplish their management goals. Removal or introductions are common forms of management. Habitat manipulation is also a management effort that can influence wildlife. All actions, however, have consequences for other populations. Thus, communities and ecosystems often are affected by impacts to 1 species, or by local habitat impacts. In this paper we examine the role of populations, communities, and ecosystems in managing vertebrates. There are many seemingly minor impacts that can change natural systems. For example, toxicants can alter species composition, thereby affecting ecosystem processes. On the other hand, major habitat destruction can create an entirely new community.

This chapter concentrates on the different management activities that influence populations, communities, and ecosystems, which in turn influence animal habitat (Figure 1-1). The needs of species and the ways in which they might be affected by human activities also are examined. Other questions addressed include the following:

1) Should we continue to manage species only, or should we focus on ecosystem management?
2) How can we manage ecosystems?
3) What information do managers need to have to effectively maintain populations and ecosystems?

My purpose is to show that any changes in species composition, numbers, habitat, or ecosystems can alter the natural system.

Chapter Preview

Community-level effects 2
Stability in the natural system 4
Populations 4
Ecosystems 9
Environmental impacts 11
Putting it all together 13

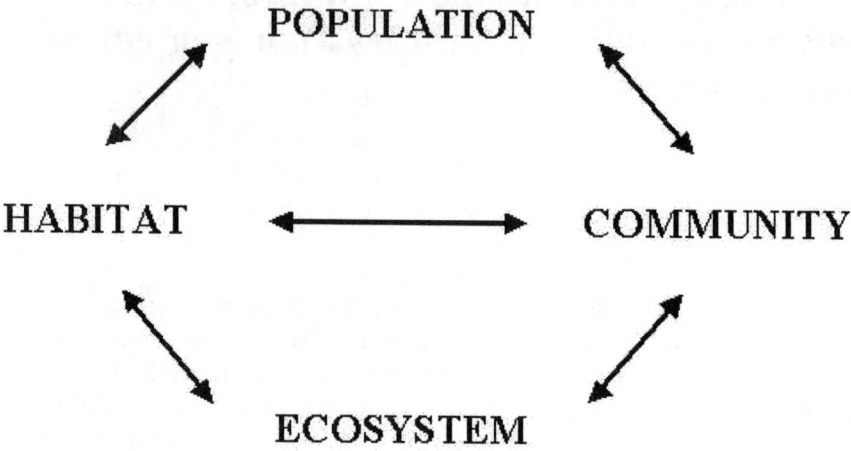

Figure 1-1 Populations, communities, and ecosystems all influence one another and all affect an animal's habitat

Community-level Effects

The gray wolf (*Canis lupus*) reintroduction is an example of how management effort affects communities of wildlife. In preparation for the wolf reintroduction, the U.S. Fish and Wildlife Service (USFWS 1994) prepared an Environmental Impact Statement (EIS). The purpose of the EIS was to identify problems that might be encountered during the reintroduction and to identify what information was needed to resolve the problems. The EIS authors also worked to determine how recovery might be achieved through the introduction. Although the EIS examined a number of components of the reintroduction effort, this paper emphasizes the impact of the wolf as a predator.

The wolf recovery plan stated that 10 packs of wolves in the Yellowstone ecosystem, each reproducing for 3 years, would be considered to be a recovered population. The plan estimated that each pack would consist of an average of 10 wolves; thus, there would be about 100 wolves in the ecosystem. There also might be a few lone wolves and some packs with fewer than 10 wolves. Authors of the EIS consulted 15 wolf "experts" who predicted that reintroduction would take about 20 years, at which time, a stable population of 110 to 150 wolves would exist in the park. There were, however, differences of opinion among the experts.

The wolf experts, with the help of modelers, evaluated the prey base available along with data from other wolf populations and predicted that each wolf would kill 7 to 28 ungulates per year, depending on prey size, weather, and alternative prey. Eight prey species were potentially available in the park:

- elk (*Cervus elaphus*),
- mule deer (*Odocoileus hemionus*),
- bison (*Bison bison*),
- pronghorn (*Antilocapra americana*),
- bighorn sheep (*Ovis canadensis*),
- moose (*Alces alces*),
- an occasional white-tailed deer (*Odocoileus virginianus*), and
- mountain goat (*Oreamnos americanus*).

Based on an estimate of 23,100 ungulates in the early 1990s, there would be 1 wolf per 330 ungulates in summer and 1 wolf per 230 ungulates in winter. It was true that there were some preferred prey. Experts believed that elk, numbering 17,500, would be the primary prey. Both wolf experts and modelers believed that the reintroduced wolf population would reduce the elk population that had increased by more than 30% in the late 1980s because of a series of mild winters. The modeling efforts predicted wolves would reduce the elk by 4% to 30%.

Mule deer, also a popular prey item, were predicted to be reduced by 20% to 30%. Like the elk, they had increased in numbers because of the mild weather and no hunting pressure in the Park. Wolves were predicted to take only an occasional individual from the other ungulate populations.

Predictions were that wolf reintroduction would result in fewer elk and mule deer; and, as a result, the amount of forage would increase and the community and ecosystem would improve. Coyote (*Canis latrans*) numbers would decrease in areas where wolves were found, but coyotes would remain in areas near humans where wolves would not likely go. Impact on other predators would be minimal. Grizzly bears (*Ursus arctos*) would not be affected because they only occasionally took elk or mule deer (USFWS 1994).

Was the reintroduction a success? That question was answered with a resounding yes. On 27 July 1998, there were 10 wolf packs and a total of 118 wolves in Yellowstone National Park (Yellowstone Park 1998). This number resulted from a release of 14 wolves in 1995 and 17 wolves in 1996. Other wolves may have been there already or may have arrived between 1995 and 1998. As predicted, introduced wolves clearly preferred elk in their diet. Only a few moose, deer, and antelope were taken. One sheep was known to have been taken. Although bison were numerous, none were taken as prey. Field data showed that each pack took an elk every 1.1 to 5 days (Phillips and Smith 1997).

Authors of the EIS believed that the social value of wolves in Yellowstone National Park would be high. People would visit Yellowstone to see wolves but would be satisfied just knowing they were present even though they did not see them. People not living in the area also would enjoy knowing wolves existed in the area. Most importantly, however, would be the changes that the introduction of another predator would bring to the plant and animal community. Additional plant biomass could be available to support animal presence. The results could affect ungulate movement, winter feeding, as well as other dynamics in Yellowstone and neighboring systems.

Stability in the Natural System

Without outside interference, most populations proceed toward, though they seldom reach, a state of self-regulation that eliminates extreme fluctuations in size through positive and negative feedback mechanisms. The growth of a prey population is positive feedback for a predator population, i.e., more prey, more food. The growth of a predator population is a negative feedback mechanism for the prey population, i.e., more predators mean more prey eaten, leaving fewer to reproduce. Some cyclic species fluctuate more than others.

Any change in a natural system produces alteration. For example, the introduction of a different species often causes a disturbance in the natural system. These introductions can harm native species and disturb their habitat. One example is the introduction of lake trout (*Salvelinus namaycush*) into Yellowstone Lake in Yellowstone National Park.

On 30 July 1994, lake trout were discovered in Yellowstone Lake, the core of the remaining undisturbed natural habitat for native cutthroat trout (*Oncorhynchus clarki bouvieri*). Data suggested that lake trout, which eat the cutthroat trout, may have been in the lake since 1989 when they probably were introduce illegally.

Biologists estimate that lake trout could reduce the number of cutthroat trout by more than 90% in the next 20 to 100 years. Since cutthroat trout spawn in the streams around the lake, they are a popular sport fish in both the lake and its tributaries. In addition, they are an important food item for grizzly bear, ospreys (*Pandion haliaetus*), white pelicans (*Pelecanus erythrorhynchos*), river otter (*Lutra canadensis*), and other animals. Unfortunately, biologists see little chance of removing all lake trout from Yellowstone Lake (Kaeding et al. 1996).

Populations

Populations frequently form the basis of management actions. Most biologists define a population as a group of organisms of a single species that interact and interbreed in a defined space and time (Anderson 1999). A species such as the North American mule deer may occupy large areas. Other populations may be found only in a very small area. Populations usually are managed by changing the numbers of individuals—removal through hunting, introduction to a new area (like the gray wolf or lake trout), or by manipulation of the habitat.

Hunting

Hunting is a form of mortality affecting a population. Managers use hunting as a tool to keep animals within the carrying capacity of the habitat. For example, if a population gets too large (e.g., elk in Yellowstone National Park), they eat most of the browse and degrade the habitat. There are times when hunted populations are kept at artificially high numbers or natural predators are reduced so people can have a "better hunt."

The impact of hunting on a population always creates questions. When hunting substitutes for other forms of mortality, it is said to be compensatory. When hunting adds to the total mortality, it is said to be additive. The basic idea underlying the compensatory mortality hypothesis is that as hunting increases, other forms of mortality such as predation by other animals, disease, and starvation decrease, keeping the survival rate of the population the same. The additive hypothesis states that as hunting increases, the size of the population decreases in a linear manner. Hunting, then, adds to the total mortality of the population.

From 1987 to 1991, the Alaska Department of Fish and Game conducted an extensive study of wolves in northwestern Alaska. Results from the study helped us understand hunting and predator-prey interactions. The Department radio-collared 86 wolves and followed them to find out what they captured and what killed them. Wolves preyed primarily on caribou (*Rangifer tarandus*), killing 6% to 7% of the herd annually. In the winter when caribou migrated, the wolves changed prey, primarily capturing resident moose. The wolves removed 11% to 14% of the moose population annually. In years of heavy snowfall, the moose had greater trouble moving. As a result, more starved or were taken more easily by wolves; this mortality of moose could be very high in some years, creating concerns that they could be eliminated in a series of severe winters. Since the moose population was established in the area approximately 40 years before the study began, biologists speculated that wolves historically left the area with the caribou during their migration.

Biologists found at the end of the study that
- 11% of the radio-collared wolves were alive,
- 28% were missing,
- 42% were killed by hunters,
- 4% starved,
- 1% was killed by another wolf,
- 1% died of old age, and
- 13% died from rabies.

These results indicated that the food available in the winter was not sufficient to support the wolf population. If hunter kills were reduced, other causes of mortality would occur (e.g., wolf kills, disease, starvation). In all likelihood, the wolf population was above the carrying capacity of the habitat (Ballard et al. 1997).

Habitat

Each animal requires habitat for food, shelter, cover, and reproduction. Groups of animals and plants together comprise the community in which many interactions occur. When we add the physical components and pathways, we have the ecosystem. Habitat is an important key to the success of an individual species. Many species require different types of habitats seasonally for different stages in their life history.

Some animals require a very specialized habitat. For example, some species of fish can thrive only in water that has a consistent temperature. On the other end of the scale, house sparrows (*Passer domesticus*) are able to live in a multitude of habitats. If we wish to maintain desirable species, we must be certain that their basic needs continue to be met. This includes preventing impacts on the quality of a habitat from any type of pollution.

It is important to keep in mind that conditions necessary for the survival of a species vary. The composition of a community is a result of the evolution of its component parts as well as recent history, including human influence on the natural system. Each species has physical and biological needs that must be satisfied in order for the species to survive and reproduce. Conditions include

- amount of water,
- solar radiation,
- temperature,
- daylight or photoperiod,
- air currents,
- types of soils and minerals, and
- barometric pressure.

Some birds, mammals, and fish require several habitats in order to maintain their populations. In fact, many kinds of animals require a variety of habitats at different times in their life cycles. Contamination of a physical attribute or any component of the habitat can adversely affect the species.

The Yuma clapper rail (*Rallus longirostris yumanensis*), a small rail species that is listed as endangered, is found along the lower Colorado River in California and Arizona. In early winter, clapper rails use marsh areas far from the upland part of the river, near open water, possibly to increase foraging efficiency. Crayfish, the primary prey item of the Yuma clapper rail, normally are found in open-water pools (Anderson and Ohmart 1985). These areas are generally newer portions of the marsh and have newly established emergent vegetation that lacks the heavy residual-vegetation characteristics of the interior portions of the marsh along this part of the river.

During the breeding season, individual rails seem to need the safety of the marsh interior to conceal nests and young; as a result, they move back into these areas and make only occasional foraging excursions out into the open water areas. The marsh interior provides more bare ground interspersion, is closer to upland vegetation, and has high marsh-plant stem density. The areas are close to open water, but the water takes the form of inland pools, rather than the large open-water areas along the river (Conway 1990). Because of their nesting habits, development along the river directly influences the Yuma clapper rail. Human use of shoreline areas and human-waste disposal directly affect the breeding habitat in terms of both lost cover and lost food sources. Water skiing and other forms of water recreation in the winter disrupt the winter habitat of the birds. Pollution, such as the placement of chemicals in waterways, as well as harvesting of crayfish, causes a decline in available food.

These examples show how the whole marsh is important to this species. Disruption of 1 biological or physical component of the ecosystem can harm the Yuma clapper rail. Several other species of rails live in the same marsh: Virginia rail (*Rallus limicola*), sora (*Porzana carolina*), and black rail (*Laterallus jamaicensis*). Each species has its own use of the habitat by partitioning areas during different seasons. Habitat disruption of 1 species has direct effects on the other species. We know the whole marsh is necessary for rail survival. If data are collected during only 1 season, misleading results can occur.

Sage grouse (*Centrocercus urophasianus*) are found in sagebrush country of the West. Early each spring, when the snow is still on the ground, the males gather early in the morning at traditional mating sites (leks). They strut each morning for about a month trying to attract and mate with females. Females then move away to find nest sites. When the brood is hatched, the female and chicks move to sites where there is an adequate protein diet to support the growing chicks. In the winter, sage grouse require areas protected from snow accumulation in order to find food, indicating that they require a number of habitats during the year (Figure 1-2). Disturbance of any of these different habitats can harm the sage grouse population. Thus, road construction, oil and gas development, grazing by domestic livestock, and heavy human recreation may adversely affect the birds.

While these examples of rails and sage grouse illustrate how some animals require several habitats each year, others may require different habitats on a seasonal basis (e.g., migratory birds, gray whales, and big game) or may require various habitats in order to complete their life histories (e.g., salmon) (Di Silvestro 1997).

Another example of the complexity of animal habitat use is represented by the marbled murrelet (*Brachyramphus marmoratus*). The murrelets are small seabirds that breed in the northern Pacific region. The species ranges from the Bering Sea

SAGE GROUSE HABITAT NEEDS

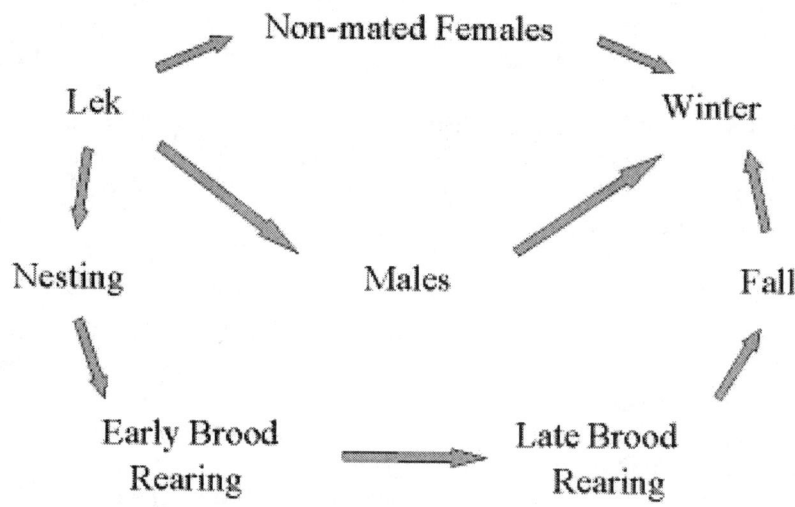

Figure 1-2 Habitats used by sage grouse vary by life stages and seasons during the year

south to the coast of northern California. During the nesting season, the birds fly to the pockets of coastal old-growth forest, where they nest on the nearly horizontal branches of trees (Meyer 1999). In Alaska, about 5% of the murrelet population nests on the ground in areas where there are no predators. The nesting birds are secretive and are often very difficult to detect, except when they are flying to the ocean to forage. In order to survive, this seabird must have 2 very distinct habitats—the open ocean and the old-growth forests. Human impact on both areas is a source of concern for the welfare of the marbled murrelet population. Cutting down old-growth forests decreases the birds' nesting sites, and oceanic pollution reduces food supply and damages their feathers. Gill nets used by fishing companies pose yet another hazard.

Sometimes people alter habitat but still want wildlife present. In order for this to happen, either special management actions are needed, or the animals must adapt to human-created environments. In Jackson, Wyoming, the elk herd is kept artificially at a high level because it is provided by humans with plots of land on which they are given food during the winter. Because urban development now encroaches on elk winter habitat, the elk would be grazing in peoples' backyards if they were not fed this way. This artificial feeding keeps the elk numbers high enough to satisfy both the hunters and the viewing public.

Reintroduced turkeys in parts of the West must have adequate nesting and brood rearing habitat as well as roost trees. When snow falls, their habitat is limited. Turkeys come to ranchers' haystacks and obtain grain to survive through the winter months. When considering reintroduction, biologists must know whether there is adequate winter support for the population. If not, the introduction is doomed to failure.

Ecosystems

When a number of populations are found in an area, they form a community of organisms. Together with the abiotic system they form the ecosystem. Throughout the world, ecosystems are influenced by climate and physical components to form deserts, tropical forests, grasslands, and other biotic and abiotic combinations.

While most wildlife management efforts continue to focus on single species, some individuals are now suggesting that ecosystem management be promoted. There are a number of ideas about ecosystem management; however, the essence of the concept is preservation of ecosystem processes (Grumbine 1994). Ecosystem management means that we would not focus on individual species and the associated management costs, but rather on preservation of the health of the ecosystem, thereby maintaining healthy populations overall (Simberloff 1998). In this way, we would sustain the processes that are part of each system.

While ecosystem management seems like a reasonable approach, there are many pitfalls and controversies. People are not able to clearly define ecosystem boundaries. This makes it difficult to focus on units, a concept that humans need. It is also apparent that ecosystem processes can continue if 1 or more species are removed. This goes against the concept of endangered-species preservation, although not necessarily against the idea of biodiversity if lost species are replaced by others. If ecosystem management truly were to operate, it would be necessary to have "measures of impact" to the ecosystem. In some cases this is possible. People can see the changes that occur in species composition and diversity as the oxygen levels of waterways are increased. But how does that translate into an evaluation of the whole ecosystem? People are important components of many ecosystems. How can we evaluate and manage systems with many different human interests?

In the U.S., we hear of impacts on the grasslands, eastern deciduous forests, and southwestern deserts because the public easily views them. However, the boreal forests of northern Canada, Scandinavia, and northern Russia are important, yet often overlooked ecosystems. The climate is slightly warmer than the northern tundra, with much more precipitation (about 28 to 102 cm annually). The soil thaws, allowing tree roots to penetrate more deeply and develop more fully. The boreal forests (taiga in northern Europe and Asia) have rather poor soil because of slow decay there. The acid produced in the decay is carried into the soil by rain or melting snow, making the soil relatively infertile for most crops. Although the growing season is short, a large number of birds migrate to these areas, and many mammals, e.g., moose, caribou, wolverine (*Gulo gulo*), and snowshoe hare (*Lepus americanus*), live in the forests. Large lakes and streams are found in many parts of the boreal forest region.

Over the past 30 years, combinations of human activities and natural events have resulted in both dramatic and subtle changes in the boreal ecosystem. Climatic warming, acid deposition, and increased exposure to ultraviolet radiation caused by ozone depletion are global stresses that may have serious environmental consequences in these regions (Schindler 1998). Thousands of fish populations and perhaps millions of invertebrate populations from the boreal waters of Canada were lost as a result of acid precipitation (Minns et al. 1990; Schindler 1998). The 3 stresses may have a synergistic impact that is difficult to assess. Extensive logging is another factor, particularly in the former Soviet Union.

Another example of a difficult ecosystem to manage is the Everglades ecosystem in southern Florida which consists of a complex mosaic of communities linked by the flow of fresh water. It is being fundamentally altered in order to support a rapidly growing human population (Harwell 1997). The essence of the Everglades is the abundance and diversity of species that live in diverse habitats across vast open spaces of the "river of grass" and associated coastal estuaries. This area is highly vulnerable to natural disturbances. Fires are common. Drought occurs at regular intervals and hurricanes at irregular intervals. The many human communities around the region all have their own water development projects that have a major impact on the natural operation and flow of nutrients that keep the system alive.

The foundation of ecosystem management comes from defining ecological sustainability goals by deciding which parts of the system are to be protected and which parts are to be sacrificed to support human development and recreation. This is a very difficult task, as we do not know all the intricacies of the Everglades system, but plans are currently underway to develop appropriate management programs. To accomplish this task, biologists are creating a model of human interactions with the environment as well as a set of ecosystem management principles. Another aspect of the project involves defining the bounds of regional ecosystems. Endpoints need to be identified and goals set. A part of the management consideration involves 2 endangered species, the Everglades snail kite (*Rostrhamus sociabilis plumbeus*) and the wood stork (*Mycteria americana*). All of these factors must be integrated in order to establish ecological sustainability goals in terms of this complex ecological system (Harwell 1997).

We can see that human impacts make it increasingly difficult to maintain the natural system. It becomes particularly difficult to integrate all the special interest concerns while keeping the natural system viable. People do not understand the limits of the ecosystem. They are interested in short-term gains and fail to look at long-term sustainability. It is difficult to provide managers with a single form of data to manage these complex systems. Maintenance of the natural ecosystem means trying to sustain the estimated 10 million plants, animals, and microbes on the earth (Pimentel et al. 1997).

Environmental Impacts

Many forms of habitat alteration are obvious. When terrestrial habitat is reduced or destroyed, an area often can no longer support the same species. Biological changes such as the introduction of exotic plants or animals can occur, making the area unfit for some desirable species. Temperature can be changed. More subtle, but still of great importance, is the introduction of chemical pollutants. Some chemicals can be introduced into the air, water, and soil either to accomplish a purpose such as insect control or as waste from an industrial process.

Throughout American history, changes in agriculture have usurped large amounts of land, particularly grasslands and nearby forests. In the past 40 years, approximately 810,000 hectares have been converted annually to agriculture. In addition, some 202,000 hectares of wetlands have been destroyed each year (Anderson 1999). All of these actions have caused major changes in the species composition of an area.

Some forms of habitat structural changes can be both detrimental and beneficial to a wildlife community. Humans want to prevent grassland and forest fires. In reality, fires create an important element in the natural landscape. Some communities, such as the chaparral in southern California, are a form of fire climax. Some plants sprout only after a fire burns through the area, allowing seeds to pop out of their pods. Unfortunately, some of the hillside chaparral communities are considered ideal places for people to build houses. Fires are often difficult to control and they destroy many homes.

Forests are often set back in succession as fire sweeps through them. Generally, grasses grow in the year following a fire, providing browse for many different species of animals. Some species of animals appear only after a fire. The Lewis woodpecker (*Melanerpes lewis*) is called a bird of fire. It likes to drill holes in the snags left standing after a fire burns through the forest. Thus, Lewis woodpeckers are not found in a forest much after 20 years following a forest fire.

Exotics

Exotics have created a major impact on our ecosystems. Some parts of the world, such as New Zealand and Australia, have had exotics introduced that expanded so quickly that much of the native fauna has been imperiled. In the U.S., a number of exotic introductions have occurred. One individual wanted to introduce all animals mentioned in Shakespeare's writings into the U.S. This resulted in the European starling (*Sturnus vulgaris*) spreading throughout most of the country. Other animals have been accidentally introduced, such as the Norway rat (*Rattus norvegicus*), which arrived on ships. Along with the rat came fleas that carried plague, now a problem in many of our mammal species. Mongoose populations introduced to control rats have virtually destroyed all ground-nesting birds on the Hawaiian Islands. Profit has been a major motive in the introduction of exotics. The nutria (*Myocaster coypus*), a large rodent found in many of our small lakes and marshes, was introduced as a source of fur in the southern part of our country. However, several escaped from their pens, and the rest were subsequently released into the wild. As a result, their range has expanded throughout a large part of the country. Nutria, unlike beaver (*Castor canadensis*), destroy dirt embankments and dikes, creating havoc for many water-dependent organisms.

During the latter half of the 19th century and the early part of the 20th, 2 species of exotic fish were introduced into the U.S. One, the carp (*Cyprinus carpio*) (considered a mistake), and the brown trout (*Salmo trutta*) (considered a success). The carp, a native of Asia, had been widely cultured in Europe and Asia for centuries as a food fish. Selective breeding in captivity had produced a carp with few bones and scales, which made it a popular food for European dinner tables. It was not realized at the time of importation that the cultured variety would revert to its bonier condition in the wild, with coarse flesh and an increased number of scales. The American public rejected this variety, and the carp ended up causing substantial destruction to the habitat of native fish. Because of their feeding habits, carp damaged wildlife habitat by destroying aquatic vegetation, the loss of which frequently caused massive plankton growth and turbid water. Habitat destruction has resulted in the expenditure of millions of dollars on largely ineffective control programs.

The brown trout, which has been successfully introduced into many streams in this country, does not appear to compete with native trout species or to harm the habitat. Generally, it inhabits waters not suited for native trout.

Heavy Metals

Mining can have a major impact not only on wildlife habitat but also on the food source of some species. A study of river otter, mink (*Mustela vison*), and raccoon (*Procyon lotor*) in the upper Clark Fork River, Montana, showed that the abundance of these species was considerably lower than that in the surrounding rivers (Szumski 1998). Riparian habitat features important to semi-aquatic mammals (e.g., cover type, amount of canopy closure, and bank characteristics) were not significantly different between the Clark Fork and the other rivers in the area. Likewise, biologists found the level of human disturbance in the riparian zone of the Clark Fork River to be similar to or less severe than surrounding rivers. However, a decline in fish populations, presumably as a result of toxic metals, reduced the aquatic prey base for all piscivorous mammals on the upper Clark Fork River. Such reductions are expected to most severely affect an obligate piscivore like the river otter. Facultative piscivores like mink and raccoons are less seriously affected.

Metal concentrations of arsenic, cadmium, copper, lead, mercury, and zinc were significantly elevated in tissues of mink collected from the Clark Fork River when compared with mink collected from reference streams. Analysis of scat and gastrointestinal tracts demonstrated that aquatic prey items made up a significant portion of the minks' diet on both the Clark Fork and reference streams. Also, whole-body concentrations of cadmium, copper, and lead were significantly elevated in fish collected from the Clark Fork River, and other whole-body metal concentrations were similar to or slightly above the national average. This information indicated that the transfer of metals through the food chain had occurred.

None of the individual metal concentrations found in mink tissues were at levels associated with clinical signs of toxicity. However, metals such as lead and mercury can produce subtle, permanent behavioral changes at exposure levels far below those associated with any clinical signs of toxicity. In addition, the toxic effects of metals such as arsenic, cadmium, lead, and mercury may be additive, producing tissue injury or biochemical changes at concentrations previously associated with no adverse effects. Furthermore, tissues collected from mink frequently demonstrated a number of additional pathologies including internal parasites (lungworm nematodes, protozoa sporocysts, and cestodes) and bacterial and viral infections.

The results of the study support the overall conclusion that the observed absence of otter on the upper Clark Fork River, and observed reductions in mink and raccoon abundance, is the result of past mining and smelting operations and present metal contamination in the upper Clark Fork River. For piscivorous mammalian species like the otter, mink, and raccoon, the reduced prey base on the river is most likely the primary cause of this decline.

Putting It All Together

Ecosystems generally achieve an equilibrium state in which different species of plants and animals can exist in a state of balance. Destruction of the vegetation or introduction of an exotic or a contaminant can disrupt this balance (Figure 1-3). If the system is large and resilient, it can establish a new equilibrium. Most systems continually work toward equilibrium, though they may never reach that point. Furthermore, the balance in a system is influenced by natural events, such as rainfall, that differ each year. Often, ranchers want to graze the same number of

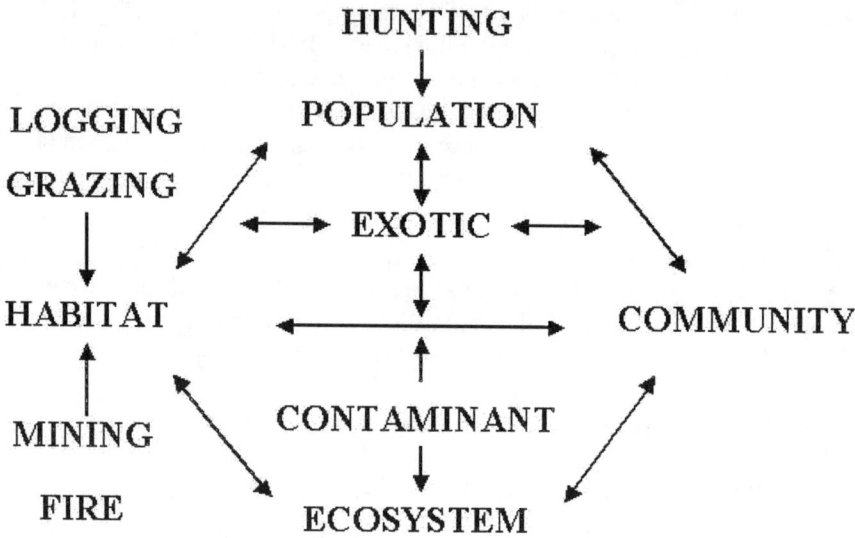

Figure 1-3 Impacts can occur at the population, community, or ecosystem level. Likewise, impacts on habitat occur. Any impact can create changes throughout the system.

cattle each year, despite changing moisture patterns. Populations like sage grouse definitely respond to moisture. In years of heavy rainfall during nesting and brood-raising, success may be low because nests are abandoned or the chicks are too cold to thermoregulate during the first few days of life. Usually animals that are adversely affected by weather during one year rebound the next year. If, however, humans create sub-optimal habitat through grazing or development, a species may be unable to respond. Several years of heavy rain or drought, in which the food supply is low during brood-rearing, may result in a population decline.

Renewable resources are an important component of human interaction in the natural system. When we carefully plan removal of wildlife or trees and allow the resource to recover, we can use this resource forever. We must maintain the community in which the resource is found.

We must look at sustainable resource management, which means looking at how the ecosystem functions. For wildlife species, we must leave their basics of food and shelter so that they can live and reproduce; otherwise, we cannot remove a portion of the population and still have them survive. Most people cannot understand that most wildlife species require several types of habitat during the annual or life cycle. We must address these realities in our resource management.

While we are inclined to look at individual species (deer, sage grouse, wolf) as either a desirable or undesirable species, we need to look more broadly at the entire system. A number of approaches can be taken to manage ecological processes. One approach is landscape ecology, which is defined as the study of how land patterns influence processes (Golley 1993). This means that the landscape is the focus of management (Simberloff 1998).

Because ecological processes are difficult to understand, adaptive management of ecosystems is being suggested. This involves project-as-experiment (Simberloff 1998). Experiments are in place and changes in the system are made according to the results of the experiment. Procedures are modified during the experiment. This may or may not be effective, as it is difficult to know how far changes should go.

Today any form of ecosystem management must consider the effects of humans. There are few, if any, ecosystems in the world that are not affected by people. As in the previously mentioned example of the Everglades, we must consider how human impacts change the processes in the system.

Simberloff (1998) discusses the concept of keystone species, whose activities govern the well-being of many other species in the system, as an alternative to single–species or ecosystem management. This concept might be valid if a true keystone species can be found in the system.

Selection of a keystone species for the prairie ecosystem serves as an example. Currently, the black-footed ferret (*Mustela nigripes*) is an endangered species. There is a good captive–breeding program underway that produces enough animals to be released into the wild in order to establish a number of populations. However, the system as a whole presents a problem. The black-tailed prairie dog (*Cynomys ludovicianus*) is down in numbers by more than 99% since 1898. Ferrets depend on prairie dogs for food. The U.S. Fish and Wildlife Service now has a petition to list the black-tailed prairie dog as threatened. Listing of the swift fox (*Vulpes velox*), ferruginous hawk (*Buteo regalis*), and burrowing owl (*Athene cunicularia*), all animals that live in the prairie ecosystem and use prairie dogs as prey, has been or is being considered. Natural fires and disease influence all prairie dog species and therefore affect the grassland ecosystem (Anderson and Williams 1997). Heavy grazing, poisoning, shooting, and habitat destruction are all contributing to the loss of this system and thus the wildlife community within it (Figure 1-3). In addition, contaminants such as oil, gas, and mineral debris are being added. The whole prairie ecosystem should be managed, and the prairie dog might be considered a keystone species. As we manage the prairie, we need to consider human use for recreation and sustainable yield in terms of cattle grazing and hunting.

We cannot effectively manage an ecosystem by working with only a single animal species. Aquatic and terrestrial systems are not separable. We must look at each system as a whole and all of the systems together. It is important to develop holistic management plans so that individual managers can know how their activities fit into the total picture of integrating all levels of natural resource planning, including human use. We must evaluate any changes made to a system. Even small changes, such as the introduction of contaminants at one point, can trickle throughout a community much as the ripples radiate from a pebble dropped into a pond. We must be prepared to make changes as we go along to permit the system to maintain its stability. This pursuit of system stability produces a balance between people and nature.

References

Anderson BW, Ohmart RD. 1985. Habitat use by clapper rails in the lower Colorado River Valley. *Condor* 87:116–126.

Anderson SH. 1999. Managing our wildlife resources (3rd Ed.). Englewood Cliffs NJ: Prentice Hall. 540 p.

Anderson SH, Williams ES. 1997. Plague in a complex of white-tailed prairie dogs and associated small mammals in Wyoming. *J Wildl Dis* 33:720–732.

Ballard WBL, Ayaares LA, Krausman PR, Reed DS, Fancy SG. 1997. Ecology of wolves in relation to a migratory caribou herd in northwest Alaska. *Wildl Monogr* 135:1–47.

Conway CJ. 1990. Seasonal changes in movement and habitat use in three sympatric rails, [Masters thesis]. University of Wyoming.

Di Silvestro R. 1997. Steelhead trout: Factors in protection. *BioScience* 47:409–414.

Golley FB. 1993. Development of landscape ecology and its relation to environmental management. In: Jensen ME, Bourgeron PS, editors. Eastside Forest Ecosystem Health Assessment, Volume 2, Ecosystem management: Principles and application. U.S. Department of Agriculture National Forest System, Washington DC. p 39–46.

Grumbine RE. 1994. What is ecosystem management? *Conserv Biol* 8:27–38.

Harwell MA. 1997. Ecosystem management of south Florida. *BioScience* 47:499–512.

Kaeding LR, Boltz GD, Carty DG. 1996. Lake trout discovered in Yellowstone Lake threatens native cutthroat trout. *Fisheries* 21:16–20.

Meyer C.B. 1999. Marbled murrelet use of landscapes and seascapes during the breeding season in California and southern Oregon, [Ph.D. dissertation]. Laramie: University of Wyoming.

Minns CK, Moore LR, Schindler DW, Jones ML. 1990. Assessing the potential extent of damage to inland lakes in eastern Canada due to acid deposition. IV. Predicting the response of species richness. *Can J Fish Aquat Sci* 47:483–492.

Phillips MK, Smith DW. 1997. Yellowstone Wolf Project Biennial Report, 1995 and 1996. Yellowstone National Park. YCR-NR-97-4.

Pimentel D, Wilson C, McCullum C, Huang R, Dwen P, Flack J, Tran Q, Saltman T, Cliff B. 1997. Economic and environmental benefits of biodiversity. *BioScience* 47:747–757.

Schindler DW. 1998. A dim future for boreal waters and landscapes. *BioScience* 48:157–164.

Simberloff D. 1998. Flagships, umbrellas, and keystones: Is single species management passe in the landscape era? *Biol Conserv* 83:247–257.

Szumski MJ. 1998. The effects of mining and related metal contamination on piscivorous mammals along the Upper Clark Fork River, Montana [Ph.D. dissertation]. Laramie WY: University of Wyoming.

[USFWS] U.S. Fish and Wildlife Service. 1994. Final environmental impact statement: The reintroduction of gray wolves to Yellowstone National Park and Central Idaho. Helena, MT.

Yellowstone National Park. 1998. Yellowstone Gray Wolf Restoration Project. Current Status Report. Unpublished report.

CHAPTER 2

Perturbations in Terrestrial Vertebrate Populations: Contaminants as a Cause

Glen A. Fox

The term "ecotoxicology" was introduced by Truhaut (1977), reflecting a growing concern about effects of environmental chemicals upon species other than man. Ecotoxicology is the science dealing with the adverse effects of toxic agents on living systems. The early focus of ecotoxicology was the identification and quantitation of chemicals in environmental media with little effort directed at establishing the environmental significance of these residues. As a multidisciplinary science, ecotoxicology recognizes that the toxicity of all contaminants originates with molecular interactions that alter essential cellular activities, resulting in a hierarchical cascade of changes in physiological processes, in the responses of whole organisms, in populations, in community composition, and in ecosystem function. In this biological continuum, processes at 1 level have their mechanistic origins in the preceding level and express their consequences in the next level. Much more effort has been expended on the study of effects on aquatic organisms than on terrestrial organisms, both in terms of understanding their significance and in the development of techniques for detecting effects. Wildlife toxicologists have focused on field investigations directed at the understanding of effects of long-term or repeated exposure to contaminants on individual organisms and their physiology (the level at which exposure can be adequately described and assessed) and, to a far lesser extent, on the implications of these effects for populations and gene pools. Cairns and Niederlehner (1996) have introduced the term "landscape ecotoxicology" for the study of effects of toxic chemicals dispersed over a large spatial scale and the interactions between physical and chemical patterns and the impairment of ecological processes. Here, the focus is on emergent and cumulative responses at the level of communities and ecosystems at the local, landscape, regional, continental, and global scales.

Chapter Preview

Theory and background   20
Modeling   25
Field observations of effects on populations   29
Lessons learned from these experiences   45
Conclusions   47

My objective is to review what we know about the effects of toxic contaminants on terrestrial vertebrate populations and to suggest how we might improve our science to facilitate the expansion of terrestrial ecotoxicology beyond the responses of tissues and organisms to the responses of populations.

Theory and Background

What is a population?

A population is a group of interbreeding, or potentially interbreeding, individuals of the same species occupying a particular space at the same time. Populations have a density, an age structure, a rate of growth and a genetic composition. The individuals that compose populations are genetically unique entities whose behavior and physiology are adaptively integrated to allow them to survive and reproduce. Resources are partitioned among individuals, and it is at the level of the individual that genetic diversity is maintained; evolutionary processes act; and behavior, reproduction, growth, and death occur and are regulated. Contaminants and other environmental factors act by their effects, direct or indirect, on individuals. The consequences of these effects are expressed in the population.

What factors affect populations and how?

According to Ricklefs (1973), "One of the objectives of population biology is to predict changes in population size as a result of observable changes in primary demographic parameters . . . whether these changes are caused by natural environmental factors or man. Implicit in this aim is to determine to what extent populations can adjust to changes in their environment and remain viable." For a variety of reasons, the population ecology of birds is better known than that of any other group of terrestrial vertebrates, and there is no reason to suppose that bird populations are limited in ways different from other terrestrial vertebrates. I will therefore base most of my discussion on population regulation and the effects of contaminants on populations on the results of field studies of birds. I will draw extensively on the works of Ian Newton, particularly from his recent monumental synthesis *Population Limitation in Birds* (Newton 1998).

Within the habitats they occupy, populations are limited by external, or environmental, factors that are the ultimate causes of population regulation and by the intrinsic or demographic features of the population whose net effects mediate the influence of external factors to determine population trends. If we are to understand what determines the average level of populations, and why this level varies temporally and spatially, we must focus our efforts on the external factors (Newton 1998).

The environmental factors that influence populations are both natural (e.g., resources, natural enemies, competing species) and anthropogenic (e.g., exploitation and contaminants). Effects of anthropogenic factors often can be mitigated or reversed (at some cost) to enhance population recovery. The intrinsic or demographic factors that affect populations are birth rate, death rate, and movements (i.e., immigration and emigration). Social systems provide the behavioral framework within which the day-to-day regulation of density and distribution occurs. "Contest competition" serves as the mechanism by which the effects of changes in resource levels are translated into changes in demography and population level. As Newton (1991) points out, "Whether dominance occurs with respect to food supplies, nest sites, mating opportunities, or predator avoidance, it always resolves to defense of the single resource—space."

Resources, natural enemies, and competing species act as density-dependent limiting mechanisms that stabilize populations and modulate population changes towards long-term stability. In contrast, weather, natural disasters, exploitation, and contaminants act as density-independent factors that tend to destabilize populations by affecting demographic parameters and thereby causing large and unpredictable fluctuations. For natural enemies, weather, exploitation, or contaminants to reduce breeding numbers below the level that would otherwise occur, at least part of the mortality they inflict must be "additive" to other mortality and not simply "compensatory," i.e., replacing other causes of mortality. Compensation is expected in populations that are resource-limited and experience strong intraspecific competition for resources. Additivity is expected in populations that are well below their resource limit and are relatively free of intraspecific competition. Like other forms of ecological disturbance, the impact of additive mortality on a population depends upon its timing and frequency, severity, predictability, scale, and synergy with other disturbances. Extreme and unpredictable events have the most impact, and exposure to contaminants is definitely unpredictable.

Unless a species can expand its range, population density will increase as habitat is lost. Whether this will affect the local, regional, and global population depends on whether the rates of emigration, mortality, and reproduction are already, or will become, density-dependent. As habitat is removed or altered, there may be no effect on numbers until at some point in the life cycle, a threshold density is reached at which major density-dependent losses begin to occur. Further habitat loss from this point onward would lead to a decrease in the carrying capacity of the remaining habitat.

Species are distributed across mosaics of habitat patches that differ in their quality and value to the individual. Dispersal (immigration and emigration) facilitates rapid changes in local densities in response to local conditions and thereby maintains population stability and genetic continuity. In widespread species, production of young is more likely to offset mortality in good patches of habitat ("source" populations) than in poor patches ("sink" populations), where mortality may exceed production and densities are maintained by continued immigration (Pulliam 1988). If the differential in production between source and sink populations is great, a small amount of source habitat can ensure the continued occupancy of a large amount of sink habitat. The balance between source and sink populations and between local extinctions and recolonizations ultimately determines whether a regional population persists.

The overall abundance of any species depends upon the extent of its geographical range, the amount of suitable habitat within that range, and the mean density achieved within the habitat. However, seldom is the whole population of any species studied, and the general applicability of a researcher's findings depends on how typical the study area is of the species' range and whether it is sufficiently large to be influenced by those factors that influence the overall population.

What are contaminants?

For the present purpose, I will define contaminants as substances that occur in the environment at least in part through human action and that, at some concentrations, have adverse effects on living organisms (i.e., pesticides, manmade chemicals, effluents and emissions or natural substances).

Three main mechanisms by which contaminants cause declines in terrestrial vertebrate populations have been identified:
- directly, by causing breeding failures or deaths;
- indirectly, by reducing food supply; and
- indirectly, by altering the physical or chemical structure of habitats making them less suitable for certain species (Newton 1998).

The characteristics of contaminants that seem most important in terms of their impact on vertebrates include
- the mode of action, use-pattern, target, and specificity of contaminants such as pesticides;
- environmental transport and fate;
- persistence in the environment; and
- a propensity to bioaccumulate.

The proof and quantitation of exposure are essential to establishing any linkage between contaminants and impacts on individuals and populations. Mineau et al. (1994) based their framework for ranking contaminants with respect to their relative importance as threats to biodiversity on 3 characteristics:
- the extent to which they are distributed through the environment at biologically effective concentrations,
- their degree of selectivity in affecting different taxa, and
- their relative persistence in the environment at biologically effective concentrations.

Mineau et al. (1994) suggest the following:
- Substances that are extensively distributed in biota, reach biologically effective concentrations in a variety of environmental media, and are relatively persistent are of greatest concern because of their potential to threaten many species or habitats over wide geographic areas.
- High selectivity increases the risk that a single or few specialized species will be lost from an ecosystem, whereas pollutants with low selectivity threaten many species irrespective of their ecological roles.
- Stable substances that persist at toxicologically effective levels for extended periods of time (e.g., polychlorinated biphenyls [PCBs], metals) or that lead to long-lasting abiotic changes (e.g., loss of acid buffering capacity of system) are of greater concern than "hit-and-run" contaminants.
- Persistent, rather than episodic, effects have the greatest potential for causing shifts in gene frequencies in populations.

Manmade chemicals play no part in the normal biochemistry of living organisms and are new forces in the population ecology of all organisms. As the pioneer raptor ecologist Tom Cade (1968) put it:

> Down through the centuries, not all the falcon trappers, egg collectors, war messengers concerned for their messenger pigeons, or misguided gunmen have been able to effect a significant reduction in the numbers of breeding falcons. But the simple laboratory trick of adding a few chlorine molecules to a hydrocarbon and the massive application of this unnatural class of chemicals to the environment can do what none of these grosser, seemingly more harmful agents could do.

What is more, dichlorodiphenyltrichloroethane (DDT) brought about the decline or extirpation of disparate populations of peregrine falcons (*Falco peregrinus*) within 10 to 20 years of its widespread use.

Do contaminants differ from natural factors, exploitation and disturbance in their effects on populations?

Because some contaminants such as pesticides are biocides by design, they are capable of altering the quantity, quality and availability of food and habitat. Like natural enemies and exploitation, some contaminants are capable of killing individuals or altering reproductive performance. Like those of parasites, the impacts of pesticides are influenced by factors that increase exposure or decrease resistance. Wobeser (1997) stresses the importance of the local environment to the incidence of various diseases and suggests that similarities in ecology may be more important than similarities in phylogeny and that species differ in their adaptability. Like hurricanes and wildfires, oil spills are a form of natural disaster, destroying both individuals and habitat in a single, brief stochastic event. Like weather, the population impacts of indirect effects of contaminants on food supply and critical habitat vary with severity and/or intensity (the relative decrease in availability or the amount altered or destroyed), duration, timing and/or frequency and scale. A food shortage will have its greatest impact when it occurs during the breeding season, killing the breeding adults and/or reducing reproductive success. When mortalities, reproductive failures, shortages of food or breeding habitat occur in successive years they may limit numbers or result in stepwise population declines and, in extreme cases, cause local extinctions. Their impact will depend on the proportion of the population or critical habitat affected, relative to the size of the local, regional, and global population.

Like adverse weather and parasites, contaminants can have sublethal effects on individuals that alter their survival and reproductive performance. However, laboratory and field studies have shown that these manmade toxicants are unique in the variety of their modes of action and the diversity of their sublethal effects, effects that are frequently unexpected and often delayed. These sublethal effects alter reproduction and growth and may result in short-, medium-, and long-term increases in vulnerability to disease, predation, and starvation resulting from effects on the immune system, stress response, energetics, and neurobehavioral function. In addition, some contaminants can damage or alter genetic material. Hence, exposure to contaminants may have serious implications for populations and gene pools.

The implications of landscape scale and life-history strategies on population effects

In any area within a species' geographical range, the landscape will consist of patches of breeding or feeding habitat, inhospitable areas, and areas that can be used as dispersal corridors. The landscape structure, the spatiotemporal distribution of the toxic events, and the dispersal characteristics and habitat-specific demographics of the species will influence the impact of a toxic event on the local population's survival (Fahrig and Freemark 1995). These authors suggest that, in general, landscape-scale propagation of effects is likely as the population becomes increasingly patchy, the toxic events become increasingly widespread or chronic, and the occurrence of separate toxic events become increasingly synchronous (See Chapter 12).

In allocating resources and time, individuals make tradeoffs among growth (the process that renders a new individual capable of reproduction), maintenance, protection, and reproduction (the addition of new individuals to the population). Allocation of energy to reproduction is a function of mortality (the loss of existing population members), population size, and the predictability of the environment. Contaminants challenge biological systems by inducing energy-expensive, stress-resisting processes (Calow 1991).

Within a species, larger individuals appear to survive better, breed earlier, and have a higher fecundity (Gibbons et al. 1981; Congdon and Tinkle 1982; Semlitch et al. 1988; Berven 1990; McKnight and Gutzke 1993; Sedinger et al. 1995; Saether et al. 1997). Amphibians and reptiles grow at a continuously decreasing rate throughout most of their lives, whereas birds and many mammals grow very quickly. In amphibians, reptiles, and fish, growth rate, body condition, and fecundity are all tightly linked. In contrast, young birds often reach adult (maximum) size by the time they fledge (i.e., within weeks) and almost certainly by the time they initiate their first reproductive attempt. This relatively brief period of very rapid growth implies a very great demand for protein and calcium for growing tissues and, in species whose parents feed their young, a marked increase in energy requirements of the parents. This intense need for protein and calcium forces species that may be predominantly seed-eaters at other times to seek out insects, mollusks, and other invertebrates.

The appropriate physiological and behavioral tradeoffs are optimized by evolution and are modified by the individual's current physiological and physical condition. These compromises constitute the "life-history strategy" by which species populations adjust their reproductive effort to yield a new generation of reproducing adults. The life-history characteristics of so called "K-" and "r-"strategists (after Pianka 1970) are listed in Table 2-1. Generally, populations of K-strategists are most sensitive to changes in adult survival, whereas r-strategists are most affected by reductions in reproductive success.

An examination of longitudinal studies of reproductive success in a wide variety of terrestrial birds and mammals revealed that
- a large fraction of all young produced die before they can breed,
- not all individuals that survive to breed will produce offspring successfully,
- successful individuals vary greatly in their productivity, and
- the most successful individuals contribute disproportionately to the next generation (Clutton-Brock 1988; Newton 1989).

Breeding lifespan is the single most important demographic determinant of lifetime reproductive success. Phenotype, but not genotype, appears to influence lifetime reproductive success. Because individuals vary greatly in their productivity, removal of a disproportionate number of successful individuals by contaminants could affect the dynamics of the local population.

Modeling

Population models have not been used extensively in evaluating the effects of contaminants on terrestrial vertebrate populations. However, many of the models that have been used for evaluating non-contaminant–related impacts could be extended to include contaminants if their effects are known (Emlen 1989; Pulliam 1994; Williams and Emlen 1994). The understanding of the dynamics of animal

Table 2-1 Attributes associated with life-history strategies and their implications for population studies

Attributes	r-strategists	K-strategists
Habitats occupied	Use temporary habitats in unpredictable environments	Make efficient use of specialized habitats
Population density	Variable, below carrying capacity	Stable, near carrying capacity
Main cause of mortality	Density-dependent factors	Density-independent factors
Reproductive effort	Prolific, may provide parental care, young develop quickly	Few offspring, provide parental care, young develop slowly
Breeding seasons	Few	Multiple to many
Non-breeding component	Some non-territory–holding floaters	Many subadults and non-territorial individuals
Population turnover	Rapid	May take decades
Recovery from population reduction	Rapid. Declines in reproduction may be followed by declines in numbers	Slow
Vulnerability	Prone to local extirpation	Especially vulnerable to human impacts
Most sensitive to	Changes in recruitment	Changes in adult survival
Example	Anurans, small lizards, passerines, galliforms, dabbling ducks, small rodents	Turtles, snakes, alligators, raptors, seabirds, seaducks, ungulates, large carnivores

populations and related ecological issues depends upon a direct analysis of life-history parameters that rely on individually marked individuals (Lebreton et al. 1992). Statistical theory and comprehensive software now exist for most types of capture–recapture studies and provide powerful ways to address efficiently many questions in vertebrate population dynamics. Matrix population models have been developed into a powerful framework for demographic analyses (McDonald and Caswell 1993). Such models allow the classification of individuals by any variable of biological interest, lead easily to sensitivity analyses that pinpoint the most ecologically significant portions of the life history, and provide an objective basis upon which to allocate field effort, and are easily extended to include stochastic variation.

Even relatively simple models can assist us in our understanding of demographic effects (Calow et al. 1997). Henny (1972) used mortality-rate schedules obtained for banded birds, age at which a species begins to breed, and the recruitment-rate schedule to examine the impact of pesticides on the rates of mortality and recruitment of 16 species representing raptors, fish-eaters, and passerines for 2 periods, 1925 to 1945 and 1946 to 1965. Grier (1980) used a deterministic life-table model and a stochastic Monte Carlo model to simulate the effects of different rates of reproduction and survival on population outcomes of the bald eagle (*Haliaeetus leucocephalis*). Both models showed that survival was far more critical to the existence of the population than was reproduction and that short-term chances of extinction depend largely on survival rate, partially on the initial size of the population, and almost not at all on reproductive rate. Higher reproduction rates simply led to larger populations among those that survived.

A landscape perspective is required to understand how species are distributed across complex mosaics of habitat patches that differ in their quality and use to the individual. Within a relatively small geographic region, most mobile animal populations inhabit a variety of different habitats, and intraspecies variation in reproductive success and survival occurs with habitat type. Spatially explicit population models, which consider both species–habitat relationships and the arrangement of the habitats in space and time and their relation to population parameters, have been used frequently in the study of individual and population responses to habitat change (Dunning et al. 1992). They also are proving useful in linking behavioral and ecological studies that examine habitat selection to population studies that consider how populations are regulated. Simulations using a model incorporating dispersal, survivorship, reproductive success, and habitat information for Bachman's sparrow (*Aimophila aestivalis*) suggested that variations in demographic variables affect population size more than variation in dispersal ability does and that changes in adult and juvenile survivorship have especially large impacts on the probability of population extinction (Pulliam et al. 1992). When this model was modified to include agricultural habitats, the breeding population decreased, although none of the simulated populations went extinct even at the highest rate of agricultural land use (Pulliam 1994). However, when pesticide applications, assumed to reduce the fledging success of pairs breeding in territories next to agricultural habitats by 50%, were incorporated into the model, an increase from 0% to 30% in the amount of agricultural habitat reduced the population size by 54% and increased the probability of extinction to 33%.

Spromberg et al. (1998) used a toxicant-dosed, single-species metapopulation model to explore the range of possible effects of persistent and degradable contaminants on a metapopulation. Important conclusions that emerged from their simulations were that
- mortality in 1 subpopulation has ecologically significant effects on unexposed populations,
- migration from contaminated to uncontaminated sites means that these uncontaminated sites cannot be used as reference sites,

- arrangement of the patches is critical to the dynamics of the system and the overall impact of a toxicant, and
- multiple discrete outcomes are possible from the same initial conditions.

Because acute or chronic exposure to endocrine-disrupting contaminants may alter reproduction, mortality, and life-stage transitions in exposed individuals or their offspring, the ability of traditional toxicity tests to predict population-level effects is confounded. McTavish et al. (1998) have proposed a framework for modeling the population-level effects of endocrine-disrupting contaminants on the reproduction, mortality, and life-stage transitions of age-structured populations. This model should prove valuable in analyzing the transgenerational impacts and delayed disturbance/recovery response mechanisms potentially associated with such exposures.

Laboratory and field observations of effects on individuals

The variety of impacts of contaminants observed in field and laboratory studies includes the following:
- immediate directed lethal toxicity to insects, rodents, plants, and fungi;
- lethal toxicity to nontarget or exposed wildlife and their habitat;
- various forms of sublethal toxicity to nontarget or exposed wildlife, including failure to attempt to breed and reproductive toxicity (eggshell thinning, embryotoxicity, teratogenicity);
- a variety of behavioral effects including decreased territorial behavior, increased aggressive behavior, altered courtship and parental behavior, decreased or abnormal foraging behavior, increased vulnerability to accidents and predators, and altered migratory behavior/orientation;
- a variety of adaptive biochemical tissue- and organ-level responses;
- a variety of chemical injuries at the tissue and organ level;
- altered energetics and thermoregulatory ability;
- altered osmoregulatory ability;
- immunotoxicity;
- alterations in endocrine function (sex steroids, thyroid and adrenal hormones, etc.); and
- genotoxicity.

Sublethal effects can influence most aspects of an individual's physiology and behavior. In almost all cases, the variance in the measured parameter in contaminant-exposed individuals is greatly increased. This chemically induced variability is superimposed on the natural variation in time and space in the measures of interest. Therefore, in order to assess the significance of the contaminant-induced effect, we require knowledge of the natural variation over time in numbers, breeding success, survival, etc.

Field Observations of Effects on Populations

I will use a limited number of examples to illustrate how well our observations match what we might predict based on our knowledge of population regulation.

Pesticides

Because of their biocidal activity, the high volumes applied, and the spatial extent of applications, pesticides are the contaminants with the greatest potential to affect terrestrial wildlife populations. Mineau et al. (1994) suggest that pesticidal impacts on biodiversity are likely to extend to the ecoregional level, where pesticide use is widespread and where there are limited refugia outside crop areas proper. In agricultural habitats, disturbances including applications of pesticides and other agrochemicals are usually frequent, synchronous, and large-scale (O'Connor and Shrubb 1986). If the frequency of toxic events is high, Fahrig and Freemark (1995) suggest that the toxic events themselves can restructure the population so that it is no longer continuous, and the ability of a species to recolonize affected areas may be constrained by the imposed spatial structure. If the population is patchy or the frequency of toxic events is high, the spatiotemporal pattern of the events can have a large impact on the long-term survival of the population. The more synchronous the toxic events in different locations, the lower the survival probability of the regional population.

Direct effects of pesticides

In forestry, insecticides are used primarily to control epidemics of defoliating lepidopteran insects, which, if left uncontrolled, would result in large decreases in the yield of marketable timber from north-temperate and boreal forests. Repeated defoliations can cause the death of large areas of forest. In North America, more than 90% of insecticide spraying in forestry is directed at budworms (*Choristoneura* spp.) and the gypsy moth (*Lymantria dispar*). Outbreaks tend to be synchronous over large regions of susceptible forest. New Brunswick has had the largest and most continuous program of forest spraying for "protection" against eastern spruce budworm, with large areas being repeatedly treated with various insecticides in most years between 1952 and 1992 (Freedman 1995). In excess of 17 million kg of chemical pesticides were applied to New Brunswick's forests between 1952 and 1992. The insecticides applied were DDT, phosphamidon, fenitrothion, aminocarb, and trichlorphon.

Between 1952 and 1965, about 5.75 million kg of DDT were applied to budworm-infected stands in New Brunswick. Between 1958 and 1963, there was a highly significant inverse correlation between the breeding success of woodcock (*Philohela minor*) and the amount of DDT applied (Wright 1965). In 1954 to 1958, the ratio of immature to adult hunter-killed woodcock in the fall in sprayed areas was 39% of that in unsprayed areas, and between 1959 and 1963, the ratio of immatures to adult females was depressed at least by 50% (Wright 1960, 1965). In 2 years, during which the composition of the fall hunter-kill in New Brunswick was compared to that in Nova Scotia, where no spraying occurred, the breeding success of New Brunswick woodcock was 48% lower (Wright 1965). Woodcock live on the forest floor, where they eat earthworms and other soil invertebrates. The impact on breeding success was not due to eggshell thinning (Dilworth et al. 1972), but no information is available on embryo or juvenile mortality. The nutritious lepidopteran larvae provide a superabundant food source for adult and nestling songbirds in infected stands and a conduit for insecticide exposure. Based on censuses of singing birds, it was estimated that as many as 376,000 ruby-crowned kinglets (*Regulus calendula*) died in New Brunswick in 1975 after operational use of the highly toxic organophosphate phosphamidon, a nonpersistent substitute for DDT (Pearce et al. 1976). Aerial application of phosphamidon also was associated with 70% decreases in forest bird numbers post-spray in Montana (Finley 1965) and Switzerland (Schneider 1966). Because of the difficulties inherent in taking a census of forest birds, population-level effects of insecticide exposure have to be rather large before they can be detected. However, measurements of cholinesterase inhibition in field-caught individuals provide a means to assess the degree of exposure and assess the potential lethal and sublethal impact of organophosphate and carbamate insecticides based on a wealth of controlled laboratory and field studies (Grue et al. 1991). Fenitrothion was the principal insecticide applied for budworm control in the forests of eastern North American through the 1970s and 1980s. In an intensive study of a marked breeding population of white-crowned sparrows (*Zonotrichia albicollis*), Busby et al. (1990) documented extensive adult mortality, territory abandonment, inability to defend a territory, and disruption of incubation and clutch desertion, resulting in a 75% reduction in productivity following an experimental fenitrothion spray that produced levels of cholinesterase inhibition similar to those measured after operational sprays.

Modern technological agriculture relies on pesticides to control weeds, arthropods, and plant diseases. Pesticides therefore affect species diversity at least in the area they are applied and beyond if the method of application is imprecise or the pesticide is mobile. Insecticides have varying degrees of selectivity, but herbicides are designed to control all plants other than the crop.

The accumulated evidence suggests that the following groups of birds are most at risk to lethal insecticide exposure:

- seed eating-species that are attracted to treated seeds and insecticide granules,
- herbivorous species that may consume large quantities of recently sprayed foliage or treated seedlings,
- insectivorous species that gorge themselves on pest insects during outbreaks, and
- scavengers and predators that feed on carcasses or disabled individuals that have been poisoned.

Historically, extensive avian mortality was associated with persistent and bioaccumulating alkyl-mercury and cyclodiene seed dressings and other organochlorines, whereas today, the pesticidal products of concern are the cholinesterase-inhibiting organophosphates and carbamates.

The incidence and extent of lethal pesticide poisoning in terrestrial vertebrates is certainly greatly underestimated. Lack of measurable effects on avian abundance is likely the result of

- movements of birds in and out of treated areas,
- a floating surplus of nonbreeding individuals that can rapidly replace killed territorial birds,
- errors inherent in census methodology, and
- an underestimation of direct mortality because of difficulties in finding sick or dead birds as a result of their low density, their propensity to hide, and scavenger activity, etc. (Mineau and Peakall 1987; Mineau and Collins 1988).

Pesticidal treatment of seed grain with concentrates of toxic chemicals to deter pests has resulted in numerous incidents of wildlife mortality. Alkyl-mercury seed dressings were associated with population declines of a variety of seed-eaters, rooks (*Corvus frugilegus*), and raptorial birds in Sweden (Otterlind and Lennerstedt 1964); dieldrin-treated seed dressings were associated with population declines of the stock dove (*Columba oenas*), kestrel (*Falco tinnunculus*), sparrowhawk (*Accipiter nisus*) (Cramp et al. 1962; Newton et al. 1992), and, to a lesser extent, the peregrine falcon (Ratcliffe 1980; Nisbet 1988) in the United Kingdom (UK). Walker and Newton (1998) analyzed the data for dieldrin concentrations in the liver of 605 individual sparrowhawks found dead in the field between 1963 and 1986 and concluded that individuals died of direct lethal toxicity and as a result of sublethal neurotoxic effects and that the latter appeared to delay population recovery. Widespread use of heptachlor-treated grain resulted in lowered reproductive success, mortality of adults, and population decline of resident western Canada geese *(Branta canadensis moffitti)* in the Columbia Basin of Oregon in the late 1970s (Blus et al. 1984). Reproductive success increased, mortality decreased, and nesting population increased after lindane (gamma-HCH) was substituted for heptachlor. When the environmentally undesirable organochlorines were replaced by less persistent organophosphorus seed treatments in the UK in the early 1970s, numerous mortalities of grey geese (*Anser* spp) were reported (Stanley and Bunyan 1979). More recently, granular carbamate and organophosphate insecticides formulated for incorporation into the seed furrow of row crops have been a problem, since a surplus of granules always seems to be available to foraging birds (Hardy et al. 1986; Mineau 1993). One or two granules of some formulations are sufficient to kill a small songbird. Granules are approximately the same size as dietary grit, and those that are silica-based are very attractive to birds. A kill of an estimated 2000 Lapland longspurs (*Calcarius lapponicus*), which landed on a carbofuran-treated canola field in Saskatchewan, illustrates the potential hazard of such products to large flocks of arctic migrants that traverse the northern prairies at canola seeding time (reviewed in Mineau 1993). Numerous mortalities have been associated with the use of granular carbofuran in corn fields.

Using radiotelemetry to locate individuals and cholinesterase determinations to monitor exposure, Blus et al. (1989) determined that many successful adult female sage grouse (*Centrocercus urophasianus*) were killed when they took their young to dimethoate-sprayed fields adjacent to sagebrush in areas where their summer range interfaced with irrigated fields of alfalfa and potatoes in southeastern Idaho. More recently, radiotelemetry led investigators to discover very large mortalities of Swainson's hawks (*Buteo swainsoni*) in Argentina, where the majority of the species winters (Woodbridge et al. 1995; Goldstein et al. 1996). The death of an estimated 20,000 individuals resulted from consumption of OP-insecticide-contaminated acridid grasshoppers captured on treated alfalfa and sunflower fields. Hawks banded on the Canadian Prairies and in northern California were present in the kills. This serendipitous discovery highlights the potential impact of the use of highly toxic cholinesterase-inhibiting pesticides like monocrotophos on neotropical migrants as the uninformed farmers of Central and South American countries make the transition from primitive, low-intensity agriculture and range-based livestock production to intensive agricultural cultivation.

Henny et al. (1999) were unable to detect any obvious declines in Swainson's hawk populations as a result of the anti-ChE insecticide-induced mortality on their wintering grounds in Argentina. Neither were they able to detect any effect on long-term numbers of either wintering or breeding bald eagles in the Fraser River Delta of British Columbia resulting from secondary anti-ChE poisoning while feeding on dead or debilitated waterfowl. The documented kills represented 2% to 5% of the species and regional population, respectively. The authors conclude that it is highly unlikely that any procedure can detect a 2% to 5% pesticide-related change in a normally fluctuating raptor population.

The persistent, bioaccumulating lipophilic residues of the pesticides DDT, dieldrin, and heptachlor are embryotoxic and have been associated with decreased reproductive success in free-living birds. When Henny (1972) examined the mortality and recruitment rates of 16 species for temporal and spatial differences during the time periods 1925 to 1945 and 1946 to 1965, he found no increases in post-fledgling mortality in any species in 1946 to 1965, but he did find evidence of lowered recruitment in 5 of 9 species where there was sufficient information. Those species showing lowered reproductive success fed primarily on fish, birds, or amphibians and reptiles and had decreased eggshell thickness (Hickey and Anderson 1968; Anderson and Hickey 1972). DDE-induced eggshell thinning and embryotoxicity have had the greatest effect on populations of birds worldwide (Hickey and Anderson 1968; Ratcliffe 1970; Anderson and Hickey 1972). Eggshell thinning in excess of 15% to 20% has been associated with increased egg breakage in a variety of avian species. The bird-eating hawks, the fish-eating eagles (*Haliaeetus* spp.), the osprey

(*Pandion haliaetus*), the brown pelican (*Pelecanus occidentalis*), and the double-crested cormorant (*Phalacrocorax auritus*) were the most affected by dichlorodiphenyldichloroethylene (DDE). A comparison of different peregrine populations from around the world showed that all populations with an average of <17% shell thinning maintained their numbers, whereas all those with >17% declined, some to the point of extinction (Peakall and Kiff 1988). Hamerstrom (1986) studied a color-marked population of harriers (*Circus cyanus*) in the Beuna Vista Marsh in central Wisconsin during 1960 through 1983. She observed an absence of "sky-dancing" (courtship behavior), a failure of about 75% of the available pairs to lay eggs, and a failure of the population to respond normally to a peak in vole abundance from 1964 through 1968, when DDT was intensively applied in her study area. In addition to these effects on reproductive behavior, 90 of 91 marked breeders were never seen again, and the eventual population recovery occurred entirely through immigration.

In 1987, the USFWS listed the white-faced ibis (*Plegadis chihi*) as a migratory nongame bird of management concern. Intensive studies of the reproductive performance of a colony at Carson Lake NV, were undertaken using the "sample egg technique" (Blus 1984) to determine if DDE, other organochlorines, selenium or mercury were affecting reproduction (Henny and Herron 1989). Henny and Herron estimated that 45% of the nesting population in 1985 and 1986 was adversely impacted by DDE and that 20% of the population's expected production was lost. This species is extremely sensitive to the eggshell-thinning effects of DDE, which was apparently accumulated while wintering in the interior agricultural region of Mexico. Selenium (from agricultural drainwater) and Hg (lost during gold and silver processing in the late 1800s) were present in the eggs but were not accumulated to concentrations sufficient to affect hatching success.

Fox (1971) examined temporal variation in the numbers of bird-eating raptors counted in fall migration at Hawk Mountain, PA (migrants from eastern Canada and the New England states) and Duluth, MN (migrants originating in an area extending from eastern edge of the Rockies to central Ontario) and concluded that the numbers of peregrines, merlins, sharp-shinned hawks (*Accipiter striatus*), and Cooper's hawks (*Accipiter cooperii*) decreased markedly during 1960 and 1967 compared with number from 1952 to 1959. Bednarz et al. (1990) used the hawk-count data from Hawk Mountain to infer population trends from 1934 to 1986 and concluded that numbers of sharp-shinned and Cooper's hawks, peregrine falcons and bald eagles crashed and recovered coincident with the history of DDT usage in North America.

Because most insecticides are neurotoxicants by design, sublethal poisoning is most frequently associated with abnormalities in territorial, parental, foraging, and predator avoidance behaviors (Busby et al. 1990; Grue et al. 1991). Patnode and White (1991) studied the fate of 212 nests, 491 eggs, and 456 nestlings of 3 altricial open-nesting species—the northern cardinal (*Cardinalis cardinalis*), brown thrasher

(*Toxostoma rufum*), and northern mockingbird (*Mimus polyglottos*)—located at the periphery of pecan orchards and peanut crops receiving multiple operational pesticide application in southern Georgia. They concluded that the daily survival rate of eggs and nestlings varied inversely ($P < 0.05$) with exposure for all species combined and that nestling weight gain was significantly reduced with increasing exposure levels in 2 of the 3 species. Newton (1998) points out that:

> Although the numbers of most bird species can recover quickly from direct pesticide impacts, if usage continues, local populations may remain permanently depressed or die out. The frequent use of a lethal but nonpersistent pesticide can have consequences for bird populations that are just as severe as the less frequent use of a persistent one.

Indirect effects of pesticides

The intensive and extensive chemical use in farming is 1 completely new element that modern agriculture has introduced into the countryside. According to O'Connor and Shrubb (1986), pesticides have 3 main ecological effects: 1) they kill plants and animals, often selectively, directly influencing the agricultural landscape and its ecology; 2) they change food and habitat resources by removing plant species and altering invertebrate fauna; and 3) they allow the farmer much greater freedom to select his crops and rotations and thereby markedly modify farmland habitat from year to year.

The problem of the deleterious indirect effects of pesticides has been extensively investigated in the UK, where the Joint Nature Conservation Committee commissioned a review of the indirect effects of pesticides on farmland birds (Campbell et al. 1997). Their findings are summarized as follows:

- There is convincing evidence that a range of both scarce and more widespread farmland breeding birds are in decline. Of 40 species of birds breeding on farmland, 24 have declined, 8 by more than 50% in the last 20 to 30 years.
- There is good evidence of both long-term and short-term declines in many invertebrate and plant species upon which these birds feed. This evidence suggests that declines in bird food are, in part, attributable to the effects of pesticides and that there are temporal associations between the trends in pesticide use and the period of rapid decline of many of the declining bird species.
- Pesticide use has reduced the abundance of insects and plants taken as food, both over the longer term and within-season. Organic farms hold higher densities of breeding and wintering birds. Numbers of seed-eating birds are higher in winters on fields with more seeds.
- Set-aside and organically managed fields support higher skylark (*Alauda arvensis*) breeding densities and productivity, relative to other fields, after

controlling for the effect of crop type, field boundary structure, and field area (Wilson et al. 1997).

- In a continuous 30-year study of the grey partridge (*Perdix perdix*), experimental manipulation of cropping patterns and pesticide applications have provided convincing evidence that chick survival has been reduced by the indirect effects of herbicides and broad spectrum insecticides (Potts and Aebischer 1995). There is some evidence (N. Brickle in prep.) to suggest that the same holds true for the corn bunting (*Miliaria calandra*).

These findings strongly suggest that modern agricultural use of pesticides affects populations of birds through habitat modification. Campbell et al. (1997) identified 3 possible indirect impacts of pesticides: 1) insecticides and some fungicides may kill invertebrates that are important food for chicks; 2) herbicides may kill the food plants of some insects, reducing the abundance of the insects; and 3) herbicides may reduce the abundance of weeds and seeds that provide a food source for birds. In addition, herbicides may eliminate essential habitat, alter habitat quality, and increase habitat fragmentation (Freemark and Boutin 1995). An examination of 31 pairs of organic and conventional farms in Denmark revealed that 20 of 35 common farmland bird species were more common on organic farms, and 15 of the 35 species exhibited a decline with increasing pesticide use (Braae et al. 1988). On the basis of total bird density, they estimated the carrying capacity on conventionally farmed land to be only 37% to 51% of the organically farmed land. As part of the same study, Hald and Reddersen (1990) showed that many of the herbivorous insects known to be important as food for birds, as well as a number of plant species important to these insects, were more abundant in the organically farmed fields. The yellowhammer (*Emberiza citrinella*) was significantly more numerous on Danish organic farms, where reproductive success was higher, than on conventional farms (Petersen and Nohr 1991). The correspondence between species affected in the UK and Denmark and those affected much earlier in Sweden (Otterlind and Lennerstedt 1964) suggests that these guilds are particularly vulnerable to habitat modifications. Grassland birds, as a guild, are declining more rapidly than any other group in Canada (Dunn and Downes 1998) and North America (Askins 1993).

The indirect effects of insecticides and herbicides on the numbers and reproductive performance of the grey partridge have been the subject of a long-term study in the UK. Partridge chick survival is a principal factor determining population changes (Southwood and Cross 1969). Partridge chicks are mainly insectivorous during the first 10 days after hatching, and their survival largely depends on the number of arthropods available in cereal crops in June (Potts 1980). The densities of 5 preferred taxa, representing about 20% of the local arthropod community, explained more than 50% of the variation in chick survival (Green 1984). Routinely used herbicides were responsible for a reduction in the density of insect-rich plant species and a 50% reduction in the numbers of preferred insects. A series of experiments in which the selectivity of the herbicide and insecticide regime applied to the preferred habitats (field margins and headlands) was varied markedly affected the densities of preferred insects and survival of partridge chicks (Rands 1985). Similar effects of agricultural landscape structure on the food resources and survival of grey partridge chicks were recently reported in Poland (Panek 1997).

Application of the herbicide glyphosate reduced structural and floral complexity of vegetation in 4-to 5-year-old clearcuts in spruce-fir forests in Maine and lowered the abundance of birds and small mammals, particularly insectivores, for 3 years post-treatment (Santillo, et al 1989a; Santillo, et al 1989b).

Sheehan et al. (1987) estimated the indirect impact of pesticides on aquatic invertebrates and vegetation in wetlands and uplands that play a dominant role in recruitment of waterfowl in the prairie pothole region of Canada. Their various scenarios, some of which are extremely conservative, indicate that the number of breeding waterfowl at risk and the possible loss of breeding waterfowl broods are in the same range as the impact from the recreational harvest of waterfowl in the 3 Prairie Provinces.

Metals

Metals are natural substances and are non-biodegradable by nature. Anthropogenic activities have dispersed them widely, greatly increased their environmental concentrations, and have particularly enriched those that are the most toxic (lead, mercury, zinc, and cadmium).

Emissions from sulfide ore smelters

For many decades, sulfide ore smelters have emitted large quantities of SO_2, arsenic and heavy metals to the atmosphere. The severity of the extensive ecological damage caused by these toxic emissions diminishes with increasing distance from the point source. The Sudbury, ON, example is one of the world's best documented cases of extensive vegetation and related ecological damage caused primarily by the emissions of toxic gasses (SO_2) (reviewed in Freedman 1995). The wildlife habitat in the impact zone of such smelters exhibits vegetation characteristics that are anomalous in comparison to adjacent areas. In the area affected by the smelter at Anaconda, MT, the large-scale arsenic and metal enrichment was translated into phytotoxic changes in the ecological landscape, including widespread loss of indigenous plant communities and gross alterations in horizontal and vertical structure, which reduced the area's ability to support diverse wildlife populations (Galbraith et al. 1995). Studies of vertebrate abundance and wildlife habitat suitability in the vicinity of the zinc smelters near Palmerton, PA revealed that the highest diversity of amphibians and birds occurred in habitat more than 5 km upwind or 5 km downwind of the smelter (Storm et al. 1993). Mean capture rates of small mammals were markedly higher at upwind and downwind sites compared with sites close to the smelters. Habitat suitability scores for 10 indicator species were lowest at sites closest to the smelters and consistent with the depauperate vertebrate community observed.

Nyholm (1994) studied the effects of heavy metals on breeding populations of free-living pied flycatchers (*Ficedula hypoleuca*) in the pollution gradient from the sulfide ore smelter at Ronnskarsverken, Sweden. He found that eggshell quality, clutch size, embryo survival, nestling survival, and nestling brain mass decreased with proximity to the smelter. In a subsequent egg-exchange experiment (Nyholm 1998), the effects on embryo and nestling survival and nestling brain mass were confirmed. Effects on eggshell quality, clutch size, and embryo survival were associated with the degree of heavy metal contamination of the female, whereas effects on nestling survival and development (i.e., brain mass) were associated with the heavy metal contamination of nestlings via their invertebrate diet.

Lead

More than 6000 metric tons of lead shot were deposited annually in North American wetlands by waterfowl hunters for several decades (Eisler 1988). Lead poisoning occurs in waterfowl that ingest spent shot during feeding and retain them as grit in the gizzard where they are eroded; then the soluble lead is absorbed from the gastrointestinal tract (Wobeser 1997). Lead adversely affects virtually all body systems by inhibiting enzymes essential in all cells, resulting in chronic disease and low-level, often undetected, mortality, particularly during the late winter. Mallards dosed with lead shot were more vulnerable to hunting, had a lower propensity to migrate, and had a higher overall mortality (Bellrose 1959). An estimated 1.6 to 2.4 million ducks per year were lost as a result of lead poisoning in the U.S. (Bellrose 1959). Frequency of occurrence of shot in the gizzards of waterfowl was lowest in geese, intermediate in dabbling ducks, and highest in diving ducks (Aythyinae). When captive adult ring-necked ducks (*Aythya collaris*) were dosed with a single No. 4 lead shot (i.e., minimal exposure) 15% died, and by the sixth and seventh week after dosing, 87% of the survivors were emaciated, and 78% were incapable of normal locomotion and swimming activities (Mautino and Bell 1986). These observations suggest that this species, and possibly other species of diving ducks (Aythyinae and Mergini), may be particularly vulnerable to lead poisoning.

The incidence of lead in European waterfowl was as high or higher than that reported for the U.S. (Pain 1992), suggesting that lead poisoning has had a significant effect on waterfowl populations wherever they are hunted intensively. Ingestion of lead fishing sinkers by mute swans (*Cygnus olor*) caused major population declines on several rivers in the UK (Birkhead and Perrins 1985; Sears 1988). Lead poisoning associated with the ingestion of fishing sinkers has recently been identified as an important cause of mortality in the common loon (*Gavia immer*) in Canada (Pokras and Chafel 1992; Scheuhammer and Norris 1996). Lead poisoning in waterfowl has been reduced and could eventually be eliminated by the introduction of nontoxic shot and fishing sinkers. The use of lead shot for sport hunting of waterfowl was banned in the U.S. in 1991 and in Canada in 1999.

The spectacled eider (*Somateria fischeri*) population nesting on the arctic coastal tundra of the Yukon-Kuskokwim Delta, AK, declined by 94% between 1971 and 1992 (Stehn et al. 1993) and was listed as threatened by the USFWS in 1993. Subsistence waterfowl hunters were still depositing lead shot into the breeding habitat of the Y-K Delta in 1997. Lead poisoning was first diagnosed in spectacled eiders found dead in the Y-K Delta in the early 1990s (Franson et al. 1995). Examination of spectacled eiders trapped on the nesting areas in the Y-K Delta during 1993 to 1996, suggested that lead ingestion occurs locally and that during the brood-rearing period 36% of adult females and 12% of their ducklings were exposed to lead when captured (Flint et al. 1997). Monitoring of marked individuals suggested that lead exposure prior to hatching their eggs reduced the subsequent survival of females by 34% and that most mortality occurred over winter (Grand et al. 1998). Lead ingestion significantly reduced the body mass of wintering canvasbacks (*Aythya valisineria*) wintering in Louisiana (Hohman et al. 1990).

It is very likely that lead poisoning from ingestion of bullet fragments present in carrion contributed to the demise of the California condor (*Gymnogyps californianus*). This highly endangered species is 1 of 2 immense vultures in the world, a relic from the Pleistocene and recent prehistoric megafauna. Lead was certainly a significant factor in the 1980s when 36% of the free-living condors had elevated blood lead levels and 3 known deaths (from a total wild population of about 15 individuals) were due to lead poisoning (Wiemeyer et al. 1988). The wild population had been declining for many years, and in the winter of 1984 to 1985 it declined from 15 to 9 individuals, and 4 of 5 known breeding pairs were lost. In 1985, all the remaining individuals were taken into captivity. The California condor is a classic *K*-strategist. After modeling the population dynamics of the species, Mertz (1971) concluded that this species has such a low reproductive rate that its population could not persist except for the extraordinary longevity of adults. Lead poisoning reduces the survival of subadults and the established breeders alike. Kiff et al. (1979) produced strong evidence to suggest that reproductively significant DDE-induced eggshell thinning occurred in this species in the 1960s and concluded that this species is among the most sensitive to this effect discovered to date. Wiemeyer et al. (1986) reported 16% thinning in eggshells of turkey vultures (*Cathartes aura*) collected within the condor's range from 1980 to 1981.

Mercury

Atmospheric deposition is associated with acidification and increased bioavailability of mercury in lakes in northeastern North America where high blood mercury levels (Evers et al. 1998) are associated with significantly lower rates of nesting and production of hatchlings by territorial pairs of common loons (Burgess et al. 1998). These effects were also reported for loons from the Indian-Wabigoon River system in northwestern Ontario in the 1970s (Barr 1986) where mercury originated from a leaking chlor-alkali facility. Mercury is also an issue downstream from old gold and silver mining and processing sites (Henny and Herron 1989) and was an issue, historically, when used as a pesticidal seed dressing.

Selenium

Selenium is an essential dietary trace element whose toxicity may be manifest at dietary levels only slightly higher than those required nutritionally. Sources of selenium in freshwater environments include releases from metallic ore smelters, fly ash from coal-fired power plants, weathering of selenium-rich geologic deposits, and most importantly, irrigation drainage water from selenium-rich agricultural soils. In the early 1980s, aquatic birds nesting on the Kesterson National Wildlife Refuge in the San Joaquin Valley of California were exposed to high concentrations of selenium in an agricultural drainage water evaporation impoundment. Observations of poor reproductive success and overt teratogenesis were observed in 3 species of dabbling ducks, American coots (*Fulica americana*), eared grebes (*Podiceps nigricollis*), black-necked stilts (*Himantopus mexicanus*), and American avocets (*Recurvirostra americana*) (Ohlendorf et al. 1986). When data for 1983 to 1985 for all species were pooled, at least 39% of 578 nests contained 1 or more nonviable eggs and 26% of 2281 fully incubated eggs were nonviable (Skorupa 1998). When the data were pooled by species, 4% to 49% of fully incubated eggs failed to hatch, and 0% to 15% contained embryos with 1 or more deformities of the eyes, beak, and limbs. Complete post-hatch juvenile mortality was observed in the coots and grebes (Ohlendorf et al. 1986). Acute poisoning of adult coots was also reported (Ohlendorf et al. 1988). Further investigations have revealed that this problem is widespread, and that species differed considerably in their sensitivity (Heinz 1996).

Oil

According to Mineau et al. (1994), bulk petroleum hydrocarbon spills may have genetic-, population-, and ecosystem-level effects on biodiversity, generally at lower than the ecoregional scale. Species with the most limited ranges are clearly at the most risk. Fahrig and Freemark (1995) predict that a single, unusual toxic event such as an oil spill in a continuous population would not propagate throughout the landscape if emigration is unrestricted by the spatial configuration of the habitat. Recolonization is expected to occur quickly from the neighboring areas. Even if the toxic event is chronic, its effects will remain local. However, if the population is patchy, the large-scale effect of the toxic event depends upon the relationship of the population in the exposed area to the dynamics of the regional population.

The water repellency and insulative properties of the plumage of birds are detrimentally affected by oil, and birds ingest oil both while trying to preen their plumage and through the water and food they ingest. Oil has been shown to be an hepatotoxin (Szaro et al. 1981; Lee et al. 1985), to cause hemolytic anemia (Leighton et al. 1983), to alter osmoregulation (Peakall et al. 1983; Hughes et al. 1990), to affect the adrenals (Gorsline and Holmes 1982; Peakall et al. 1983), to alter ovarian endocrine function (Cavanaugh and Holmes 1987), and to decrease male fertility (Holmes and Cavanaugh 1990). Ingestion of oil has been shown to alter energy metabolism and physiology (Butler et al. 1986; Culik et al. 1991). Several field studies of experimentally exposed individuals have been conducted that have resulted in reduced reproductive success (Fry et al. 1986; Butler et al. 1988). Anderson et al. (1996) reported that an unexpectedly large proportion of oiled brown pelicans, which appeared to be successfully rehabilitated, died of unknown causes within a year of their release, presumably as a result of immunosuppression and/or an altered stress response. In contrast, Underhill et al. (1999) found that the majority of African penguins (*Spheniscus demersus*) that were rehabilitated after the *Apollo Sea* oil spill were sighted again on breeding colonies in successive years. This difference likely reflects differences in the composition of the oils to which they were exposed. Nevertheless, we would expect large-scale direct mortality of fouled individuals plus sublethal effects on reproduction and survival following an oil spill.

Oil spills are single events, but if they are followed by other disasters in successive years, they could contribute to reduced recruitment or depletion of potential immigrants. The majority of vulnerable species are K-strategists, with large numbers of subadult and non-breeding individuals. On 24 March 1989, the supertanker *Exxon Valdez* ran aground in Prince William Sound, releasing some 44 million liters of crude oil and contaminating roughly 2100 km of shoreline. Some 35,000 carcasses were recovered, representing 90 species; 74% of the carcasses were murres (*Uria* spp.). There was a second wave of mortality of different species, most in poor body condition, 6 months later (Piatt et al. 1990), which may represent victims of chronic, sublethal toxicity. An estimated 250,000 birds were killed (Piatt and Ford 1996). Despite the obvious acute impact on murres, investigators were unable to detect changes in numbers, phenology, or egg or chick production or to distinguish between long-term effects of the spill and a natural response of the murres to long-term changes in their marine environment. Any effects of the spill on murre colony attendance were relatively short-term and may reflect recruitment of formerly non-breeding individuals at spill-affected colonies or immigration from unaffected colonies (Boersma et al. 1995; Erikson 1995; Piatt and Anderson 1996). No attempt was made to assess the impact of the spill on genetic diversity in remaining populations of murres.

Oakley and Kuletz (1996), who conducted an on-going study of the population, reproduction, and foraging of pigeon guillemots (*Cepphus columba*) on an island 30 km from the grounding site at the time of the spill, detected a decline in numbers that lasted for 3 years after the spill, but no effects on reproduction or foraging were observed. Monitoring of the size and productivity of 24 colonies (10 oiled, 14 not) of black-legged kittiwakes (*Rissa tridactyla*) in Prince William Sound between 1984 and 1994 revealed no changes in numbers in response to the spill, but productivity at oiled colonies was markedly reduced from 1989 to 1994 (Irons 1996). The black oystercatcher (*Haematopus bachmani*) feeds on intertidal invertebrates. According to Sharp et al. (1996) and Andres (1997) the proportion of non-breeding pairs was higher on oiled sites in 1989, and eggs were smaller. Oystercatcher chick mortality was higher on oiled sites than on the non-oiled site and was positively correlated with the amount of oil present in foraging territories. In 1990, oystercatcher chick production remained lower on oiled territories and failed completely on bioremediated shorelines. Nesting surveys of bald eagles in 1989 and 1990 in coastal south-central Alaska revealed a large-scale reproductive failure only in western Prince William Sound and only in 1990 (Bernatowicz et al. 1996).

Wiens et al. (1996) examined the effect of the oil spill on the marine bird community based on survey counts between June 1989 and 1991 from 10 study bays differing in the magnitude of initial oiling. Species richness, species diversity, and species occurrence were all decreased, especially in heavily oiled bays, but the differences had disappeared by 1991. Species richness of several guilds of birds feeding on or close to the shoreline was negatively related to initial oiling level until mid-1990. Of these guilds, the richness of a guild of winter visitant and resident species showed the greatest negative association with initial oiling. However, the richness of guilds of solitary or colonial species that dive or feed on fish showed no significant relationships with oiling at any time.

Persistent bioaccumulative organochlorine contaminants

Fish-eating wildlife in the Great Lakes are chronically exposed to a complex mixture of organochlorine contaminants (DDE, PCBs, HCB, TCDD, mirex), the composition of which has varied spatially and temporally (Suns et al. 1993; Bishop et al. 1996; Pekarik and Weseloh 1998; Ryckman et al. 1998). Generally, concentrations peaked in the late 1960s and early 1970s, decreased until the mid-1980s, and, in some locations, leveled off in the late 1980s and 1990s (Pekarik and Weseloh 1998). A variety of reproductive and physiological effects have been observed in fish-eating birds (reviewed by Grasman et al. 1998) and reptiles and amphibians (Bishop and Gendron 1998).

Once an abundant shoreline resident, nesting bald eagles are no longer found on Lake Ontario, and those nesting on the Michigan shorelines of Lakes Erie, Huron, and Michigan have productivity less than that required to maintain a stable population (Bowerman et al. 1998). Deformities have been found in several eaglets. Eagles nesting on the Michigan shorelines of Lakes Huron and Michigan, and possibly Erie, appear to be a "sink" population. Although DDE-induced eggshell thinning and embryo lethality was probably responsible for their population crash (Wiemeyer et al. 1984; Nisbet 1989), current investigations have focused on the contribution of PCBs (Colborn 1991; Bowerman et al. 1995).

Like the bald eagle, the piscivorous herring gull (*Larus argentatus*) is a year-round resident on the Great Lakes. Studies in eastern Lake Ontario in the early 1970s suggested that reproductive success was, at best, 10% of normal (Gilbertson 1974; Gilman et al. 1977). Eggshell thinning was less important than embryo death and egg disappearance. Incubation behavior was abnormal, allowing for excessive overheating and chilling and predation of eggs. Impaired growth, congenital malformations, enlarged livers, subcutaneous edema, and hepatic porphyria were observed in young that hatched (Peakall and Fox 1987; Gilbertson et al. 1991). Herring gull reproduction in Lake Ontario began to recover by 1977 and was normal by 1981, corresponding to the period in which residues of HCB, PCBs, and 2,3,7,8-TCDD declined most rapidly. Today we detect several toxicant-induced biochemical abnormalities in adults and suppression of T-cell–mediated immunity in fledglings (Fox 1993; Grasman et al. 1996).

Reproduction was unsuccessful in migratory Forster's terns (*Sterna forsteri*) nesting in Green Bay in Lake Michigan in 1983, as a result of nest abandonment, egg disappearance, and hatching failure. Abnormal incubation behavior was documented (Kubiak et al. 1989), and chicks that hatched were smaller, had enlarged livers, and some were deformed (Hoffman et al. 1987). Kubiak et al. (1989) associated the reproductive failure with coplanar PCB contamination. When the same colonies were studied in 1988, reproduction was normal and was accompanied by a 67% decrease in total PCBs and a 42% decrease in calculated TCDD-equivalents (Harris et al. 1993). However, a significant proportion of apparently normal chicks hatched in 1988 exhibited unexplained weight loss at 14 to 20 days of age, which resulted in severe emaciation and death. Ludwig et al. (1993) provide convincing evidence of an association between TCDD-EQs and chick wasting, deformities, and egg viability in Caspian terns (*Sterna caspia*) in the years following a 100-year flood event in the Saginaw River/Bay ecosystem.

Unlike herring gulls, the populations of the double-crested cormorant crashed in the 1950s and 1960s throughout the Great Lakes as a result of egg breakage from DDE-associated eggshell thinning (Anderson and Hickey 1972; Weseloh et al. 1983). Overall, the population within the Great Lakes declined by more than 80%, paralleling a similar population crash in this species (Gress et al. 1973) and the closely related brown pelican off the coast of southern California (Anderson et al. 1975). Cormorants and pelicans are particularly vulnerable to eggshell thinning because they incubate their eggs by standing on them. Since the late 1970s, eggshell thickness has increased markedly, and the Great Lakes cormorant population is currently increasing by 20% to 30% per year (Price and Weseloh 1986; Weseloh et al. 1995). Currently, eggshell thinning is no longer a problem, but DDE may still influence embryo survival (Custer et al. 1999). However, there is conflicting evidence as to the relative importance of DDE (Custer et al. 1999) and PCBs (Tillitt et al. 1992; Ludwig et al. 1996) in the chemical etiology of embryotoxicity. The causes of crossed bills and other deformities (Fox et al. 1991), which appear more common in cormorants than other species, is not well understood at this time (Kuiken et al. 1999). Although there are associations between the incidence of deformities and PCBs and TCDD-EQs (Ludwig et al. 1996), studies conducted within a single colony (Larson et al. 1996) and laboratory studies in which cormorant eggs are injected with PCB 126, TCDD, or an extract derived from cormorant eggs collected from the colony with the highest incidence of deformities (Powell et al. 1997) have not confirmed this association.

Caspian terns did not appear to suffer from poor reproductive success associated with eggshell thinning. However, Ludwig et al. (1993, 1996) have shown this species to be sensitive to contaminant-associated reproductive effects. In a survey conducted in 1990, plasma PCB concentrations in adult Caspian terns from Saginaw and Green Bays were greater than those in plasma from terns in colonies in Georgian Bay and the North Channel of Lake Huron (Mora et al. 1993). In addition, these investigators reported a significant negative correlation between mean plasma PCB concentrations by region and the percent of nesting terns banded as nestlings returning to their natal region to breed. Studies conducted from 1992 to 1994 found that T-cell function of pre–fledglings, as measured by the PHA skin test, decreased as total PCB and DDE exposure increased (Grasman et al. 1996). At the most contaminated sites (Saginaw Bay and eastern Lake Ontario) the PHA skin test response was 30% lower than at the least contaminated sites. These findings suggest a possible physiological mechanism for the regional source–sink dynamics in recruitment reported by (Mora et al. 1993).

Genotoxic effects of contaminants

There has been relatively little published on the chemical- or radiation-induced changes in the genetic material of wild terrestrial vertebrates (reviewed by Hebert and Murdoch-Luiker 1996). Changes may be direct alterations in genes and gene expression or selective effects of pollutants on gene frequencies (Anderson et al. 1994; Hebert and Murdoch-Luiker 1996).

One-third of the individual sea otters (*Enhydra lutris*) sampled 1.5 years after exposure to the *Exxon Valdez* oil spill showed evidence of clastogenic damage in the leukocytes (Bickham et al. 1998). On-going investigations suggest that the rate of induction of microsatellite germline mutations in Herring gulls increases with proximity to steel-smelting and coking facilities (Yauk et al. 2000).

Clastogenic damage was found in the erythrocytes of slider turtles (*Trachemys scripta*) populating a reservoir contaminated with ^{137}Cs and ^{90}Sr (Lamb et al. 1991). Increased DNA strand breakage was found in the hepatocytes of slider turtles and snapping turtles (*Chelydra serpentina*) collected from a nearby reservoir contaminated with similar low-level radioactive wastes (Meyers-Schone et al. 1993).

Ellegren et al. (1997) used both genetic and morphological data to show that barn swallows (*Hirundo rustica*) breeding in the contaminated region around Chernobyl,

Ukraine, in 1991 and 1996 had suffered elevated levels of both expressed (increased frequency of partial albinism) and non-coding (microsatellite) loci from the 1986 release of radioactive material from the nuclear power plant. These genotoxic effects seem to have resulted in a loss of fitness among individuals in the breeding population and also may have been associated with a significant decline in breeding population size between 1986 and 1996 (Ellegren et al. 1997).

Acid rain

Fahrig and Freemark (1995) predict that toxic events that occur over an area that is as large or larger than the metapopulation (such as acid precipitation or global warming) will pose the most serious threat to species survival. The immediate effect is reduction in numbers across all local populations, some of which, in combination with other stochastic factors, are likely to lead to local extinctions. Because the regional population is reduced, the probability of colonization is reduced. If the habitat is continuous, the limiting factor is the reproductive rate of the species. The more chronic the event, the higher the probability of regional extinction. Similarly, Mineau et al. (1994) rated acidic deposition as a major ecoregional threat to biodiversity for both populations and ecosystems.

The atmospheric pollutants SO_2 and NO acidify precipitation, which lowers the pH of poorly buffered soils, streams, and lakes from 6–7 to 4–5 or less. This in turn affects the mobility and bioavailability of toxic metals, while decreasing the availability of calcium and magnesium. Different species are eliminated from the ecosystem as pH decreases, depending on their pH tolerance and susceptibility to metal toxicity. Therefore, there are declines in food quality and alterations in species composition that are not necessarily accompanied by decreased total biomass. Some lakes become fishless as a result of recruitment failure, thereby reducing the competition for the remaining acid-tolerant arthropods. These changes in the availability of preferred foods result in alterations in the breeding distributions of waterfowl species (Alvo et al. 1988; McNicol and Wayland 1992). Observations of lower duckling survival of ring-necked ducks in acid-sensitive wetlands in east-central Maine led McAuley and Longcore (1988) to suggest that lower recruitment coupled with a high annual harvest rate may have contributed to a population decline in the Atlantic Flyway.

The distribution (Omerod et al. 1985) and breeding performance of dippers (*Cinclus cinclus*) inhabiting streams in the uplands of Wales were altered by stream acidity through effects on insect abundance and calcium availability (Ormerod et al. 1991). The reduced availability of calcium-rich foods (Blancher and McNicol 1991) and the change in types of food available may play a role in the reduced reproductive success (Blancher and McNichol 1988) of tree swallows (*Tachycineta bicolor*) at acidic wetlands in Ontario.

On poor, acidified soils in the Netherlands, Graveland et al. (1994) report an increasing number of great tits (*Parus major*) and other forest passerines that produce eggs with thin, porous shells, desert their clutches, or occupy empty nests. This is apparently the result of a decrease in the availability of calcium-rich snail shells on the forest floor which is caused by, acid deposition. Soil acidity limited the distribution and reproduction of red-backed salamanders (*Plethodon cinerous*) in a hardwood-hemlock forest in New York (Wyman and Hawksley-Lescault 1987). These authors suggest that altered salamander distribution may fundamentally change the forest floor decomposer food web.

Lessons Learned from These Experiences

Although there are relatively few examples of well-documented contaminant-induced population declines or limitations, numerous field studies provide ample evidence that contaminants have the potential to contribute to declines in terrestrial vertebrate populations directly, by causing breeding failures or death, or indirectly, by locally reducing food supply or crucial habitat or by altering the physical structure and flow of nutrients within the landscape on a regional or global scale. Newton (1998) suggests that during the last 50 years, many bird species over large areas have evidently been held by direct pesticide impacts well below the level that contemporary landscapes would otherwise support. Species differ in their vulnerability to exposure and the effects of contaminants (Table 2-2). The discrepancy between what we are now learning concerning the declining numbers of various terrestrial vertebrates, the spatiotemporal correspondence of these declines with massive inputs of contaminants, and our knowledge of the effects of these chemicals on individuals suggests that our inability to detect population-level impacts in many of our studies is an artifact of our science. Several fundamental problems emerge from this assessment of the impacts of contaminants on populations:

- Effects of contaminants need to be distinguished from those of other ecological factors. Contaminant effects occur concomitantly with habitat depletion and other problems. We measure the effect of the integrated cumulative stress on any endpoint we examine. The significance of the contaminant-induced stress must be assessed in relation to the remaining assimilative capacity of the population (see Marschall and Crowder 1996 for a possible approach). Investigations to separate out the role of contaminants, or a particular contaminant, or to detect significant subtle sublethal effects are labor-intensive and involve long-term commitment of resources. As Lewis and Malecki (1984) point out, a multitude of variables must be considered when quantitatively assessing population impacts of contaminants, including the number of individuals of each sex and age class suffering lethal and sublethal effects, data on population abundance and distribution, age and sex composition, proportion of non-breeding adults, reproductive potential,

Table 2-2 Factors contributing to the vulnerability of various species to the effects of contaminants

Biochemical or physiological factors	Behavioral or life-style factors
Route of exposure	Insectivory, piscivory, carnivory, and scavenging
Metabolic rate (species size and amount of food consumed)	Presence in agricultural or forested areas where pesticides are applied
Ability to detoxify and excrete or sequester versus to activate or store	Living at air–water or land–water interface where floating oil accumulates
Target organ sensitivity	Very specialized breeding habitat or nest sites requirements
	Gregarious behavior at some point in life cycle
	A geographically restricted breeding, migration, or wintering area
	Unique behaviors

rates of emigration and immigration, post-fledging survival, relationship of natality and mortality to density, and the influence of other environmental stressors on population numbers. Most methodological biases consistently underestimate exposure and impact. Biomarkers of exposure such as cholinesterase (ChE) or aminolevulinic acid dehydratase (ALAD) inhibition can be used to more precisely link exposure with specific effects in individuals. Biomarkers may also be used to measure the interactive effects of contaminants (Walker 1998).

- The need has arisen for established definitions and criteria to identify impacts, to assess their significance, and to determine whether biological recovery has occurred. We must balance the relative importance of statistical probability against statistical power and biological significance in assessing the impact we attach to findings. In order to be protective of populations, we must focus on minimizing Type II errors and consider the weight of evidence.
- Population effects would ideally be assessed at a spatially relevant scale, relative to the population at risk and to the species population. What is the size of the population exposed, and does it represent a local or regional population or the entire population of that species? How many individuals in the population were killed? Most studies that have shown local population effects have made no effort to translate this to the larger population. Many investigations are restricted to plots smaller than those recommended for the census methods they employ, methods that have inherent methodological problems. It may not be possible for an investigator to determine the relationship of his study population to the larger population of the species, and in this case we might refer to the study population as an "evolutionarily significant unit."
- Studies that focus entirely on the number of individuals in a population do not provide us with any information about the origin, health, or genetic

make-up of the individuals of which it is comprised. Seldom are the dynamics of the study populations understood (i.e., immigration and source–sink relationships, life-history strategy and effects of local habitat quality). Studies should involve marked individuals wherever possible and be conducted over multiple breeding seasons. We have almost no information on survival of individuals, but this is the most important demographic parameter in K-strategists. Frequently there is a lack of good baseline data on population size and reproductive performance prior to exposure or a lack of sufficient reference populations. Few studies are conducted for long enough to document recovery.

- The effects of contaminants may be indirect, delayed, unexpected, and widespread. These surprises often result from interspecies differences in sensitivity to particular chemicals and vulnerability in general (i.e., not all species are created equal). Contaminants capable of disrupting endocrine function may alter reproductive, mortality, and life-stage transition rates, either in the exposed individuals or in their offspring. In order to detect surprises, we need to monitor population numbers, reproductive performance, contaminant concentrations, etc., in at least 2 generations and to archive appropriate samples. To understand the impact of stochastic events such as oil spills, a plan of action must be in place that can be initiated immediately after a spill occurs.
- Assessment of exposure is frequently the weakest component in the linkage between contaminants and observed effects. Economic and biological concerns usually drive exposure assessment methodologies. This false economy limits the types of statistical approaches that can be used and often weakens the interpretation and impact of the findings. Nondestructive biomarkers can be used as surrogate measures of exposure to lead and ChE-inhibiting agents, or contaminant concentrations can be measured in blood or feathers. For studies of the impact of egg-borne contaminants on reproduction, the "sample egg method" has proven very valuable, particularly in studies of colonially nesting species (Blus 1984; Henny and Herron 1989; Skorupa 1998).

Conclusions

- Avian ecologists now recognize contaminants as important agents in bird population declines, influencing the distribution and abundance patterns on both local and widespread scales. This may well be the case for other less-studied terrestrial vertebrates.
- We need to improve the design of field studies to detect population effects. The key parameters measured in field studies should always be chosen with the species' life-history strategy in mind. The vast majority of studies con-

ducted to date have not had sufficient statistical power to eliminate the possibility that biologically significant changes have occurred. The use of prospective power analysis to guide research design will greatly increase our probability of *correctly* detecting biologically significant trends or effects (Peterman 1990; Steidl et al. 1997). When our routine statistical tests reject the null hypothesis, this is not a *de facto* proof that the hypothesis is true. Conversely, when we accept the null hypothesis, we must not assert, either explicitly or implicitly, that there was "no effect" unless our study had high power (Peterman 1990). Rather, we should routinely use retrospective power analysis to provide information about the validity of hypotheses accepted by effectively estimating the number of samples or effect size that would have been necessary to reject the hypothesis under the specific conditions of the study. The costs of type II errors can be high (Peterman 1990; Taylor and Gerrodette 1993) since small changes in vital rates can have significant impacts on populations over time and/or space.

- We must come to some agreement on the minimum biological response that we will consider biologically significant in order to use power analysis in our decision making. We must establish some definitions and criteria in terms of what a population is and what a significant (or acceptable) population effect is. Some knowledge of the population at risk is needed before interpretation is possible. What is the importance of the local study population in terms of the global population, numerically and genetically, versus local value in terms of ecological services? What about species and populations in decline? Population sustainability and biodiversity are now important to state and federal governments. The Convention on Biodiversity, the European Community Directive on the Conservation of Wild Birds, the Comprehensive Environmental Response, Compensation, and Liability Act, and the Migratory Birds Convention Act should be consulted and a consensus-building exercise undertaken. We should also assign a significance to sublethal impacts on energetics, immune function, and behavior that often are secondarily lethal. It is increasingly apparent that many populations are "living on the edge," and impacts of this type may push them over the threshold. We must remember that there is no social safety net in nature.

- Models that merge toxicological, demographic, and ecological data hold considerable promise for improving our ability to accurately assess risks of contaminants to populations. We can make much greater and better use of models to interpret the spatial, demographic, and genetic significance of observed effects to populations and gene pools. Models are available that allow us to estimate the transgenerational impacts of endocrine-disrupting contaminants.

- Field manipulations should be used where possible to tease out the contribution of contaminants to the cumulative stress on the population and strengthen the cause-effect linkage. Suitably chosen biomarkers in blood or contaminant concentrations in blood or feathers can be used to assess exposure in individuals in a nondestructive manner, while other biomarkers can provide mechanistic information to strengthen the cause-effect linkage. The individual's performance and survival can be followed via radiotelemetry and colormarking.

- A role for studies exists at the individual level that focuses on growth, survival, reproduction, and genetic diversity, all of which are crucial to the continuity of the population. These provide measures that can be incorporated into demographic and other population models. In addition, we should look for effects on energetics, immune function, and behavior using function or "challenge" tests where possible.
- We must become more and more concerned about the quality of remaining habitats as their quantity declines. Through phytotoxicity, some contaminants alter primary production, reduce the complexity and alter the structure of the habitat, and encourage the colonization of exotic species. These changes result in the elimination of nest-sites and niches, reduction of ecological services, and changes in nutritive quality. The habitat requirements of most species change seasonally (Chapter 1).
- Long-term, intensive, multidisciplinary studies are essential to understanding the significance of contaminant impacts on populations. Because many factors influence population size, linkages must be made between ecology, population dynamics, and toxicology. The ongoing work by the Game Conservancy in the UK, focusing on the population regulation of the grey partridge and other birds in an agricultural setting, has taught us much about the subtle and long-term indirect impacts of modern, nonpersistent pesticide use. Similarly, Newton's long-term, intensive studies of the population ecology of the sparrowhawk have greatly increased our understanding of the impacts of persistent, bioaccumulative pesticides on predatory species. Much knowledge could be gained from long-term intensive multidisciplinary studies focusing on declining amphibian populations, declining guilds of forest birds, and other r-strategists. Virtually no work has been done on the impacts of contaminants on shorebirds, a group whose lifestyle would surely make it vulnerable to the effects of metals and lipophilic organochlorines.

References

Alvo R, Hussell DJT, Berrill M. 1988. The breeding success of common loons (*Gavia immer*) in relation to alkalinity and other lake characteristics in Ontario. *Can J Zool* 66:746–752.

Anderson DW, Gress F, Fry DM. 1996. Survival and dispersal of oiled brown pelicans after rehabilitation and release. *Marine Pollut Bull* 32:711–718.

Anderson DW, Hickey JJ. 1972. Eggshell changes in certain North American birds. *Proc Int Ornithol Congr* 15:514–540.

Anderson DW, Jehl JR Jr., Risebrough RW, Woods LA Jr., Deweese LR, Edgecomb WG. 1975. Brown pelicans: Improved reproduction off the southern California coast. *Science* 190:806–808.

Anderson S, Sadinski W, Shugart L, Brussard P, Depledge M, Ford T, Hose J, Stegeman J, Suk W, Wirgin I, Wogan G. 1994. Genetic and molecular ecotoxicology: A research framework. *Environ Health Perspect* 102 (Suppl 12):3–8.

Andres BA. 1997. The *Exxon Valdez* oil spill disrupted the breeding of black oystercatchers. *J Wildl Manage* 61:1322–1328.

Askins RA. 1993. Population trends in grassland, shrubland, and forest birds in eastern North America. In: Power DM, editor. Vol. 11, Current ornithology. New York NY: Plenum. p 1–34.

Barr JF. 1986. Population dynamics of the common loon (*Gavia immer*) associated with mercury-contaminated waters in northwestern Ontario. Occasional Paper 56. Ottawa, ON. Canadian Wildlife Service, Environment Canada. 23 p.

Bednarz JC, Klem Jr D, Goodrich LJ, Senner SE. 1990. Migration counts of raptors at Hawk Mountain, Pennsylvania, as indicators of population trends, 1934–1986. *Auk* 107:96–109.

Bellrose Jr FC. 1959. Lead poisoning as a mortality factor in waterfowl populations. *Ill. Nat Hist Surv Bull* 27:235–288.

Bernatowicz JA, Schempf PF, Bowman TD. 1996. Bald eagle productivity in south-central Alaska in 1989 and 1990 after the *Exxon Valdez* oil spill. *Am Fish Soc Symp* 18:785–797.

Berven KA. 1990. Factors affecting population fluctuations in larval and adult stages of the wood frog (*Rana sylvatica*). *Ecology* 71:1599–1608.

Bickham JW, Mazet JA, Blake J, Smolen MJ, Lou Y, Ballachey BE. 1998. Flow cytometric determination of genotoxic effects of exposure to petroleum in mink and sea otters. *Ecotoxicology* 7:191–199.

Birkhead M, Perrins C. 1985. The breeding biology of the mute swan *Cygnus olor* on the River Thames with special reference to lead poisoning. *Biol Conserv* 32:1–11.

Bishop CA, Gendron AD. 1998. Reptiles and amphibians: Shy and sensitive vertebrates of the Great Lakes basin and St. Lawrence River. *Environ Monit Assess* 53:225–244.

Bishop CA, Ng P, Norstrom RJ, Brooks RJ, Pettit KE. 1996. Temporal and geographic variation of organochlorine residues in eggs of the common snapping turtle (*Chelydra serpentina serpentina*) (1981–1991) and comparisons to trends in the herring gull (*Larus argentatus*) in the Great Lakes basin in Ontario, Canada. *Arch Environ Contam Toxicol* 31:512–524.

Blancher PJ, McNicol DK. 1988. Breeding biology of tree swallows in relation to wetland acidity. *Can J Zool* 66:842–849.

Blancher PJ, McNicol DK. 1991. Tree swallow diet in relation to wetland acidity. *Can J Zool* 69:2629–2637.

Blus LJ. 1984. DDE in birds eggs: Comparison of two methods for estimating critical levels. *Wilson Bull* 96:268–276.

Blus LJ, Henny CJ, Lenhart DJ, Kaiser TE. 1984. Effects of heptaclor- and lindane-treated seed on Canada geese. *J Wildl Manage* 48:1097–1111.

Blus LJ, Staley CS, Henny CJ, Pendleton GW, Craig TH, Craig EH, Halford DK. 1989. Effects of organophosphate insecticides on sage grouse in southeastern Idaho. *J Wildl Manage* 53:1139–1146.

Boersma PD, Parrish JK, Kettle AB. 1995. Common murre abundance, phenology, and productivity on the Barren Islands, Alaska: The Exxon Valdez oil spill and long-term environmental change.

In: Wells PG, Butlerand JN, Hughes JS, editors. Exxon Valdez oil spill: Fate and effects in Alaskan waters. Philadelphia PA: ASTM. STP 1219. p 820–853.

Bowerman WW, Best DA, Grubb TG, Zimmerman GM, Giesy JP. 1998. Trends of contaminants and effects in bald eagles of the Great Lakes region. *Environ Monit Assess* 53:197–212.

Bowerman WW, Giesy JP, Best DA, Kramer VJ. 1995. A review of factors affecting productivity of bald eagles in the Great Lakes region; implications for recovery. *Environ Health Perspect* 103 (Suppl. 4):51–59.

Braae L, Nohr H, Petersen BS. 1988. Bird fauna in organically and conventionally farmed areas: A comparative study of bird fauna and the effects of pesticides. Environmental Project No. 102. Danish Ministry of the Environment, Copenhagen. [English translation]

Burgess NM, Evers DC, Kaplan JD, Duggan M, Kerekes JJ. 1998. Mercury and reproductive success of common loons breeding in the Maritimes. In: Burgess NM, Beauchamp S, Brun G, Clair T, Vaidya O, Roberts C, Rutherford L, Tordon R, editors. Mercury in Atlantic Canada: A progress report. Sackville NB: Environment Canada. p 104–109.

Busby DG, White LM, Pearce PA. 1990. Effects of aerial spraying of fenitrothion on breeding white-crowned sparrows. *J Appl Ecol* 27:743–755.

Butler RG, Harfenist A, Leighton FA, Peakall DB. 1988. Impact of sublethal oil and emulsion exposure on the reproductive success of Leach's storm-petrels: Short and long-term effects. *J Appl Ecol* 25:125–143.

Butler RG, Peakall DB, Leighton FA, Borthwick J, Harmon RS. 1986. Effects of crude oil exposure on standard metabolic rate of Leach's storm-petrel. *Condor* 88:248–249.

Cade TJ. 1968. The gyrfalcon and falconry. *Living Bird* 7:237–240.

Cairns J and Niederlehner BR. 1996. Developing a field of landscape ecotoxicology. *Ecol Applic* 6:790–796.

Calow P. 1991. Physiological costs of combating chemical toxicants: Ecological implications. *Comp. Biochem Physiol* 100C:3–6.

Calow P, Sibly RM, Forbes V. 1997. Risk assessment on the basis of simplified life-history scenarios. *Environ Toxicol Chem* 16:1983–1989.

Campbell LH, Cooke AS, editors. 1997. The Indirect effects of pesticides on birds. Peterborough, UK. Joint Nature Conservation Committee. 18 p.

Cavanaugh KP, Holmes WN. 1987. Effects of ingested petroleum on the development of ovarian endocrine function in photostimulated mallard ducks (*Anas platyrhynchos*). *Arch Environ Contam Toxicol* 16:247–253.

Clutton-Brock TH, editor. 1988. Reproductive success. Chicago IL: University of Chicago.

Colborn T. 1991. Epidemiology of Great Lakes bald eagles. *J Toxicol Environ Health* 33:395–453.

Congdon JD, Tinkle DW. 1982. Reproductive energetics of the painted turtle (*Chrysemys picta*). *Herpetologica* 38:228–237.

Cramp S, Conder PJ, Ash JS. 1962. Deaths of birds and mammals from toxic chemicals, January-June 1961. 2nd report of the Joint Committee of the British Trust of Ornithology and the Royal Society for the Protection of Birds on Toxic Chemicals, in collaboration with the Game Research Association. Sandy UK: The Royal Society for the Protection of Birds. 25 p.

Culik BM, Wilson RP, Woakes AT, Sanudo FW. 1991. Oil pollution of Antarctic penguins: Effects on energy metabolism and physiology. *Marine Pollut Bull* 22:388–391.

Custer TW, Custer CM, Hines RK, Gutreuter S, Stromborg KL, Allen PD, Melancon MJ. 1999. Organochlorine contaminants and reproductive success of double-crested cormorants from Green Bay, Wisconsin, USA. *Environ Toxicol Chem* 18:1209–1217.

Dilworth TG, Keith JA, Pearce PA, Reynolds LM. 1972. DDE and eggshell thickness in New Brunswick woodcock. *J Wildl Manage* 36:1186–1193.

Dunn EH, Downes CM. 1998. Monitoring Canada's songbirds: Status and trends. *Bird Trends* 6:2–11.

Dunning JB, Danielson BJ, Pulliam HR. 1992. Ecological processes that affect populations in complex landscapes. *Oikos* 65:169–175.

Eisler R. 1988. Lead hazards to fish, wildlife, and invertebrates: A synoptic review. Washington DC: U.S. Fish and Wildlife Service. Biol. Rep. 81.

Ellegren H, Lindgren G, Primmer CR, Moller AP. 1997. Fitness loss and germline mutations in barn swallows breeding in Chernobyl. *Nature* 389:593–596.

Emlen JM. 1989. Terrestrial population models for ecological risk assessment: A state-of-the-art review. *Environ Toxicol Chem* 8:831–842.

Erikson DE. 1995. Surveys of murre colony attendance in the northern Gulf of Alaska following the Exxon Valdez oil spill. In: Wells PG, Butler JN, Hughes JS, editors. *Exxon Valdez* oil spill: Fate and effects in Alaskan waters. Philadelphia PA: ASTM. STP 1219. p 780–819.

Evers DC, Kaplan JD, Meyers MW, Reaman PS, Braselton WE, Major A, Burgess N, Scheuhammer AM. 1998. Geographic trend in mercury measured in common loon feathers and blood. *Environ Toxicol Chem* 17:173–183.

Fahrig L, Freemark K. 1995. Landscape-scale effects of toxic events for ecological risk assessment. In: Cairns J Jr, Niederlehner BR, editors. Ecological toxicity testing. Boca Raton FL: Lewis. p 193–208.

Finley RB. 1965. Adverse effects on birds of phosphamidon applied to a Montana forest. *J Wildl Manage* 29:580–591.

Flint PL, Petersen MR, Grand JB. 1997. Exposure of spectacled eiders and other diving ducks to lead in western Alaska. *Can J Zool* 75:439–443.

Fox GA. 1971. Recent changes in the reproductive success of the pigeon hawk. *J Wildl Manage* 35:122–128.

Fox GA. 1993. What have biomarkers told us about the effects of contaminants on the health of fish-eating birds in the Great Lakes? The theory and a literature view. *J Great Lakes Res* 19:722–736.

Fox GA, Collins B, Hayakawa E, Weseloh DV, Ludwig JP, Kubiak TJ, Erdman TC. 1991. Reproductive outcomes of colonial fish-eating birds: A biomarker for developmental toxicants in Great Lakes food chains. II. Spatial variation in the occurrence and prevalence of bill defects in young double-crested cormorants in the Great Lakes, 1979–1987. *J Great Lakes Res* 17:158–167.

Franson JC, Petersen MR, Meteyer CU, Smith MR. 1995. Lead poisoning of spectacled eiders (*Somateria fischeri*) and of a common eider (*Somateria mollissima*) in Alaska. *J Wildl Dis* 31:268–271.

Freedman B. 1995. Environmental ecology. 2nd Ed. The ecological effects of pollution, disturbance, and other stresses. San Diego CA: Academic Press. 610 p.

Freemark K, Boutin C. 1995. Impacts of agricultural herbicide use on terrestrial wildlife in temperate landscapes: A review with special reference to North America. *Agric Ecosyst Environ* 52:67–91.

Fry DM, Swenson J, Addiego LA, Grau CR, Kang A. 1986. Reduced reproduction in the wedge-tailed shearwater exposed to weathered Santa Barbara crude oil. *Arch Environ Contam Toxicol* 15: 453–463.

Galbraith H, LeJeune K, Lipton J. 1995. Metal and arsenic impacts to soils, vegetation communities and wildlife habitat in southwest Montana uplands contaminated by smelter emissions: I. Field evaluation. *Environ Toxicol Chem* 14:1895–1903.

Gibbons JW, Semlitsch RD, Green JL, Schubauer JP. 1981. Variation in age and size at maturity of the slider turtle (*Pseudemys scripta*). *Am Natur* 117:841–845.

Gilbertson M. 1974. Pollutants in breeding herring gulls in the lower Great Lakes. *Can Field Nat* 88:273–280.

Gilbertson M, Kubiak T, Ludwig J, Fox G. 1991. Great Lakes embryo mortality, edema, and deformities syndrome (GLEMEDS) in colonial fish-eating birds: Similarity to chick-edema disease. *J Toxicol Environ Health* 33:455–520.

Gilman AP, Fox GA, Peakall DB, Teeple SM, Carrol TR, Haymes GT. 1977. Reproductive parameters and egg contaminant levels of Great Lakes herring gulls. *J Wildl Manage* 41:458–468.

Goldstein MI, Woodbridge B, Zaccagnini ME, Canavelli SB, Lanusse A. 1996. An assessment of mortality of Swainson's hawks on wintering grounds in Argentina. *J Rapt Res* 30:106–107.

Gorsline J, Holmes WN. 1982. Suppression of adrenocortical activity in mallard ducks exposed to petroleum-contaminated food. *Arch Environ Contam Toxicol* 11:497–502.

Grand JB, Flint PL, Petersen MR, Moran CL. 1998. Effect of lead poisoning on spectacled eider survival rates. *J Wildl Manage* 62:1103–1109.

Grasman KA, Fox GA, Scanlon PF, Ludwig JP. 1996. Organochlorine-associated immunosuppression in prefledgling Caspian terns and herring gulls from the Great Lakes: An ecoepidemiological study. *Environ Health Perspect* 104 (Suppl 4):829–842.

Grasman KA, Scanlon PF, Fox GA. 1998. Reproductive and physiological effects of environmental contaminants in fish-eating birds of the Great lakes: A review of historical trends. *Environ Monit Assess* 53:117–145.

Graveland J, van der Wal R, van Balen JH, van Noordwijk AJ. 1994. Poor reproduction in forest passerines from decline of snail abundance on acidified soils. *Nature* 368:446–448.

Green RE. 1984. The feeding ecology and survival of partridge chicks (*Alectoris rufa* and *Perdix perdix*) on arable farmland in East Anglia. *J Appl Ecol* 21:817–830.

Gress F, Risebrough RW, Anderson DW, Kiff LF, Jehl JR Jr. 1973. Reproductive failures of double-crested cormorants in southern California and Baja California. *Wilson Bull* 85:197–208.

Grier JW. 1980. Modeling approaches to bald eagle population dynamics. *Wildl Soc Bull* 8:316–322.

Grue CE, Hart ADM, Mineau P. 1991. Biological consequences of depressed brain cholinesterase activity in wildlife. In: Mineau P, editor. Cholinesterase-inhibiting insecticides. Their impact on wildlife and the environment. Amsterdam: Elsevier. p 151–209.

Hald AB, Reddersen J. 1990. Wildlife indicators - insects and wild plants. Danish Ministry of Environment, Copenhagen. [in Danish with English summary].

Hamerstrom F. 1986. Harrier, hawk of the marshes. Washington DC: Smithsonian Institution. 171 p.

Hardy AR, Fletcher MR, Stanley PI. 1986. Pesticides and wildlife: Twenty years of vertebrate wildlife incident investigations by MAFF. *State Vet J* 40:182–192.

Harris HJ, Erdman TC, Ankney GT, Lodge KB. 1993. Measures of reproductive success and polychlorinated biphenyl residues in eggs and chicks of Forster's Terns on Green Bay, Lake Michigan, Wisconsin - 1988. *Arch Environ Contam Toxicol* 25:304–314.

Hebert PDN, Murdoch-Luiker M. 1996. Genetic effects of contaminant exposure - towards an assessment of impacts on wildlife populations. *Sci Tot Environ* 191:23–58.

Heinz GW. 1996. Selenium in birds. In: Beyer WN, Heinz GH, Redmon-Norwood AW, editors. Environmental contaminants in wildlife: Interpreting tissue concentrations. Boca Raton FL: SETAC Press. p 447-458.

Henny CH. 1972. An analysis of the population dynamics of selected avian species with special reference to changes during the modern pesticide era. Washington DC: U.S. Department of Interior, Bureau of Sport Fisheries and Wildlife. Wildl. Res. Rep. 1.

Henny CJ, Herron GB. 1989. DDE, selenium, mercury, and white-faced ibis reproduction at Carson Lake, Nevada. *J Wildl Manage* 53:1032–1045.

Henny CJ, Mineau P, Elliott JE, Woodbridge B. 1999. Raptor poisonings and current insecticide use: What do isolated kill reports mean to populations? In: Adams NJ, Slotow RH, editors. Proc. 22 Int. Ornithol. Congr., Durban. Johannesburg: Birdlife South Africa. p 1020-1032.

Hickey JJ, Anderson DW. 1968. Chlorinated hydrocarbons and eggshell changes in raptorial and fish-eating birds. *Science* 162:271-273.

Hoffman DJ, Rattner BA, Sileo L, Docherty D, Kubiak TJ. 1987. Embryotoxicity, teratogenicity, and aryl hydrocarbon hydroxylase activity in Forster's terns on Green Bay, Lake Michigan. *Environ Res* 42:176-184.

Hohman WL, Pritchert RD, Pace RM, Woolington DW, Helm R. 1990. Influence of ingested lead on body mass of wintering canvasbacks. *J Wildl Manage* 54:211-215.

Holmes WN, Cavanaugh KP. 1990. Some evidence for an effect of ingested petroleum on the fertility of the mallard drake (*Anas platyrhynchos*). *Arch Environ Contam Toxicol.* 19:898-901.

Hughes, MR, Kasserra C, Thomas BR. 1990. Effect of externally applied bunker fuel on body mass and temperature, plasma concentration, and water flux of glaucous-winged gulls, *Larus glaucescens*. *Can J Zool* 68:716-721.

Irons, DB. 1996. Size and productivity of black-legged kittiwake colonies in Prince William Sound before and after the *Exxon Valdez* oil spill. *Am Fish Soc Symp* 18:738-747.

Kiff LF, Peakall DB, Wilbur SR. 1979. Recent changes in California condor eggshells. *Condor* 81:166-172.

Kubiak TJ, Harris HJ, Smith LM, Schwartz TR, Stalling DL, Trick JA, Sileo L, Docherty DE, Erdman TC. 1989. Microcontaminants and reproductive impairment of the Forster's tern on Green Bay, Lake Michigan - 1983. *Arch Environ Contam Toxicol* 18:706-727.

Kuiken T, Fox GA, Danesik KL. 1999. Bill malformations in double-crested cormorants with low exposure to organochlorines. *Environ Toxicol Chem* 18:2908-2913.

Lamb T, Bickham JW, Gibbons JW, Smolen MJ, McDowell S. 1991. Genetic damage in a population of slider turtles (*Trachemys scripta*) inhabiting a radioactive reservoir. *Arch Environ Contam Toxicol* 20:138-142.

Larson JM, Karasov WH, Sileo L, Stromborg KL, Hanbidge BA, Giesy JP, Jones PD, Tillitt DE, Verbrugge DA. 1996. Reproductive success, developmental anomalies, and environmental contaminants in double-crested cormorants (*Phalacrocorax auritus*). *Environ Toxicol Chem* 15:553-559.

Lebreton JD, Burnham KP, Clobert J, Anderson DR. 1992. Modeling survival and testing biological hypotheses using marked animals; a unified approach with case studies. *Ecol Monographs* 62:67-118.

Lee Y-Z, Leighton FA, Peakall DB, Norstrom RJ, O'Brien PJ, Payne JF, Rahimtula AD. 1985. Effects of ingestion of Hibernia and Prudhoe Bay crude oils on hepatic and renal mixed function oxidase in nestling herring gulls (*Larus argentatus*). *Environ Res* 36:248-255.

Leighton FA, Peakall DB, Butler RG. 1983. Heinz-body hemolytic anemia from ingestion of crude oil: A primary toxic effect in marine birds. *Science* 220:871-873.

Lewis SJ, Malecki RA. 1984. Effects of egg oiling on larid productivity and population dynamics. *Auk* 104:584-592.

Ludwig JP, Auman HJ, Kurita H, Ludwig ME, Campbell LM, Giesy JP, Tillitt DE, Jones P, Yamashita N, Tanabe S, Tatsukawa R. 1993. Caspian tern reproduction in the Saginaw Bay ecosystem following a 100-year flood event. *J Great Lakes Res* 19:96-108.

Ludwig JP, Kurita-Matsuba H, Auman HJ, Ludwig ME, Summer CL, Giesy JP, Tillitt DE, Jones PD. 1996. Deformities, PCBs and TCDD-equivalents in double-crested cormorants (*Phalacrocorax auritus*) and Caspian terns (*Hydroprogne caspia*) of the upper Great Lakes 1986–1991: Testing a cause-effect hypothesis. *J Great Lakes Res* 22:172–197.

McAuley DG, Longcore JR. 1988. Survival of juvenile ring-necked ducks on wetlands of different pH. *J Wildl Manage* 52:169–176.

McDonald DP, Caswell H. 1993. Matrix methods for avian demography. In: Power DM, editor. Vol. 10. Current ornithology. New York NY: Plenum. p 307–321.

McKnight CM, Gutzke WHN. 1993. Effects of embryonic environment and of hatchling housing conditions on growth of young snapping turtles (*Chelydra serpentina*). *Copeia* 93:475–482.

McNicol DK, Wayland M. 1992. Distribution of waterfowl broods in Sudbury area lakes in relation to fish, macroinvertebrates, and water chemistry. *Can J Fish Aquat Sci* 49 (Suppl. 1):122–133.

McTavish K, Stech H, Stay F. 1998. A modeling framework for exploring the population-level effects of endocrine disruptors. *Environ Toxicol Chem* 17:58–67.

Marschall EA, Crowder LB. 1996. Assessing population responses to multiple anthropogenic effects: A case study with brook trout. *Ecol Applic* 6:152–167.

Mautino M, Bell JU. 1986. Experimental lead toxicity in the ring-necked duck. *Environ Res* 41:538–545.

Mertz DB. 1971. The mathematical demography of the California condor population. *Am Nat* 105:437–453.

Meyers-Schone L, Shugart LR, Beauchamp JJ, Walton BT. 1993. Comparison of two freshwater turtle species as monitors of radionuclide and chemical contamination: DNA damage and residue analysis. *Environ Toxicol Chem* 12:1487–1496.

Mineau P. 1993. The hazard of carbofuran to birds and other vertebrate wildlife. No. 177. Technical Report Series. Ottawa ON: Canadian Wildlife Service, Headquarters.

Mineau P, Collins BT. 1988. Avian mortality in agro-ecosystems 2. Methods of detection. In: Greaves MP, Smith BD, Greig-Smith PW, editors. Environmental effects of pesticides. Thornton Heath UK, British Crop Protection Council. BCPC Mono. No. 40, p 13–27.

Mineau P, Peakall DB. 1987. An evaluation of avian impact assessment techniques following broad-scale forest insecticide sprays. *Environ Toxicol Chem* 6:781–791.

Mineau P, Scheuhammer AM, Clark T. 1994. Effects of environmental pollutants on biodiversity. In: Biodiversity Science Assessment Team. Biodiversity in Canada: A science assessment for Environment Canada. Ottawa ON: Environment Canada. p 165–179.

Mora MA, Auman HJ, Ludwig JP, Giesy JP, Verbrugge DA, Ludwig ME. 1993. Polychlorinated biphenyls and chlorinated insecticides in plasma of Caspian terns: Relationships with age, productivity, and colony site tenacity in the Great Lakes. *Arch Environ Contam Toxicol* 24:320–331.

Newton I, editor. 1989. Lifetime reproduction in birds. London UK: Academic Press. 479 p.

Newton I. 1991. Concluding remarks. In: Perrins CM, Lebreton J-D, Hirons GJM, editors. Bird population studies: Relevance to conservation and management. Oxford UK: Oxford University. p 637–654.

Newton I. 1998. Population limitation in birds. San Diego CA: Academic Press. 597 p.

Newton I, Wyllie I, Asher A. 1992. Mortality from pesticides aldrin and dieldrin in British sparrowhawks and kestrels. *Ecotoxicology* 1:31–44.

Nisbet ICT. 1988. The relative importance of DDE and dieldrin in the decline of peregrine falcon populations. In: Cade TJ, Enderson JH, Thelander CG, White CM, editors. Peregrine Falcon populations, their management and recovery. Boise ID: The Peregrine Fund. p 351–375.

Nisbet ICT. 1989. Organochlorines, reproductive impairment and declines in bald eagles *Haliaeetus leucocephalus* populations: Mechanisms and dose-response relationships. In: Meyburg B-U,

Chancellor RD, editors. Raptors in the Modern World. Berlin: World Working Group on Birds of Prey. p 483–489.

Nyholm NEI. 1994. Heavy metal tissue levels, impact on breeding and nestling development in natural populations of pied flycatcher (Aves) in the pollution gradient from a smelter. In: Donker MH, Eijsackers H, Heimbach F, editors. Ecology of soil organisms. Chelsea MI: Lewis. p 373–382.

Nyholm NEI. 1998. Influence of heavy metal exposure during different phases of ontogeny on the development of pied flycatchers, *Ficedula hypoleuca*, in natural populations. *Arch Environ Contam Toxicol* 35:632–637.

Oakley KL, Kuletz K. 1996. Population, reproduction, and foraging of pigeon guillemots at Naked Island, Alaska, before and after the *Exxon Valdez* oil spill. *Am Fish Soc Symp* 18:759–769.

O'Connor RJ, Shrubb M. 1986. Farming and birds. Cambridge UK: Cambridge University.

Ohlendorf HM, Hoffman DJ, Saiki MK., Aldrich TW. 1986. Embryonic mortality and abnormalities of aquatic birds: Apparent impacts of selenium from irrigation drainwater. *Sci Total Environ* 52:49–63.

Ohlendorf HM, Kilness AW, Simmons JL, Stroud RK, Hoffman DJ, Moore JF. 1988. Selenium toxicosis in wild aquatic birds. *J Toxicol Environ Health* 24:67–92.

Ormerod SJ, O'Halloran J, Gribbin SD, Tyler SJ. 1991. The ecology of dippers *Cinclus cinclus* in relation to stream acidity in upland Wales: Breeding performance, calcium physiology and nestling growth. *J Appl Ecol* 28:419–433.

Ormerod SJ, Tyler SJ, Lewis JMS. 1985. Is the breeding distribution of dippers influenced by stream acidity? *Bird Study* 32:32–39.

Otterlind G, Lennerstedt I. 1964. Avifauna and pesticides in Sweden (English summary). *Var Fagelvarld* 23:412–415.

Pain DJ. 1992. Lead poisoning in waterfowl. In: Pain DJ, editor. Lead poisoning in waterfowl. Proceedings IWRB Workshop, Brussels, Belgium, 1991. Slimbridge UK: International Waterfowl and Wetland Research Bureau. Spec. Publ. 16. p 7–13.

Panek M. 1997. The effect of agricultural landscape structure on food resources and survival of grey partridge *Perdix perdix* chicks in Poland. *J Appl Ecol* 34:787–792.

Patnode KA, White DH. 1991. Effects of pesticides on songbird productivity in conjunction with pecan cultivation in southern Georgia: A multiple-exposure experimental design. *Environ Toxicol Chem* 10:1479–1486.

Peakall DB, Fox GA. 1987. Toxicological investigations of pollutant-related effects in Great Lakes gulls. *Environ Health Perspect* 71:187–193.

Peakall DB, Kiff LF. 1988. DDE contamination in peregrines and American kestrels and its effect on reproduction. In: Cade TJ, Enderson JH, Thelander CG, White CM, editors. Peregrine Falcon populations, their management and recovery. Boise ID: The Peregrine Fund. p 337–350.

Peakall DB, Miller DS, Kinter WB. 1983. Toxicity of crude oils and their fractions to nesting herring gulls - 1. Physiological and biochemical effects. *Marine Environ Res* 8:63–71.

Pearce PA, Peakall DB, Erskine AJ. 1976. Impact on forest birds of the 1975 spruce budworm spray operation in New Brunswick. CWS Progr. Notes 62. Ottawa ON: Canadian Wildlife Service. 7 p.

Pekarik C, Weseloh DV. 1998. Organochlorine contaminants in herring gull eggs from the Great Lakes, 1974–1995: Change point regression analysis and short-term regression. *Environ Monit Assess* 73:77–115.

Peterman RM. 1990. Statistical power analysis can improve fisheries research and management. *Can J Fish Aquat Sci* 47:2–15.

Petersen BS, Nohr H. 1991. Breeding success in the yellowhammer *Emberiza citrinella* on conventional and organic farms. Unpublished report prepared by Ornis Consultants for the Danish Ministry of Environment, Copenhagen. [in Danish with English summary]. p 2–24.

Pianka ER. 1970. On *r*- and *K*-selection. *Am Nat* 104:592–597.

Piatt JF, Anderson P. 1996. Response of common murres to the *Exxon Valdez* oil spill and long-term changes in the Gulf of Alaska marine ecosystem. *Am Fish Soc Symp* 18:720–737.

Piatt JF, Ford RG. 1996. How many seabirds were killed by the *Exxon Valdez* oil spill? *Am Fish Soc Symp* 18:712–719.

Piatt JF, Lensink CJ, Butler W, Kendziorek M, Nysewander DR. 1990. Immediate impact of the '*Exxon Valdez*' oil spill on marine birds. *Auk* 107:387–397.

Pokras MA, Chafel R. 1992. Lead toxicosis from ingested fishing sinkers in adult common loons (*Gavia immer*) in New England. *J Zoo Wildl Med* 23:92–97.

Potts GR. 1980. The effects of modern agriculture, nest predation and game management on the population ecology of partridges (*Perdix perdix* and *Alectoris rufa*). *Adv Ecol Res* 11:1–82.

Powell DC, Aulerich RJ, Meadows JC, Tillitt DE, Powell JF, Restum JC, Stromborg KL, Giesy JP, Bursian SJ. 1997. Effects of 3,3',4,4', 5-pentachlorobiphenyl (PCB 126), 2,3,7,8–tetrachlorodibenzo-*p*-dioxin (TCDD), or an extract derived from field-collected cormorant eggs injected into double-crested cormorant (*Phalacrocorax auritus*) eggs. *Environ Toxicol Chem* 16:1450–1455.

Price IM, Weseloh DV. 1986. Increased numbers and productivity of double-crested cormorants, *Phalacrocorax auritus* on Lake Ontario. *Can Field-Nat* 100:474–482.

Pulliam HR. 1988. Sources, sinks and population regulation. *Am Nat* 132:652–661.

Pulliam HR, Dunning JB, Liu J. 1992. Population dynamics in complex landscapes: A case study. *Ecol Applic* 2:165–177.

Pulliam HR. 1994. Incorporating concepts from population and behavioral ecology into models of exposure to toxins and risk assessment. In: Kendall RJ, Latcher TE Jr., editors. Wildlife toxicology and population modeling. Boca Raton FL: Lewis. p 13–26.

Rands MRW. 1985. Pesticide use on cereals and the survival of grey partridge chicks: A field experiment. *J Appl Ecol* 22:49–54.

Ratcliffe DA. 1970. Changes attributable to pesticides in egg breakage frequency and eggshell thickness in some British birds. *J Appl Ecol* 7:67–115.

Ratcliffe D. 1980. The Peregrine Falcon. Vermillion SD: Buteo Books. 416 p.

Ricklefs RE. 1973. Fecundity, mortality, and avian demography. In: Farner DS editor. Breeding biology of birds. Washington DC: National Academy of Sciences. p 366–435.

Ryckman DP, Weseloh DV, Hamr P, Fox GA, Collins B, Ewins PJ, Norstrom RJ. 1998. Spatial and temporal trends in organochlorine contamination and bill deformities in double-crested cormorants (*Phalacrocorax auritus*) from the Canadian Great Lakes. *Environ Monit Assess* 53:169–195.

Saether B-E, Lorentsen S-H, Tveraa T, Andersen R, Pedersen HC. 1997. Size-dependent variation in reproductive success of a long-lived seabird, the Antarctic petrel *(Thalassoica antarctica)*. *Auk* 114:333–340.

Santillo DJ, Brown PW, Leslie Jr DM. 1989(a). Response of songbirds to glyphosate-induced habitat changes in clearcuts. *J Wildl Manage* 53:64–71.

Santillo DJ, Leslie Jr DM, Brown PW. 1989(b). Response of small mammals and habitat to glyphosate application on clearcuts. *J Wildl Manage* 53:164–172.

Scheuhammer AM, Norris SL. 1996. The ecotoxicology of lead shot and lead fishing weights. *Ecotoxicology* 5:279–295.

Schneider F. 1966. Some pesticide-wildlife problems in Switzerland. *J Appl Ecol* 3 (suppl.):15–20.

Sears J. 1988. Regional and seasonal variations in lead poisoning in the mute swan *Cygnus olor* in relation to the distribution of lead and lead weights, in the Thames area, England. *Biol Conserv* 46:115–134.

Sedinger JS, Flint PL, Lindberg MS. 1995. Environmental influence on life-history traits: Growth, survival, and fecundity in black brant (*Branta bernicla*). *Ecology* 76:2404–2414.

Semlitsch RD, Scott DE, Pechmann JHK. 1988. Time and size at metamorphosis related to adult fitness in *Ambystoma talpoideum*. *Ecology* 69:184–192.

Sharp BE, Cody M, Turner R. 1996. Effects of the *Exxon Valdez* oil spill on the black oystercatcher. *Am Fish Soc Symp* 18:748–758.

Sheehan PJ, Baril A, Mineau P, Smith DK, Harfenist A, Marshall WK. 1987. The impact of pesticides on the ecology of prairie nesting ducks. Technical Report Series No. 19. Ottawa ON: Canadian Wildlife Service, Headquarters.

Skorupa JP. 1998. Selenium poisoning of fish and wildlife in nature: Lessons from twelve real-world examples. In: Frankenberger WTJr, Engberg RA, editors. Environmental chemistry of selenium. New York NY: Marcel Dekker. p 315–354.

Southwood TRE, Cross DJ. 1969. The ecology of the partridge. III. Breeding success and the abundance of insects in natural habitats. *J Anim Ecol* 38:497–509.

Spromberg JA, John BM, Landis WG. 1998. Metapopulation dynamics: Indirect effects and multiple distinct outcomes in ecological risk assessment. *Environ Toxicol Chem* 17:1640–1649.

Stanley PI, Bunyan PJ. 1979. Hazards to wintering geese and other wildlife from use of dieldrin, chlorfenvinphos and carbophenothion as wheat seed treatments. *Proc R Soc Lond* B 205:31–45.

Stehn RA, Dau CP, Conant B, Butler Jr WI. 1993. Decline of spectacled eiders nesting in western Alaska. *Arctic* 46:264–277.

Steidl RJ, Hayes JP, Schauber E. 1997. Statistical power analysis in wildlife research. *J Wildl Manage* 61:270–279.

Storm GL, Yahner RH, Bellis ED. 1993. Vertebrate abundance and wildlife habitat suitability near the Palmerton zinc smelters, Pennsylvania. *Arch Environ Contam. Toxicol* 25:428–437.

Suns K, Hitchin G, Toner D. 1993. Spatial and temporal trends of organochlorine contaminants in spottail shiners from selected sites in the Great Lakes (1975 –1990). *J Great Lakes Res* 19:703–714.

Szaro RC, Hensler G, Heinz GH. 1981. Effects of chronic ingestion of no. 2 fuel oil on mallard ducklings. *J Toxicol Environ Health* 7:789–799.

Taylor BL, Gerrodette T. 1993. The uses of statistical power in conservation biology: The vaquita and northern spotted owl. *Conserv Biol* 7:489–500.

Tillitt, DE, Ankley GT, Giesy JP, Ludwig JP, Kurita-Matsuba H, Weseloh DV, Ross PS, Bishop C, Silco L, Stromborg KL, Larson J, Kubiak TJ. 1992. Polychlorinated biphenyl residues and egg mortality in double-crested cormorants from the Great Lakes. *Environ Toxicol Chem* 11:1281–1288.

Truhaut R. 1977. Ecotoxicology: Objectives, principles and perspectives. *Ecotoxicol Environ Saf* 1:151–173.

Underhill LG, Bartlett PA, Baumann L, Crawfor RJM, Dyer BM, Gildenhuys A, Nel DC, Oatley TB, Thornton M, Upfold L, Williams AJ, Whittington PA, Wolfaardt AC. 1999. Mortality and survival of African penguins *Spheniscus demersus* involved in the *Apollo Sea* oil spill: An evaluation of rehabilitation efforts. *Ibis* 141:29–37.

Walker CH. 1998. The use of biomarkers to measure the interactive effects of chemicals. *Ecotoxicol. Environ Safety* 40:65–70.

Walker CH, Newton I. 1998. Effects of cyclodiene insecticides on the sparrowhawk (Accipiter nisus) in Britain - a reappraisal of the evidence. *Ecotoxicology* 7:185–189.

Weseloh DV, Teeple SM, Gilbertson M. 1983. Double-crested cormorants of the Great Lakes: Egg laying parameters, reproductive failure, and contaminant residues in eggs, Lake Huron 1972–73. *Can J Zool* 61:427–463.

Weseloh DV, Ewins PJ, Struger J, Mineau P, Bishop CA, Postupalsky S, Ludwig JP. 1995. Double-crested cormorants of the Great Lakes: Changes in population size, breeding distribution and reproductive output between 1913 and 1991. *Colonial Waterbirds* 18 (Spec. Publ.1):48–59.

Wiemeyer SN, Jurek RM, Moore JF. 1986. Environmental contaminants in surrogates, foods, and feathers of California condors (*Gymnogyps californianus*). *Environ. Monit. Assess* 6:91–111.

Wiemeyer SN, Lamont TG, Bunck CM, Sindelar CR, Gramlich FJ., Fraser JD, Byrd MA. 1984. Organochlorine pesticide, polychlorobiphenyl, and mercury residues in bald eagle eggs —1967–1979 — and their relationships to shell thinning and reproduction. *Arch Environ Contam Toxicol* 13:529-549.

Wiemeyer SN, Scott JM, Anderson MP, Bloom PH, Stafford CJ. 1988. Environmental contaminants in California condors. *J Wildl Manage* 52:238–247.

Wiens JA, Crist TO, Day RH, Murphy SM, Hayward GD. 1996. Effects of the *Exxon Valdez* oil spill on marine bird communities in Prince William Sound, Alaska. *Ecol Applic* 6:829–841.

Williams B, Emlen J. 1994. Population models as research tools: An empirical perspective. In: Kendall RJ, Latcher Jr TE, editors. Wildlife toxicology and population modeling. Boca Raton FL: Lewis. p 501–508.

Wobeser GA. 1997. Diseases of wild waterfowl. 2nd Ed. New York NY: Plenum. 324 p.

Woodbridge B, Finely KK, Seager ST. 1995. An investigation of the Swainson's hawk in Argentina. *J Raptor Res* 29:202–204.

Wright BS. 1960. Woodcock reproduction in DDT-sprayed areas of New Brunswick. *J Wildl Manage* 24:419–420.

Wright BS. 1965. Some effects of heptachlor and DDT on New Brunswick woodcocks. *J Wildl Manage* 29:172–185.

Wyman RL, Hawksley-Lescault DS. 1987. Soil acidity affects distribution, behavior, and physiology of the salamander *Plethodon cinereus*. *Ecology* 68:1819–1827.

Yauk CL, Fox GA, McCarry BE, Quinn JS. 2000. Living near steel industries: Induced minisatellite germline mutations in herring gulls (*Larus argentatus*). *Mut Res*. In press.

CHAPTER 3

Contaminant-Effect Endpoints in Terrestrial Vertebrates At and Above the Individual Level

Barnett A. Rattner, Jonathan B. Cohen, Nancy H. Golden

Use of biochemical, physiological, anatomical, reproductive and behavioral characteristics of wild terrestrial vertebrates to assess contaminant exposure and effects has become commonplace over the past 3 decades. At the level of the individual organism, response patterns have been associated with and sometimes causally linked to contaminant exposure. However, such responses at the organismal level are rarely associated with or causally linked to effects at the population level. Although the ultimate goal of ecotoxicology is the protection of populations, communities, and ecosystems, most of the existing science and regulatory legislation focus on the level of the individual. Consequently, much of this overview concentrates on contaminant effects at the organismal level, with some extrapolation to higher-level effects.

In this chapter, we review the state of the science for the evaluation of biotic endpoints used to assess contaminant exposure and effects at or above the level of the individual. In addition, we describe extant contaminant concentration thresholds, guidelines, or standards (toxicant criteria) in environmental matrices (e.g., water, soil, sediment, foods) that have been developed to protect wild terrestrial vertebrates. Suggestions are provided to develop and embellish the use and value of such endpoints and criteria for extrapolation of effects to higher levels of biological organization. Increasing focus on populations, communities, and ecosystems is needed to develop biologically meaningful regulatory guidelines that will protect natural resources.

Chapter Preview

Biotic endpoints 62
Biomarkers, bioassays, and bioindicators of contaminant exposure and effect in
 terrestrial vertebrates 66
Contaminant effects on communities and ecosystems 78
Toxicant criteria 79
Conclusions and recommendations 83

Biotic Endpoints

Definitions and history

Using ecological risk assessment terminology, biological characteristics of plants or animals that may be affected by a stressor can be "measures of exposure," "measures of effect," or even "measures of ecosystem and receptor characteristics" (USEPA 1998a). A particular endpoint may fall under any of these categories, depending on the context of the study. "Assessment endpoints" are explicit expressions of an environmental value to be protected and are the ultimate focus in risk characterization that link the aforementioned measures of effect to the risk management process (Suter 1990; USEPA 1992; USEPA 1998a). For example, decline in the population status (an assessment endpoint) of the peregrine falcon (*Falco peregrinus*), brown pelican (*Pelecanus occidentalis*), bald eagle (*Haliaeetus leucocephalus*), and osprey (*Pandion haliaetus*) was evaluated through observations of eggshell thinning and hatching success (measures of effect) in the wild and through observation of surrogate captive mallards (*Anas platyrhynchos*) and American kestrels (*Falco sparverius*) fed p,p'-DDT and p,p'-DDE (Heath et al. 1969; Wiemeyer and Porter 1970). Thus, measures of effect (egg production and hatching success) at the level of the individual were used in part to predict effects on an assessment endpoint (population viability of raptorial and piscivorous birds).

Measures of exposure and effect may be quantified at a variety of levels of biological organization (e.g., molecular, cellular, organ system, organismal, population, and even biotic community). There have been efforts to classify toxicological measures with terminology linking them to level of biological organization (van Gestel and van Brummelen 1996). In their classification scheme, measurements of biochemical and physiological processes are "biomarker" responses; survival, growth, reproduction, and behavior of individual organisms fall in the realm of "ecotoxicity tests and bioassays"; population-level effects are referred to as "bioindicator" responses; and changes at the community and ecosystem level are categorized as "ecological indicators." Such definitions are the subject of considerable debate, as others use the term "biomarker" more broadly to encompass "any biological response to an environmental chemical at or below the level of the individual demonstrating a departure from the normal status" (Peakall and Walker 1996; Peakall and Fairbrother 1998). This definition encompasses biochemical, physiological, histological, morphological, and behavioral measurements. This is the most generally accepted definition and will be used in this chapter. Ideally, all such response endpoints are contaminant dose– and exposure time–dependent phenomena.

Originally termed "bioeffects" monitors, cytochemical and biochemical indices of stress were developed in the mid-1970s to document deleterious effects of pollutants in aquatic organisms (reviewed by Fossi et al. 1994). Perhaps more slowly to evolve was the use of biochemical and physiological indicators in free-ranging terrestrial vertebrates. Investigations during this era focused on the development and use of plasma, whole blood, and tissue enzymes to document contaminant exposure and effects based upon enzyme release associated with tissue damage (e.g., transaminases, dehydrogenases) and contaminant-specific enzyme induction (e.g., cytochrome P450, reviewed by Rattner et al. 1989) or inhibition (e.g., acetylcholinesterase, delta-aminolevulinic acid dehydratase; Dieter et al. 1976; reviewed by Westlake et al. 1983 and Fossi et al. 1994). This effort in terrestrial vertebrates was clearly biochemically and physiologically oriented, to the extent that recognition of organ system and organismal responses was overlooked by scientists focused on subcellular toxicological effects. A number of "state-of-the-science" symposia and workshops in the late 1980s demonstrated that "bioeffects" monitoring had considerable application to ecological hazard and risk assessment, and as a result, funding for such research increased. In the 1990s, considerable interest and progress emerged among ecotoxicologists in the development and use of sublethal, nondestructive, and even noninvasive biomarkers in terrestrial vertebrates. This interest probably occurred in response to the animal-rights movement, bioethics, and better recognition of the constraints associated with monitoring efforts in threatened and endangered species.

Evaluative criteria

A variety of methodological, statistical, and logistical characteristics has been utilized to compare and rank the validity and usefulness of various measures of exposure and effect (i.e., biomarkers, bioindicators, bioassays, and ecotoxicity tests) for monitoring at or above the level of the individual. At the 1989 Pellston Workshop on Biomarkers (23–28 July, Keystone, Colorado), a list of criteria (i.e., General Indicators, Relative Sensitivity, Biological Specificity, Chemical Specificity, Clarity of Interpretation, Time to Manifestation of Endpoint, Permanence of Response, Inherent Variability, Linkage to Higher Level Effects, Applicability to Field Conditions, Validation in the Field, Methodological Considerations, and Status of Method's Utility) was provided to the participants in an effort to evaluate the strengths and weaknesses of various biological endpoints (Huggett et al. 1992). Definitions of these criteria are nearly self-evident and have recently been summarized by Peakall and Fairbrother (1998). Description of literally hundreds of biomarkers, their potential value, and research needs are presented descriptively in the workshop proceedings (Huggett et al. 1992). However, this landmark reference text did not focus much on endpoints at and above the whole organismal level (e.g., behavior, growth, reproduction, and survival). Coincidently, another reference text (Peakall 1992) appeared in print at the same time, with greater emphasis on terrestrial wildlife. Attempts to rank the usefulness and value of the myriad of biomarkers systematically have been rather limited (Peakall 1992; Fossi and Leonzi 1994) but would be an extremely valuable undertaking.

Relationship of endpoints to different levels of biological organization

A suite of biomarkers and higher-level measures of effect can provide evidence to statistically test hypotheses about linkages between toxic chemical exposure and population, community, and ecosystem effects. Such measures of effect have an advantage over exposure assessments (i.e., detecting the presence of a contaminant in a tissue sample) because demonstration of an effect or response documents that the organism has been meaningfully exposed and potentially harmed (McCarthy and Shugart 1990; Mayer et al. 1992; Peakall and Fairbrother 1998). A highly cited Venn diagram (Adams et al. 1989; Figure 3-1) illustrates the relationship between responses at different levels of biological organization and the relevance and time scale of the response. Response data at each level of biological organization provide information for the interpretation of the relationship between exposure and adverse effects (McCarthy and Shugart 1990). In general, measures of effect at the lowest level of biological organization (e.g., molecular, cellular) provide information on sensitive and specific responses to particular toxicants, but their relationship to

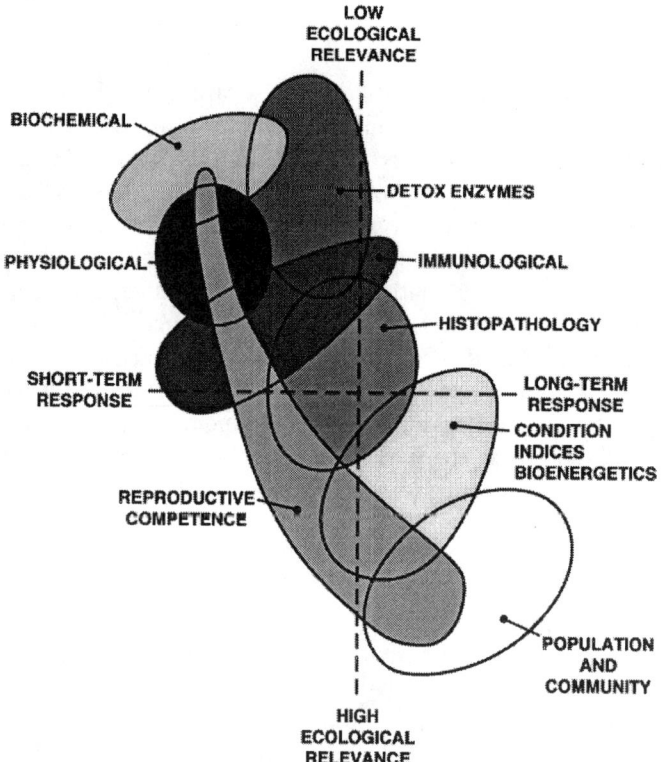

Figure 3-1 Relationship between responses at various levels of biological organization and their ecotoxicological relevance and response time-scales. (Adapted from Marine Environmental Research 28, Adams et al. The use of bioindicators for assessing the effects of pollutant stress in fish, p. 459–464, 1989, with permission from Elsevier Science.)

effects at higher levels of biological organization (e.g., populations) are usually weak and rarely demonstrate causality. Conversely, responses at higher levels of organization (e.g., abundance, species diversity) cannot by themselves demonstrate causality of contaminants. It is thought that biomarkers at the molecular and cellular levels respond quickly to contaminant exposure and thus constitute excellent "early warning" systems, while a latent period may be required before changes occur at the population or community level (Adams et al. 1989; McCarthy and Shugart 1990).

Indicator and sentinel species

One of the challenges in the use of biological endpoints in ecotoxicology and hazard assessment is the selection of appropriate sentinel species (Lower and Kendall 1990; LeBlanc and Bain 1997; Sheffield et al. 1998). As pointed out recently by Peakall and Fairbrother (1998), selection of "indicator" or "sentinel species" differs considerably between toxicologists and ecologists. The separate concept of "keystone species," put forth by Paine (1966), suggests that removal of 1 species from an ecosystem can result in the collapse of many other species in that ecosystem. Despite some supporting evidence in aquatic tide-pool communities, this concept has garnered little support from ecotoxicologists, particularly for terrestrial systems (Peakall and Fairbrother 1998). Another concept, namely the "most sensitive species," was put forth on the grounds that if such a species were protected, related less-sensitive species would also be protected (Peakall and Fairbrother 1998). There is some supporting evidence for this doctrine (e.g., daphnia more sensitive than fish and fish more sensitive than algae; Kenaga 1981), but there are numerous exceptions to such predictions, particularly in the terrestrial realm. Furthermore, despite controlled exposure studies with a myriad of toxicants, an ecological disaster usually must ensue before the most sensitive species can be identified.

Another approach to the selection of indicator or sentinel species depends on whether the goal at hand is biomonitoring of contaminated ecosystems or conservation of endangered species. As a logical extension of the Oil Vulnerability Index for marine birds (King and Sanger 1979), we are currently developing Utility and Vulnerability Indices that are based on algorithms that integrate exposure potential, relative sensitivity, and suitability for monitoring various terrestrial vertebrates residing in or near estuaries (Golden et al. 1998). The most "useful species" for biomonitoring should not necessarily be the most valued from a conservation standpoint (e.g., population could be small, species may be threatened). For Atlantic coast estuaries, our ranking scheme based on broad historical data revealed that the double-crested cormorant (*Phalacrocorax auritus*) may be the most useful (highest Utility Index score) species for biomonitoring the effects of p,p'-DDE on avian eggshell thinning, whereas, historically, the brown pelican (highest Vulnerability Index score) is more sensitive to the adverse effects of p,p'-DDE.

Alternative toxicological tests

In recent years, there has been considerable interest in developing "alternative methods" that reduce the number of animals, refine animal testing procedures to make them less stressful, or replace animals all together with in vitro systems for toxicological tests and assessment (Walker et al. 1998). A number of such in vitro assays have been or are undergoing development and assessment (e.g., Microtox, Genotox), but, to date, extrapolation to effects in vertebrate species appears to be limited to aquatic forms (Walker et al. 1998; Kaiser 1998). In vivo extrapolation can be precarious among different classes of terrestrial vertebrates. For example, acute LD50 toxicity extrapolation among mammals produces reasonable agreement on the basis of body weight raised to the 0.6–0.7 power; however, scaling factors for birds can range up to 1.55 (Mineau et al. 1996). Use of mammalian scaling factors for extrapolating acute toxicity in birds could greatly under-protect this vertebrate class.

Biomarkers, Bioassays, and Bioindicators of Contaminant Exposure and Effect in Terrestrial Vertebrates

Of the hundreds of measures of effect that have been used in laboratory studies, few have gained widespread application in terrestrial vertebrate "field" investigations. In the molecular realm, these include
- inhibition of acetylcholinesterase and other esterases,
- adduct formation (DNA and hemoglobin),
- cytochrome P450 induction,
- inhibition of delta-aminolevulinic acid dehydratase,
- oxidative stress (shifts in ratios of oxidized:reduced glutathione concentration and associated enzymes),
- alterations in concentrations of hormones, metallothionein, and porphyrins, and
- increased activities of plasma/serum enzymes released as a consequence of cellular damage.

At the cellular and tissue level, substantial literature exists describing contaminant-linked histopathological, cytogenetic, and cytotoxic endpoints, and there is emerging literature on immunological endpoints. A few remarkable organ system and whole-animal endpoints (e.g., brain asymmetry, gross anomalies such as bird deformity, teratogenesis, eggshell abnormalities, and behavior) have been linked to contaminant exposure. Many of these endpoints are sensitive, precise, and specific "indicators of exposure" to various contaminants. Still fewer endpoints have well-documented association with population-level effects in terrestrial vertebrates, and these will be discussed in detail.

Measures of exposure

A number of biochemical biomarkers are now being used by ecotoxicologists for rapid screening of contaminant exposure in wildlife to support natural resource management studies. Prime examples that have attained widespread use include acetylcholinesterase inhibition to detect exposure to organophosphorus and carbamate pesticides, cytochrome P450 induction to monitor exposure to polyhalogenated aromatic hydrocarbons (PHAHs) (Rattner et al. 1989; Melancon 1996), and porphyrins to assess exposure to PHAHs and some metals (Fox et al. 1988; Pain 1989). Delta-aminolevulinic acid dehydratase to monitor lead exposure is another biochemical endpoint that has received widespread use (Dieter et al. 1976; Pain 1989), so much so that 50% inhibition in whole blood fulfills wildlife injury criteria under Natural Resource Damage Assessment regulations (USDOI 1986). In contrast, field application and validation of other biochemical (e.g., metallothionein, measures of oxidative stress), genetic (e.g., chromosomal aberration assays, flow cytometric analysis of nuclear DNA, adduct formation, and molecular genetic biomarkers), immunological, and anatomical (e.g., brain asymmetry) endpoints have occurred more slowly. This lag is due to a number of factors including complexity and expense of assay methods, inherent variability, ambiguity in response specificity to contaminants, and even shortfalls in research funding. Measures of oxidative stress (Hoffman et al. 1998), genetic damage (Shugart et al. 1992; Bickham 1994) and immunotoxicity (Grasman et al. 1996; Luebke et al. 1997) show particular promise, partly because of their sensitivity. Despite the remarkable interest in endocrine-disrupting effects of contaminants, only a few endpoints have been developed and applied to field monitoring in terrestrial vertebrates. In and of themselves, in vivo hormone concentrations and receptor measurements in free-ranging terrestrial vertebrates are of limited value as contaminant exposure biomarkers, but they appear to be useful when coupled with anatomical, genetic, behavioral, or reproductive endpoints that integrate effects at the molecular level (e.g., ratios of estrogen and testosterone in combination with reproductive system morphology) (Fox 1992; Guillette et al. 1994; Fry 1995; Kendall et al. 1998). It is also apparent that for some individual contaminants and mixtures, suites of endpoints coupled with residue determinations (e.g., serum enzymes, histological evaluation, and lead determinations) may provide more meaningful exposure assessments (McCarthy 1990).

Measures associated with population-level effects

Endpoints most closely associated with population-level and higher-order effects are those linked to recruitment and survival. Although contaminant-induced reproductive effects and mortality are well-documented (Cooke 1973; Blus et al. 1974; Blus et al. 1979; Grue et al. 1983), the paucity of contaminant-related data on immigration, emigration, and age of reproductive competence precludes discussion. Habitat quality, food availability, predation, and disease are other critical ecological factors that regulate populations, communities, and ecosystems and, in many instances, have been shown to be affected by pollution and toxicant exposure (Kendall et al. 1996). However, from an ecotoxicological perspective, these factors' effects on populations are often "indirect," in contrast to those endpoints that are the "direct" manifestation of contaminant-induced impairment of reproduction or lethal poisoning of individuals. Following is a discussion of several measures of effect that have been linked to, and in some instances used to document, local and even global population-level effects in terrestrial vertebrates. It should be noted that our ability to verify population-level effects from the numerous studies that report individual effects is tenuous, and our success in separating contaminant-induced population effects from the effects of habitat alteration is limited.

Acetylcholinesterase inhibition

Acetylcholinesterase (AChE) is responsible for hydrolyzing acetylcholine into choline and acetic acid (O'Brien 1967), and its inhibition by organophosphorus and carbamate compounds or their metabolites is linked directly with the mechanism of toxic action (e.g., irreversible or reversible binding to the esteratic site and potentiation of cholinergic effects) (Ballantyne and Marrs 1992). Despite the short environmental half-lives of anticholinesterase pesticides compared with organochlorine compounds, some of these chemicals are extremely toxic for short periods after application. These compounds or their activated metabolites inhibit AChE and disrupt neurotransmitter processes in the central nervous system and normal neural functioning of the sensory, integrative, and neuromuscular systems. In exposed birds, alterations in behavior, endocrine function, thermoregulation, reproduction and tolerance to non-contaminant environmental stressors have been reported (reviewed by Grue et al. 1997). Death results from respiratory failure that is due to constricted airways, decreased ventilation associated with paralysis of intercostal muscles in birds and of the diaphragm in mammals, and the direct suppression of the respiratory centers in the brain.

As a measure of exposure to these compounds, AChE and nonspecific cholinesterase (ChE) activities in blood and tissues emerged as a diagnostic tool in the biomedical area. Subsequently, quantification of this enzyme was applied to laboratory and field studies in wildlife to assess exposure to organophosphorus and carbamate insecticides (Ludke et al. 1975, Hill and Fleming 1982; Grue et al. 1983; Greig-Smith 1991), although a few other contaminants, including mercury, have been reported to affect activity (Dieter 1974; Grandjean and Nielsen 1979). The biochemical methods commonly used to quantify AChE and ChE activity (differing in substrate specificity) in wildlife have been reviewed in detail (Fairbrother et al. 1991). Most often used is a modification of the Ellman colorimetric assay in which the rate of metabolism of acetyl-, butyryl- or propionyl-thiocholine iodide is monitored through the formation of a yellow color upon reaction of thiocholine with 5,5-dithiobis-(2-nitrobenzoic acid). A recent advent for free-ranging wildlife has been the use of reactivation agents (e.g., pyridine-2-aldoxine methochloride) in a split sample to detect a potential increase in activity associated with anticholinesterase exposure. This assay can be adapted to high throughput auto-analyzers and micro-well plate readers (Fairbrother et al. 1991). This biomarker is a well-accepted indicator of anticholinesterase exposure, with 50% inhibition of brain AChE activity as the accepted threshold associated with mortality and sublethal effects reported with as little as 20% inhibition (Grue et al. 1991). Physiological "reference norms" of brain AChE activity have been established for several species of birds and mammals (Westlake et al. 1983; Hill 1988). Brain AChE activity has been used as legal proof of the cause of mortality in wildlife die-offs following organophosphorus and carbamate pesticide application, misapplication, and intentional poisoning (Hill and Fleming 1982; Grue et al. 1983). Blood, plasma, and serum ChE activity have been used to monitor exposure in a number of circumstances, although degrees of inhibition in these matrices are not well-correlated with physiological and behavioral effects. The biological variability associated with age, different physiological states, stress, and disease can be substantial (Rattner and Fairbrother 1991) and should be taken into account in hazard assessments, ideally through the use of concurrent control samples. Some attempts have been made to extrapolate AChE inhibition to survival and adverse reproductive effects in free-ranging wildlife. Unintentional poisonings associated with anticholinesterase pesticides can involve a few to several hundred individuals per incident (Grue et al. 1983; Greig-Smith 1991). Despite evidence of some large die-offs supposedly involving millions of forest birds (Pearce et al. 1976) and thousands of highly conspicuous Swainson's hawks (*Buteo swainsoni*) (Goldstein et al. 1999), there is a paucity of data demonstrating population-level effects.

Teratogenesis and morphological aberrations

The occurrence of a low frequency (0% to 2%) of teratogenic and morphological aberrations is common and natural in wild populations (Hoffman et al. 1988; Ouellet et al. 1997). However, when the baseline frequency rises, it can reflect a decline in the general health of a population, which may be attributed to a number of causes, including toxicant exposure. The appearance of malformations can occur during embryonic development (teratogenesis) or at a later point in life, and in severe circumstances can lead to death. There is considerable evidence of pathological lesions, and at least some evidence of external morphologic aberrations, occurring in response to contaminant exposure in free-ranging terrestrial vertebrates.

Studies on the mechanisms of teratogenesis, pathological lesions, and morphological aberrations as a result of contaminant exposure have principally utilized domesticated laboratory rodents and in vitro systems focusing upon
- mutations,
- chromosomal breaks,
- altered mitosis,
- altered nucleic acid integrity or function,

- reduced supplies of precursors or substrates,
- decreased energy supplies,
- altered membrane characteristics,
- osmolar imbalance, or
- enzyme inhibition (Rogers and Kavlock 1996).

If not corrected by cellular repair mechanisms, these dysfunctions can manifest pathogenetic responses such as
- diminished cell proliferation,
- cell death,
- altered cellular communication,
- reduced biosynthesis,
- inhibition of morphogenetic movements, and
- mechanical disruption of developing structures (Rogers and Kavlock 1996).

Depending on the timing of exposure and the severity of the lesion, the resulting morphological deformities can lead to death of the embryo or organism, which in a few instances has been documented to threaten reproductive success of populations.

Determination of teratogenesis and morphological aberrations can be performed by simple examination in cases involving overt external lesions. Other histological malformations require necropsy and microscopic inspection for related lesions (Wobeser 1981). Effects of contaminants on skeletal growth can be determined by examination of stained embryonic preparations for evidence of defects such as shortening, contortion, or incomplete ossification (Karnofsky 1965; Hoffman 1994).

Evidence of contaminant-linked morphological aberrations has been found for all classes of terrestrial vertebrates. Controlled laboratory exposure to numerous classes of toxicants (organochlorine and organophosphorus insecticides, herbicides, and heavy metals) has produced a wide array of defects. A number of these responses have been documented in free-ranging populations (Hoffman 1990, 1994; McBee and Bickham 1990; Bantle et al. 1991; Fox 1992). While many defects produced in the laboratory are in response to chemical concentrations much greater than those typically found in the field, a few epidemic-like situations have been documented in recent decades. For example, Great Lakes Embryo Mortality, Edema, Growth Retardation, and Deformities Syndrome (GLEMEDS), which incorporates a suite of lesions (e.g., bill deformities, club feet, missing eyes, defective feathering, edema, porphyria, liver enlargement, necrosis, and lipidosis) was found to be related to the presence of dioxin-like compounds (principally coplanar polychlorinated biphenyl [PCB] congeners) and has affected several species of colonial fish-eating birds since the 1970s (Gilbertson et al. 1991). Widespread selenium toxicosis was observed during the 1983 breeding season at the Kesterson Reservoir in Merced County, California, where 22% to 65% of the nests of several aquatic bird species exposed to agricultural drainage water contained embryos with skeletal deformities and/or missing or abnormal eyes, beaks, wings, legs, and feet (Hoffman et al. 1988). In the St. Lawrence River Valley in Quebec in 1993, 12% of green frogs (*Rana clamitans*), northern leopard frogs (*Rana pipiens*), American toads (*Bufo americanus*), and bullfrogs (*Rana catesbeiana*) exposed to agricultural runoff were identified with hind-limb deformities (Ouellet et al. 1997). In a 1994 study, penis size was found to be reduced in length by an average of 24% in alligators (*Alligator mississippiensis*) inhabiting Lake Apopka, Florida, where elevated concentrations of p,p'-DDE (and perhaps other organochlorine pesticides or metabolites) in eggs and juvenile adipose tissue had previously been identified (Heinz et al. 1991; Guillette et al. 1996).

Histological lesions are a more common response to contaminant exposure and have been characterized for a number of toxicants, including lead and oil. Signs that are diagnostic of lead poisoning include both gross lesions (e.g., wasting of breast muscle, reduction of visceral fat, impaction of the esophagus or proventriculus, distended gallbladder, bile-stained gizzard) and microscopic lesions (e.g., formation of intranuclear inclusion bodies in kidney, necrosis of hepatocytes in liver, necrosis of myocardial muscle, changes in size and shape of erythrocytes) (Friend 1987; Fairbrother et al. 1996). Ingestion of oil, directly from the environment or by preening after external exposure, has been shown in several bird species to cause anemia, resulting from destructive oxidation of red blood cell membranes and proteins (Leighton 1991; Fairbrother et al. 1996). Though this lesion is not unique to oil exposure, diagnosis of this type of contamination usually can be confirmed by evidence of external oiling (Fairbrother et al. 1996).

Measurement of the frequency of teratogenic or morphological aberrations can be an indicator of the general health of a population, though its usefulness may be limited by several confounding factors. Wild populations normally exhibit a low frequency of aberrations; thus, if the baseline frequency is unknown, an abnormally "elevated" incidence will be difficult to determine. In addition, while detection of an elevated frequency of aberrations is cause for concern, their absence is not necessarily proof of a "healthy" population. Finally, an elevated incidence of aberrations may be indicative of a problem, but it does not imply contaminant exposure. Malformations attributed to contaminant exposure may be indistinguishable from those caused by genetic defects, parasites, infectious disease, poor nutrition, or physical trauma. The ambiguity resulting from these shortcomings, along with possible interspecific variability in responses to contaminants, can confound the interpretation of lesions. However, the overt nature of many teratogenic and morphological aberrations has made them useful measures of effect in the field and advocates their continued development for this purpose.

Eggshell thinning

Eggshell thinning by chlorinated hydrocarbons has been observed in several species of raptorial and fish-eating birds. The consequences of cracking and crushing thin-shelled eggs are quite serious, leading to impairment of reproduction and population decline. Ratcliffe (1967) first reported thinner eggshells of raptors sampled in 1967 compared with museum specimens predating the organochlorine pesticide era. Since this report, shell thinning has been correlated with p,p'-DDE residues or exposure and to a lesser degree exposure to other DDT metabolites, dieldrin, chloredecone, lindane, PCBs, mercury, and perhaps aluminum (Cooke 1973; Lundholm 1987; Blus 1996; Peakall 1996; Wiemeyer 1996).

The mechanism by which organochlorines thin eggshells in birds continues to be studied in detail (Lundholm 1987, 1997). In ducks, the rate of calcium translocation from the mucosal cells to the shell gland cavity is reduced by p,p'-DDE. In vivo p,p'-DDE administration and in vitro p,p'-DDE incubation studies revealed dose-dependent inhibition of calcium uptake and calcium-magnesium ATPase activity in homogenates, as well as subcellular fractions of the shell-gland mucosa. However, recent data suggest that p,p'-DDE interferes with the signal that triggers ion-activated ATPase, possibly by inhibiting prostaglandin synthetase and reducing the concentration of prostaglandin E_2 (Lundholm 1997). The mechanism by which some metals thin eggshells is less well-studied but may be related to the metals' affinity for calcium-binding sites on calmodulin, thereby altering calcium ATPase-mediated translocation of calcium (Lundholm and Mathason 1986).

Eggshell thickness may be determined directly, by averaging several measurements at the equator of a dried egg with a micrometer or indirectly, by using a thickness index (Thickness Index = weight of shell in mg/length of shell in mm × breadth of shell in mm; [Ratcliffe 1967]). Measurements may be compared to concurrent controls or historic museum specimens collected before the modern pesticide era. Eggshell breaking strength has been compared with thickness measurements and indices and appears to be a more sensitive and highly quantifiable measure of eggshell quality (Carlisle et al. 1986; Bennett et al. 1988). Moriarty et al. (1986) have hypothesized that p,p'-DDE affects the shape of the egg rather than the thickness, although this has not been rigorously studied.

Measurement of eggshell thickness is an accurate, precise, and sensitive indicator of exposure to some chlorinated hydrocarbons in raptorial and piscivorous species of birds (Blus 1996; Peakall 1996; Wiemeyer 1996). Other avian species, including ducks, gulls, and quail, are less sensitive or insensitive to eggshell-thinning effects of these compounds (Blus 1996; Peakall and Fairbrother 1998). Only limited data are available for reptiles, and these suggest that p,p'-DDE does not thin eggshells of American crocodiles (*Crocodylus acutus*) and alligators (Hall et al. 1979; Heinz et al. 1991).

Nutritional plane and various environmental stressors (e.g., elevated ambient temperature) can cause eggshell thinning in domestic fowl (Roland et al. 1973; Smith 1974). The stage of incubation in which the sample is collected is another potentially confounding factor, as thinning occurs naturally late in incubation when the embryo appears to draw upon the shell as a source of calcium. It is one of the rare biomarkers that has been used globally and can be linked directly to adverse population-level effects in many species of birds (Blus 1996; Peakall 1996; Wiemeyer 1996).

Reproductive success

The use of reproductive success as a measure of contaminant effect is valuable because it is a critical parameter of population dynamics (Kushlan 1993), although population size may be affected at any given time and in any given species by age of sexual maturity, life span, brood number and size, immigration, and emigration (Heinz 1998). Toxicants may directly alter reproduction by causing malformation or death of developing embryos, impair hatching, or adversely affect growth and survival of young, all of which constitute very sensitive life-stages (Hoffman et al. 1990; Fry 1995). Hatching or brooding success may be indirectly affected if toxic chemicals interfere with parental behavior, such as nest attentiveness or care of young (Fox et al. 1978; Fry 1995). Furthermore, contaminant effects on parental health, fertility, or fecundity may reduce reproductive performance as well (Kushlan 1993).

Several techniques have been developed that attempt to associate reproductive effects in free-ranging vertebrates with contaminant exposure. The most common method of evaluating potential effects of contaminants on reproduction is direct observation of hatching or fledging success in conjunction with contaminant analysis in eggs or organs. Comparisons may be made between sites sprayed with agrichemicals and untreated sites, or obviously polluted waste sites versus so-called "reference" sites. In some studies, the eggs or adult birds are "artificially" exposed to the contaminant of interest by the investigator (Grue and Shipley 1984; Fry et al. 1986). In all cases, the studies are conducted by marking nests and making frequent visits to monitor the fate of eggs and young. Sample size is sometimes enhanced by adding nest boxes if the species of interest is a cavity nester, and thus attracting more animals to the study site (Brewer and Hummel 1990). The nest-box technique has the advantage of low disturbance and predation associated with cavity-nesting species (Peakall 1992).

Less passive than nest marking is the "sample-egg technique," in which 1 random egg is removed from a nest for contaminant analysis. The sample egg is presumed to be representative of exposure of sibling eggs, and the remainder of the eggs are monitored for hatching and fledging success (e.g., Blus et al. 1974; Custer et al. 1983; Ohlendorf et al. 1986; Hothem et al. 1995; Blus et al. 1997). This technique has been used to compare the significance of contaminant effects on reproduction to other factors, such as predation. Specific reproductive parameters (e.g., eggshell thickness) or biomarkers of toxicant exposure (e.g., cytochrome P450) may be measured in the sample eggs. A combination of the nest-box and sample-egg techniques has been used to increase sample size and to add more control to sample-egg studies (White and Hoffman 1995).

A more manipulative approach used to study the effects of contaminants on avian reproduction is the "egg transfer" technique. Eggs are taken from a nesting colony at a polluted site and placed in a nest at a clean site, and vice versa (Wiemeyer et al. 1975; Kubiak et al. 1989). Subsequent observations are made of nest success and are accompanied by analysis of contaminants in eggs and tissues. This technique can be used to distinguish direct and indirect effects of contaminants on reproduction. For example, eggs from a polluted colony that fail to hatch when moved to a clean colony can be inferred to be directly affected by intrinsic contaminants. Eggs from a clean colony that fail to hatch, or to yield fledglings, when moved to a polluted colony may be affected indirectly via altered parental behavior. Artificial incubation of eggs from polluted and reference sites serves to further separate intrinsic from extrinsic effects of pollution on hatchability (Kubiak et al. 1989). In a variation on the egg transfer technique, young birds may be transplanted from a clean site to establish a colony at a site where pollution is suspected and then are allowed to mature and attempt to breed for a generation or more. Subsequent analysis of breeding success and contaminant residues may then be conducted (Blus et al. 1979).

Little information has been accumulated on reproductive effects of toxic chemicals in free-ranging terrestrial vertebrates other than birds. Despite extensive use of small mammals and mink (*Mustela vison*) in controlled toxicological laboratory studies (McBee and Bickham 1990; Talmage and Walton 1991; Wren 1991), remarkably few studies have examined reproductive endpoints (pregnancy rate, litter size, in utero losses, numbers of stillborn fetuses) in wild populations (Jackson 1952; Rowley et al. 1983; Thalken and Young 1983; Jett et al. 1986; Clark et al. 1996). In general, measurement of such endpoints is an ancillary component of small mammal (*Microtus* and *Peromyscus* spp.) demographic studies at contaminated and reference sites or at agricultural sites prior to and after pesticide applications.

Still fewer studies have been conducted on reproductive effects of contaminants in free-ranging reptiles and amphibians (Hall and Henry 1992). However, a decline in a population of American alligators at Lake Apopka, Florida, was linked to the endocrine-disrupting properties of certain environmental contaminants. Although the specific toxicants have yet to be unequivocally identified, the population decline was associated with gonadal aberrations, depressed androgen levels, and increased estrogen levels and occurred following the release of dicofol and sulfuric acid from a chemical plant on the lake (Guillette et al. 1994; Matter et al. 1998). Recent alarm of a worldwide decline in amphibian populations (Borchelt 1990) hopefully will spur research on potential reproductive effects of contaminants in these species.

All of these techniques face difficulties. Some parameters of reproductive success in free-ranging terrestrial vertebrates may be affected not only by a wide array of individual contaminants but also by combinations of compounds and even complex mixtures that may act additively or synergistically (Hoffman et al. 1990). Thus, in terms of causality, reproductive effects may have a low degree of contaminant specificity. Results also may be confounded by non-contaminant factors such as predation, changes in food supply or disturbance of nesting adults (Peakall 1992; Kushlan 1993). Reproductive success itself can be difficult to measure because short-term variability is to be expected. Furthermore, it may be several months or longer before reproductive effects of contaminants are manifest, which makes field studies costly and time-consuming. Finally, there appears to be considerable variability among terrestrial vertebrate species in terms of sensitivity to reproductive toxicants.

Survival and population-level effects

Perhaps the most insidious action of some contaminants is their potential to affect survival of a species. For acutely toxic compounds, such as anticholinesterase pesticides and petroleum crude oil, large numbers of dead and dying individuals may be found following a misapplication or an accidental spill. Under such circumstances, the linkage to exposure either is readily apparent or can be established through forensic investigation. However, for toxicants that gradually evoke toxicity, the protracted time course of effects may make the outcome on population dynamics, including survival, more difficult to link to contaminant exposure than to other biotic (e.g., disease, predation) and abiotic factors (e.g., weather) that can directly, indirectly, or interactively affect survival. Thus, this category of endpoints can reflect direct lethal effects on individuals or the integrated effects of a toxicant on fertility and fecundity.

Methodologies for estimating population abundance and survival are complex and have been reviewed in detail (Lancia et al. 1994; Johnson 1994). Indices including such diverse techniques as visual sightings, audible sound or song, scat counts in a plot, and individuals caught per trap night are believed to be correlated with abundance, although they are not rigorous measures of this parameter. When all individuals can be observed, more formal population-estimation methods can quantify abundance and density of wildlife through complete census (e.g., aerial photography) and census of individuals on sample plots (Lancia et al. 1994). In practice, all individuals in a population usually cannot be observed, but estimates of abundance and density can be obtained through various count methods (e.g., line transects) and capture methods (e.g., mark-recapture, capture removal) that also can provide survival estimates (Lancia et al. 1994). In addition, numerous survival models encompassing unimpeded and density-dependent growth, birth, and death rates (life tables) and interacting species models have been developed for estimating survival rates of terrestrial vertebrates, although their application in ecotoxicology has been somewhat limited.

Possibly the most rigorous technology for the study of pesticide effects on survival of free-ranging birds involves the use of radiotransmitters to monitor individuals at various intervals to verify survival. Survival rates have been estimated in birds that are equipped with radiotransmitters and exposed naturally to organophosphorus pesticides in sprayed fields. Studies of adult radio-tagged sage grouse (*Centrocercus urophasianus*) (Blus et al. 1989) and mature northern bobwhite (*Colinus virginianus*) (White et al. 1990) demonstrated that survival probability was related to anticholinesterase pesticide exposure, although one such study in mallard hens (Brewer et al. 1988) was somewhat complicated by a high incidence of predation. Survival rates of post-fledging, radio-tagged starlings (*Sturnus vulgaris*) have been examined in fields sprayed with methyl parathion and were found to be significantly lower than these of starlings in control fields sprayed with xylene-water (Whitten et al. 1998). Other studies involving free-ranging, radio-tagged northern bobwhites, eastern cottontails (*Sylvilagus floridanus*), and great horned owls (*Bubo virginianus*) documented little (i.e., blood ChE inhibition) or no effect of organophosphorus pesticides, with no measurable effect on survival (Tank et al. 1993; Buck et al. 1996). An alternate approach entailed monitoring the survival of trapped bobwhite and mourning doves (*Zenaida macroura*) that were fitted with radiotransmitters and gavaged with graded doses of an organophosphorus or carbamate pesticide (Buerger et al. 1991; Brewer et al. 1996; Hawkes et al. 1996). A related methodology utilized dosed starlings marked with highly conspicuous patagial tags (Stromborg et al. 1988). Because exposure can be well-controlled in this paradigm, sublethal pesticide effects on physiological function, behavior, and susceptibility to predation can be investigated to determine overall consequences of exposure on survival. All such studies require substantial sample size and are labor-intensive and costly.

Numerous investigations have quantified hatching and fledging success of marked bird nests, and survival estimates (daily survival rate) are often derived statistically from such reproductive studies in passerines and wading birds (e.g., Patnode and White 1991; Fluetsch and Sparling 1994; Hothem et al. 1995). As described previously, contaminant exposure may be documented in a sample egg or by using a spray deposit card situated near a nest. Again, the drawbacks of such studies include the need for a large sample size and their labor-intensive nature.

A number of small mammal studies have been conducted in pesticide-treated fields using mark-recapture techniques, but dramatic effects on survival are rarely documented (Robel et al. 1972; Jett et al. 1986; Clark et al. 1996). In 1 study, survivorship of voles (*Microtus pennsylvanicus*), based upon age estimates derived from lens-weight growth curves, was reduced dramatically in individuals trapped at the highly polluted Love Canal area of Niagara Falls, New York (Rowley et al. 1983).

Survival differences within wild populations may arise as a result of the natural resistance of certain individuals to a toxicant. Early research suggested the development of inheritable resistance to endrin in pine voles (*Microtus pinetorum*) (Webb and Horsfall 1967). The development of resistance to a pesticide is a form of "artificial selection" for specific genotypes. Pesticides may therefore reduce genetic variation in wild populations, thus decreasing long-term viability in the face of environmental change (Gillespie and Guttman 1988). However, contaminant-related alteration of the genetic makeup of wild populations has received little attention in terrestrial vertebrates. In 1 of the few studies performed, Graf et al. (1976) found no difference in genetic variation at the esterase-2 and esterase-5 loci in house mice (*Mus musculus*) populations between fields sprayed with the carbamate insecticide Sevin and control fields. However, the investigators pointed out that an examination of 2 loci in a single species, and only 1 toxicant, should not rule out the possibility of the effects of contaminants on population genetics.

Impaired survival can be documented less formally by searches for dead or moribund individuals using casual or systematic methods following a spray event or spill (Greig-Smith 1994; Ford et al. 1996). Such mortality-incident information can reveal hazards of an agrichemical not identified during registration, particularly under a wide range of environmental conditions (Greig-Smith 1994). National reporting systems have been established in a few countries including the United Kingdom, France, and The Netherlands, and a fledgling program exists in the U.S. (i.e., Ecological Incident Information System of the U.S. Environmental Protection Agency). This approach provides carcasses that can be necropsied and chemically analyzed to determine cause of death. However, there are several disadvantages, including

- occurrence of death outside the search area,
- rapid scavenging of a carcass,
- overestimation or underestimation of carcass search efficiency,
- limited opportunity to obtain control data, and
- weak power to demonstrate absence of effects (Balcomb 1986; Edwards 1990; Greig-Smith 1994).

Casualty data are occasionally misused when massive mortality is projected on the basis of observations of an inadequate number of small plots or transects.

Anecdotal observations that are suggestive of contaminant-related declines in abundance of a species given by hunters and trappers, the lay public, and even the media may have only modest reliability but often serve as the impetus for investigations that document ecotoxicological catastrophes (Carson 1962). Clearly, such observations by amateur ornithologists helped to elucidate the widespread nature and severity of organochlorine pesticide ecotoxicity, although in other instances, local changes in habitat, food abundance, disease, and predation may be the cause of such observations.

Contaminant Effects on Communities and Ecosystems

Although many scientists have examined contaminant exposure and biomagnification in ecosystems, very little research has been conducted on community- and ecosystem-level responses to contaminants in terrestrial vertebrates. This deficiency in the state of our knowledge presumably reflects the difficulty in conducting replicated trials over vast spatial scales as well as the associated financial challenges of such investigations. Oftentimes, community and ecosystem investigations are initiated following a real or perceived ecological disaster. Such post-hoc studies may lack the robust statistical nature of a well-designed, purposefully-manipulated, controlled study. Finally, measures of community and ecosystem effects integrate a myriad of contaminant and non-contaminant factors, which are often far removed from the mechanistic action of a toxicant (Figure 3-1).

Communities

Avian and small mammal assemblages have been examined following controlled pesticide applications. Research on species richness of bird communities before and after insecticides or herbicide application has often documented little or no effect (Cooper and Whitmore 1988; Freedman et al. 1988; Hanowski et al. 1997). Small mammals were eliminated when their plant or arthropod prey base was removed by herbicide application (Tew et al. 1992). In other instances, populations and communities remained unchanged (Clark et al. 1996) or even increased when the herbicide treatment enhanced certain plant cover or invertebrate prey (reviewed by Clark et al. 1996).

Ecosystem-wide contamination events such as acid precipitation and oil spills have provided opportunities to study community-level effects. Elmberg et al. (1994) compared species richness and diversity of waterfowl among lakes subject to varying levels of acidity in precipitation. The only relationship they were able to show was a lower relative abundance of fish-eating species in lakes with lower pH. Furthermore, adverse effects of the *Exxon Valdez* oil spill on species diversity of marine bird communities were demonstrated for certain foraging guilds (Wiens et al. 1996). Studies of rodent assemblages in petroleum-waste-disturbed, tall-grass prairie ecosystems documented the prominence of house mice and cotton rats (*Sigmodon hispidus*) on contaminated sites compared with reference areas, but also they revealed that such alterations were somewhat inconsistent among the 12 waste sites examined (see Chapter 13).

Ecosystems

With the exception of well-documented "descriptions" of the ecosystem-level effects of DDT and its metabolites (Carson 1962), relatively few "investigations" have been conducted on the ecosystem scale. Some studies examined the effects of pesticides on small mammal population sizes, reproductive success, and community composition in semi-enclosed grasslands, in conjunction with studies of plant primary productivity and arthropod abundance (Barrett and Darnell 1967; Barrett 1968). However, most ecosystem-level contaminant studies center around incidents of extreme damage to the environment. A detailed investigation along a spatial gradient starting at a historic zinc-smelting site in Palmerton, Pennsylvania, documented heavy metal exposure and damage to soil; vegetation; and invertebrate, amphibian, avian, and mammalian communities (Beyer and Storm 1995). Another large-scale ecosystem investigation has documented the adverse effects of metalloids leached by irrigation of agricultural fields in the Central Valley of California. Selenium in this irrigation drain water has been shown to bioaccumulate in the food chain and to exceed toxic thresholds in birds (particularly embryos) and possibly other classes of vertebrates (Ohlendorf and Hothem 1995). Acid deposition also has been the focus of study for ecosystem-scale impacts of contaminants. Acidification of wetlands and forests has been shown to adversely affect the health of aquatic macrophytes and terrestrial vegetation and to reduce reproductive success and species richness in aquatic invertebrates, plankton, amphibians, and terrestrial birds (Sparling 1995). Oil spills are another highly conspicuous ecosystem-wide contaminant event with well-studied effects. The *Exxon Valdez* oil spill caused significant damage to communities of intertidal invertebrates, algae, fish, birds and marine mammals (Spies et al. 1996).

At and above the population level, our ability to separate contaminant-induced effects from those related to extensive habitat alterations are weak at best. Nevertheless, the recovery of community structure and ecosystem functioning following toxicological disasters (e.g., "the great natural experiment" that followed the ban of DDT; Stickel 1981) can teach us much about resilience and its limits in biological systems.

Toxicant Criteria

In the U.S., numerous statutes provide contaminant-exposure and concentration thresholds for the protection of human health, but there are only limited criteria in statutes for the protection of wild terrestrial vertebrates. Although the Toxic Substances Control Act has "environmental" mandates, historically it has failed to focus upon wildlife. The Department of the Interior has statutory authorities under the Migratory Bird Treaty Act, the Endangered Species Act, and the Fish and Wildlife Coordination Act that address "take" and "habitat destruction," yet these acts do not have mandates that set standards for concentrations of pesticides, metals, and other toxicants in environmental matrices for the protection of terrestrial vertebrates. However, several noteworthy acts and regulations (e.g., Federal Insecticide, Fungicide, and Rodenticide Act; Natural Resource Damage Assessment Regulations; Great Lakes Water Quality Protection Act) have been promulgated to set standards minimizing the hazard of toxic substances to wildlife and to assess damages for injury to wildlife. Aside from the aforementioned mandates, there are many "unofficial" scientific thresholds, critical concentrations and diagnostic guidelines used by natural resources managers that are indicative of exposure and adverse effects in wild terrestrial vertebrates.

Federal Insecticide, Fungicide, and Rodenticide Act

Enacted in 1947 and amended in 1988, the Federal Insecticide, Fungicide, and Rodenticide Act (FIFRA) provides federal control of pesticide distribution, sale, and use in the U.S. The U.S. Environmental Protection Agency (USEPA) was given authority to oversee the registration of pesticides. At present, toxicity-testing data for the mallard and bobwhite are required to generate hazard information to determine "unreasonable-adverse-effect" levels. Using the quotient method, the hazard of a pesticide to wild birds and mammals in an unrestricted-use situation is presumed to be acceptable if the estimated environmental concentration (EEC) of a pesticide is less than one-fifth of the LC50, and if the EEC is less than the chronic no-effect level (USEPA 1994). For granular products, the amount of toxicant per unit area is deemed by USEPA to be a more appropriate measure, and the acceptable hazard level is considered to be less than one-fifth of the LD50 per square foot. If it becomes apparent that a registered pesticide poses an imminent hazard to the environment, its registration may be suspended immediately.

Natural Resource Damage Assessment

The U.S. Department of the Interior and the Department of Commerce have set forth standardized and cost-effective procedures for assessing damages related to the loss of beneficial use of biota and other natural resources (e.g., air, water, soil, sediment) resulting from the release of hazardous substances or the discharge of oil into the environment (USDOI 1986 with numerous portions amended thereafter; USDOC 1996). These actions are separate from the response and cleanup activities following a spill. For small releases or spills (damage claims < $100,000) in coastal and marine environments and the Great Lakes, standardized methodologies (Type A procedures) that require minimal field work and a computer model may be used to perform damage assessments. For larger releases and spills (Type B procedures), the injury determination phase is expanded to include 1) determination of toxicant concentrations in environmental matrices (surface and ground water, sediments, geologic resources, and air) and 2) determination of whether or not the concentrations of the toxicants

- are sufficient to cause injury to biological resources (i.e., scientifically rigorous determinations linking toxicant exposure to disease, behavioral abnormalities, cancer, genetic mutations, physiological malfunction, physical deformities, or death);
- exceed actions or tolerance levels in edible portions of the organism as established by the Food, Drug and Cosmetic Act; or

- exceed levels established by a state health agency that has issued directives to limit or ban the consumption of an organism.

Notably, several biological injury endpoints in the regulations (e.g., brain AChE activity, physical deformation, histopathological lesions, eggshell thinning, altered avian reproduction, and death) that are to be linked to population-level effects are quantified for assessing damages and restoration options. The most important portion of that legislation, however, is the use of these data to determine appropriate restoration, either to be carried out or paid for by the responsible party.

For oil spills, damage assessment may be initiated coincident with cleanup efforts and is focused on determination of appropriate restoration for observed losses (injury). For other toxicant releases, damage assessment may sometimes be initiated after remedial action has been completed but more often is done in conjunction with remedial work in order for the government to reach a universal settlement with the responsible party for both remediation and restoration costs. On a case-by-case basis, concentrations of toxicants in an environmental matrix may be established as minimally hazardous to terrestrial vertebrates and used as part of cleanup or remedial action. If injury to natural resources persists after these actions have been taken, assessment procedures may be used to pursue damages from the responsible party for use in restoration.

Great Lakes water-quality criteria

Recently, the USEPA published the Final Water Quality Guidance documents for the Great Lakes System (USEPA 1995), which protect wildlife. The guidance sets forth a bioaccumulation methodology for dissolved concentrations of metals and organic pollutants that might adversely affect water or aquatic organisms consumed by wildlife in the Great Lakes. The USEPA, in conjunction with the U.S. Fish and Wildlife Service, defined criteria by which acceptable levels of dissolved contaminants in Great Lakes water could be assessed. These criteria were used to set "Criteria Maximum Concentrations" (CMC) for the protection of wildlife species.

The USEPA used a 2-tiered process to determine ambient concentrations of 4 chemicals (DDT and its metabolites, mercury including methylmercury, PCBs, and 2,3,7,8-tetrachlorodibenzo-*p*-dioxin) which, if exceeded, could result in detrimental effects to terrestrial wildlife (Table 3-1). The 2 tiers were essentially equations by which CMCs could be calculated from available toxicity data. Tier 1 methodologies were utilized when laboratory toxicity data for avian and mammalian species were extensive and readily available. Tier 2 methodologies were used when less-complete data were available. Tier 1 methodologies focused on impacts of contaminants on reproduction and wildlife populations rather than impacts on individual health. The CMCs for wildlife were based on data for river otter (*Lontra canadensis*), mink, belted kingfisher (*Ceryle alcyon*), osprey and bald eagle (USEPA 1993), species that are considered to be representative of consumers likely to receive significant

Table 3-1 USEPA guidelines for maximum concentrations in water of 4 chemicals which, if exceeded, could cause detrimental reproductive and population effects in terrestrial wildlife in the Great Lakes System[1,2]

Contaminant	Dissolved concentration
DDT	1.10×10^{-5} µg/L
Polychlorinated biphenyls	7.40×10^{-5} µg/L
2,3,7,8-terachlorodibenzo-p-dioxin	3.10×10^{-9} µg/L
Mercury (including methylmercury)	1.30×10^{-3} µg/L

[1] Based on data for mink, otter, bald eagle, osprey, and kingfisher.
[2] Source: USEPA 1995

exposure to contaminants because of their feeding habits and trophic-level positions.

The USEPA guidelines attempted to relate CMCs to "no-observed-effect levels" in laboratory studies. Because the 5 representative species are not commonly used in laboratory studies, CMCs were calculated from toxicity data obtained from experiments with other species. Several uncertainty factors were used to extrapolate CMCs for the 5 representative wildlife species from laboratory studies on other species:

- an interspecies uncertainty factor,
- an uncertainty factor for extrapolating chronic or sub-chronic effects from data on acute effects,
- an uncertainty factor for extrapolating no-observed-effect levels when only data for "low-observed-effect levels" were available, and
- an uncertainty factor for intraspecies variability in susceptibility to harmful effects of chemicals.

When Tier 2 methodologies were necessary, a further uncertainty factor was included to account for interspecies variability in susceptibility. The average weight of the wildlife species, its average daily water and food consumption, and a bioaccumulation factor based on the trophic level of the species were also considered when calculating the CMCs.

The guidelines contained an admission that criteria to protect wildlife from the synergistic effects of pollutants in a mixture were lacking because of the absence of sufficient laboratory whole-effluent-toxicity data. The guidelines recommended considering the effects of pollutants in a mixture to be additive in lieu of whole effluent toxicity data (USEPA 1993).

Thresholds, critical concentrations, and diagnostic guidelines

The hazards to terrestrial vertebrates posed by most environmental contaminants in abiotic matrices (water, sediment, soil, air) have not been rigorously studied, with the exception of some organochlorine and anticholinesterase pesticides, PCBs, dioxins, lead, mercury, and cyanide. The reader is referred to summary reviews and information published in the Contaminant Hazard Review Series of the Department of the Interior (Eisler 1998) that addresses more than 30 contaminants commonly encountered by wild terrestrial vertebrates and to a compilation of state, federal, and international guidelines for evaluating soil contamination (Beyer 1990). Many other criteria documents of broad scope, often focused toward human health and domestic animals, have been published by the USEPA, Agency for Toxic Substances and Disease Registry, and the National Academy of Science. Risk assessments addressing ingestion and dermal exposure to terrestrial wildlife are often conducted using such literature, although so little is known about the significance of respiratory exposure of wild terrestrial vertebrates that this exposure route is often ignored (Newman and Schreiber 1988; Driver et al. 1991).

The practical question "How much of a chemical must be in the tissues of a wild animal to cause harm?" (Beyer et al. 1996), or to document "significant exposure" has been the topic of hundreds of scientific investigations and is reviewed in 2 recently published texts (Beyer et al. 1996; Fairbrother et al. 1996). In the U.S., there are about 87,000 substances and compounds in commerce whose registration is covered by the Toxic Substances Control Act (TSCA), FIFRA, Food Quality Protection Act, and the Food and Drug Act (USEPA 1998b), and, of these, there are inadequate basic toxicological data for 93% of the 3000 high-production-volume chemicals (> 1 million pounds produced or imported into the U.S. annually) routinely used by man in everyday life (Environmental Defense Fund 1997). Thus, it is not surprising that exposure and adverse effect thresholds, critical concentrations and diagnostic guidelines for terrestrial vertebrates (principally birds) have been well-established for remarkably few contaminants (Table 3-2).

Conclusions and Recommendations

Ecotoxicology, as introduced by Truhaut in 1969, is often defined as the study of harmful effects of chemicals on ecosystems. To many, the ultimate concern in ecotoxicology is the effects of contaminants at and above the population level. Unfortunately, as recently pointed out by Walker (1998), current risk assessment practices do not adequately address this problem, but development of new biomarker or measure-of-effect strategies could help satisfy this deficiency. Future directions might include

- use of combinations or suites of endpoints coupled with residue determinations to provide more meaningful measures of exposure and effect;

Table 3-2 Principal contaminants for which critical concentrations and diagnostic guidelines have been established for wild terrestrial vertebrates

Contaminants	
Organochlorine pesticides	Polychlorinated biphenyls
Chlordecone	Total Aroclors
p,p'-DDE	Coplanar congeners
Dicofol	Dioxins
Dieldrin	Dibenzofurans
Endrin	Petroleum hydrocarbons
Heptachlor epoxide	Crude oils
Hexachlorobenzene	Oil fractions
Hexachlorocyclohexane	Metals, metalloids and trace elements
Methoxychlor	Cadmium
Mirex	Fluoride
Oxychlordane	Lead
Toxaphene	Mercury
Organophosphorus and carbamate pesticides	Selenium

- widespread application of biochemical reactivation techniques and use of modulators for assays of AChE, delta-aminolevulinic acid dehydratase, and cytochrome P450 to minimize the need for concurrent controls;
- further development of sublethal, nondestructive and noninvasive biomarkers to facilitate monitoring of exposure and higher-order effects;
- continued field application and validation of promising behavioral, genetic, oxidative stress, and immunological measures of exposure and effect;
- initiation of much-needed large-scale contaminant exposure and effect studies that evaluate recruitment and survival of free-ranging wildlife; and
- extrapolation and validation of existing contaminant thresholds, critical concentrations and diagnostic guidelines for aquatic forms, domestic animals, and humans for use in wild terrestrial vertebrates.

Ecotoxicologists have been able to quantify and identify the mechanisms of some deleterious contaminant effects on wildlife. However, the science has not advanced to a stage that permits prediction of adverse effects of contaminants on populations so that regulatory agencies can protect populations at risk before impacts occur. As a result, many of our existing natural resource statutes impose penalties for take or damage to individual organisms, but do not deal with injury to populations or higher levels of biological organization. Our challenge for the future is the interpretation and extrapolation of responses to contaminants at low levels of biological organization to actions that protect populations, communities and ecosystems from harm.

References

Adams SM, Shepard KL, Greely Jr MS, Ryon MG, Jimenez BD, Shugart LR, McCarthy JF, Hinton DE. 1989. The use of bioindicators for assessing pollutant stress in fish. *Mar Environ Res* 28:459–464.

Balcomb R. 1986. Songbird carcasses disappear rapidly from agricultural fields. *Auk* 103:817–820.

Ballantyne B, Marrs TC. 1992. Clinical and experimental toxicology of organophosphates and carbamates. London: Butterworth. 641 p.

Bantle JA, Dumont JN, Finch RA, Linder G. 1991. Atlas of abnormalities: A guide for the performance of FETAX. Stillwater OK: Oklahoma State Publications Department. 68 p.

Barrett GW, Darnell RM. 1967. Effects of dimethoate on small mammal populations. *Am Midl Nat* 77:164–175.

Barrett GW. 1968. The effects of an acute insecticide stress on a semi-enclosed grassland ecosystem. *Ecol* 49:1019–1035.

Bennett JK, Ringer RK, Bennett RS, Williams BA, Humphrey, PE. 1988. Comparison of breaking strength and shell thickness as evaluators of avian eggshell quality. *Environ Toxicol Chem* 7:351–357.

Beyer WN. 1990. Evaluating soil contamination. Washington DC: U.S. Fish and Wildlife Service. *Biological Report* 90(2). 25 p.

Beyer WN, Storm G. 1995. Ecotoxicological damage from zinc smelting at Palmerton, Pennsylvania. In: Hoffman DJ, Rattner BA, Burton Jr AG, Cairns Jr J, editors. Handbook of ecotoxicology. Boca Raton FL: Lewis. p. 596–608.

Beyer WN, Heinz GH, Redmon-Norwood AW. 1996. Environmental contaminants in wildlife: Interpreting tissue concentrations. SETAC Special Publication Series. Boca Raton FL: Lewis. 494 p.

Bickham JW. 1994. Genotoxic responses in blood detected by cytogenetic and cytometric assays. In: Fossi MC, Leonzio C, editors. Nondestructive biomarkers in vertebrates. Boca Raton FL: Lewis. p. 147–157.

Blus LJ, Neely Jr BS, Belisle AA, Prouty RN. 1974. Organochlorine residues in brown pelican eggs: Relation to reproductive success. *Environ Pollut* 7:81–91.

Blus L, Cromartie E, McNease L, Joanen T. 1979. Brown pelican: Population status, reproductive success, and organochlorine residues in Louisiana, 1971–1976. *Bull Environ Contam Toxicol* 22:128–135.

Blus LJ. 1996. DDT, DDD, and DDE in birds. In: Beyer WN, Heinz GH, Redmon-Norwood AW, editors. Environmental contaminants in wildlife: Interpreting tissue concentration. SETAC Special Publication Series. Boca Raton FL: Lewis. p 49–71.

Blus LJ, Rattner BA, Melancon MJ, Henny CJ. 1997. Reproduction of black-crowned night-herons related to predation and contaminants in Oregon and Washington, USA. *Colonial Waterbirds* 20:185–197.

Blus LJ, Staley CS, Henny CJ, Pendleton GW, Craig TH, Craig EH, Halford DK. 1989. Effects of organophosphorus insecticides on sage grouse in southeastern Idaho. *J Wildl Manage* 53:1139–1146.

Borchelt R. 1990. Frogs, toads, and other amphibians in distress. *National Research Council News Report* 40:2–5.

Brewer LW, Driver CJ, Kendall RJ, Zenier C, Lacher Jr TE. 1988. Effects of methyl parathion in ducks and duck broods. *Environ Toxicol Chem* 7:375–379.

Brewer LW, Hummel RA. 1990. What is measurable in wildlife toxicology: Field assessment. In: Kendall RJ, Lacher Jr TE, editors. Wildlife toxicology and population modeling. SETAC Special Publication Series. Boca Raton FL: Lewis. p. 69–76.

Brewer RA, Carlock LL, Hooper MJ, Brewer LW, Cobb III GP, Kendall RJ. 1996. Toxicity, survivability, and activity patterns of northern bobwhite quail dosed with the insecticide turbufos. *Environ Toxicol Chem* 15:750–753.

Buck JA, Brewer LW, Hooper MJ, Cobb GP, Kendall RJ. 1996. Monitoring great horned owls for pesticide exposure in southcentral Iowa. *J Wildl Manage* 60: 321–331.

Buerger TT, Kendall RJ, Mueller BS, DeVos T, Williams BA. 1991. Effects of methyl parathion on northern bobwhite survivability. *Environ Toxicol Chem* 10:527–532.

Carlisle JC, Lamb DW, Toll PA. 1986. Breaking strength: An alternative indicator of toxic effects on avian eggshell quality. *Environ Toxicol Chem* 5:887–889.

Carson RL. 1962. Silent spring. Boston MA: Houghton Mifflin. 368 p.

Clark Jr DR, Moulton CA, Hines JE, Hoffman DJ. 1996. Small mammal populations in Maryland meadows during four years of herbicide (Brominal®) applications. *Environ Toxicol Chem* 15:1544–1550.

Cooke AS. 1973. Shell thinning in avian eggs by environmental pollutants. *Environ Pollut* 4:85–152.

Cooper RJ, Whitmore RC. 1988. Effect of dimilin application on a Central Appalachian forest bird community. *Trans Northeast Sect Wildl Soc* 45:60.

Custer TW, Hensler GL, Kaiser TE. 1983. Clutch size, reproductive success, and organochlorine contaminants in Atlantic Coast black-crowned night-herons. *Auk* 100: 699–710.

Dieter MP. 1974. Plasma enzyme activities in Coturnix quail fed graded doses of DDE, polychlorinated biphenyl, malathion and mercuric chloride. *Toxicol Appl Pharmacol* 27:86–98.

Dieter MP, Perry MC, Mulhern BM. 1976. Lead and PCB's in canvasback ducks: Relationship between enzyme levels and residues in blood. *Arch Environ Contam Toxicol* 5:1–13.

Driver CJ, Ligotke MW, Van Voris P, McVeety BD, Drown DB. 1991. Routes of uptake and their relative contribution to the toxicological response of northern bobwhite (*Colinus virginianus*) to an organophosphate pesticide. *Environ Toxicol Chem* 10:21–33.

Edwards PJ. 1990. Assessment of survival methods used in wildlife trials. In: Somerville L, Walker CH, editors. Pesticide effects on terrestrial wildlife. New York: Taylor and Francis. p. 129–142.

Eisler R. 1998. Contaminant hazard reviews. [CD-ROM] Washington DC: U.S. Geological Survey.

Elmberg J, Sjoeberg K, Nummi P, Poeysae H. 1994. Patterns of lake acidity and waterfowl communities. *Hydrobiologia* 279–280:201–206.

[EDF] Environmental Defense Fund. 1997. Toxic ignorance: The continuing absence of basic health testing data for top-selling chemicals in the U.S. New York: EDF. 65 p.

Fairbrother A, Marden BT, Bennett JK, Hooper MJ. 1991. Methods used in determination of cholinesterase activity. In: Mineau P, editor. Cholinesterase-inhibiting insecticides: Their impact on wildlife and the environment. Vol 2 Chemicals in agriculture. New York: Elsevier. p. 35–71.

Fairbrother A, Locke LN, Hoff GL. 1996. Noninfectious diseases of wildlife. 2nd Edition. Ames Iowa: Iowa State University Press. 219 p.

Freedman B, Poirier AM, Morash R, Scott F. 1988. Effects of the herbicide 2,4,5-T on habitat and abundance of breeding birds and small mammals of a conifer clearcut in Nova Scotia. *Can Field-Nat* 102:6–11.

Fluetsch KM, Sparling DW. 1994. Avian nesting success and diversity in conventionally and organically managed apple orchards. *Environ Toxicol Chem* 13:1651–1659.

Ford RG, Bonnell ML, Varoujean DH, Page GW, Carter HR, Sharp BE, Heinemann D, Casey JL. 1996. Total direct mortality of seabirds from the Exxon Valdez oil spill. *Amer Fisheries Soc Sym* 18:684–711.

Fossi MC, Leonzio C. 1994. Nondestructive biomarkers in vertebrates. Boca Raton FL: Lewis. 345 p.

Fossi MC, Leonzio C, Peakall DB. 1994. The use of nondestructive biomarkers in the hazard assessments of vertebrate populations. In: Fossi MC, Leonzio C, editors. Nondestructive biomarkers in vertebrates. Boca Raton FL: Lewis. p 3–34.

Fox GA, Gilman AP, Peakall DB, Anderka FW. 1978. Behavioural abnormalities of nesting Lake Ontario herring gulls. *J Wildl Manage* 42:477–483.

Fox GA, Kennedy SW, Norstrom RJ, Wigfield DC. 1988. Porphyria in herring gulls: A biochemical response to chemical contamination of Great Lakes food chains. *Environ Toxicol Chem* 7:831–839.

Fox GA. 1992. Epidemiological and pathobiological evidence of contaminant-induced alterations in sexual development in free-living wildlife. In: Colborn T, Clement C, editors. Chemically-induced alterations in sexual and functional development: The wildlife/human connection. New Jersey: Princeton Scientific Publishing Co. p. 147–158.

Friend M. 1987. Lead poisoning. In: Friend M, editor. Field guide to wildlife diseases. Washington DC: U.S. Fish and Wildlife Service, Resource Publication 167. 225 p.

Fry DM, Swenson J, Addiego LA, Grau CR, Kang A. 1986. Reduced reproduction of wedge-tailed shearwaters exposed to weathered Santa Barbara crude oil. *Arch Environ Contam Toxicol* 15:453–463.

Fry DM. 1995. Reproductive effects in birds exposed to pesticides and industrial chemicals. *Environ Health Perspect* 103:165–171.

Gilbertson M, Kubiak T, Luwig J, Fox G. 1991. Great Lakes embryo mortality, edema, and deformities syndrome (GLEMEDS) in colonial fish-eating birds: Similarity to chick edema disease. *J Toxicol Environ Health* 33:455–520.

Gillespie RB, Guttman SI. 1988. Relationships between genetic structure of electrophoretically-determined allozymes in fish populations and exposure to contaminants. *Ohio J Sci* 88:46.

Golden NH, Pearson JL, Ottinger MA, Rattner BA, Erwin RM. 1998. Biological and ecotoxicological characteristics of terrestrial vertebrates residing in estuaries. Society of Environmental Toxicology and Chemistry (SETAC) 19th Annual Meeting; 15–19 Nov 1998; PWA 196. Charlotte NC: SETAC.

Goldstein MI, Lacher TE, Jr, Woodbridge B, Bechard MJ, Canavelli SB, Zaccagnini ME, Cobb GP, Tribolet R, Hooper MJ. 1999. Monocrotophos-induced mass mortality of Swainson's hawks in Argentina, 1995–1996. *Ecotoxicol* 8:201–214.

Graf G, Guttman S, Barrett G. 1976. The effects of an insecticide stress on genetic composition and population dynamics of a feral population of *Mus musculus*. *Comp Biochem Physiol* 55C:103–110.

Grandjean P, Nielsen T. 1979. Organolead compounds: Environmental health aspects. *Residue Rev* 72:97–148.

Grasman KA, Fox GA, Scanlon PF, Ludwig JP. 1996. Organochlorine-associated immunosuppression in prefledgling Caspian terns and herring gulls from the Great Lakes: An ecoepidemiological study. *Environ Health Perspect* 104:829–842.

Greig-Smith PW. 1991. Use of cholinesterase measurements in surveillance of wildlife poisoning in farmland. In: Mineau P, editor. Cholinesterase-inhibiting insecticides: Their impact on wildlife and the environment. Vol 2 Chemicals in agriculture. New York: Elsevier. p 128–150.

Greig-Smith PW. 1994. Understanding the impact of pesticides on wild birds by monitoring incidents of poisoning. In Kendall RJ, Lacher Jr TE, editors. Wildlife toxicology and population modeling: Integrated studies of agroecosystems. Chelsea MI: Lewis. p. 310–319.

Grue CE, Fleming WJ, Busby DG, Hill EF. 1983. Assessing hazards of organophosphorus insecticides to wildlife. *Trans N Amer Wildl Nat Resour Conf* 48:200–220.

Grue CE, Gibert PL, Seeley ME. 1997. Neurophysiological and behavior changes on non-target wildlife exposed to organophosphorus and carbamate pesticides: Thermoregulation, food consumption, and reproduction. *Amer Zool* 37:369–388.

Grue CE, Hart ADM, Mineau P. 1991. Biological consequences of depressed brain cholinesterase activity in wildlife. In: Mineau P, editor. Cholinesterase-inhibiting insecticides: Their impact on wildlife and the environment. Vol 2 Chemicals in agriculture. New York: Elsevier. p. 152–210.

Grue CE, Shipley BK. 1984. Sensitivity of nestling and adult starlings to dicrotophos, an organophosphate pesticide. *Environ Res* 35:454–465.

Guillette Jr LJ, Gross TS, Masson GR, Matter JM, Percival HF, Woodward AR. 1994. Developmental abnormalities of the gonad and abnormal sex hormone concentrations in juvenile alligators from contaminated and control lakes in Florida. *Environ Health Perspect* 102:680–688.

Guillette Jr LJ, Pickford DB, Crain DA, Rooney AA, Percival HF. 1996. Reduction in penis size and plasma testosterone concentrations in juvenile alligators living in a contaminated environment. *Gen Comp Endocrin* 101:32–42.

Hall RJ, Henry PFP. 1992. Assessing effects of pesticides on amphibians and reptiles: Status and needs. *Herpetol J* 2:65–71.

Hall RJ, Kaiser TE, Robertson WB Jr, Patty PC. 1979. Organochlorine residues in eggs of the endangered American crocodile (*Crocodylus acutus*). *Bull Environ Contam Toxicol* 23:87–90.

Hanowski JM, Niemi GJ, Lima AR, Regal RR. 1997. Response of breeding birds to mosquito control treatments of wetlands. *Wetlands* 17:485–492.

Hawkes AW, Brewer LW, Hobson JF, Hooper MJ, Kendall RJ. 1996. Survival and cover-seeking response of northern bobwhites and mourning doves dosed with aldicarb. *Environ Toxicol Chem* 15:1538–1543.

Heath RG, Spann JW, Kreitzer JF. 1969. Marked DDE impairment of mallard reproduction in controlled studies. *Nature* 224:47–48.

Heinz, GH. 1998. Contaminant effects on Great Lakes fish-eating birds: A population perspective. In: Kendall RJ, Dickerson RL, Giesy JP, Sulk WP, editors. Principles and processes for evaluating endocrine disruption in wildlife. Pensacola FL: SETAC. p. 141–154.

Heinz GH, Percival HF, Jennings ML. 1991. Contaminants in American alligator eggs from Lake Apopka, Lake Griffin, and Lake Okeechobee, Florida. *Environ Monit Assess* 16:277–285.

Hill EF. 1988. Brain cholinesterase activity of apparently normal wild birds. *J Wildl Dis* 24:51–61.

Hill EF, Fleming WJ. 1982. Anticholinesterase poisoning of birds: Field monitoring and diagnosis of acute poisoning. *Environ Toxicol Chem* 1:27–38.

Hoffman DJ, Ohlendorf HM, Aldrich TW. 1988. Selenium teratogenesis in natural populations of aquatic birds in central California. *Arch Environ Contam Toxicol* 17:519–525.

Hoffman DJ. 1990. Embryotoxicity and teratogenicity of environmental contaminants to bird eggs. *Rev Environ Contam Toxicol* 115:39–89.

Hoffman DJ, Rattner BA, Hall RJ. 1990. Wildlife toxicology. *Environ Sci Tech* 24:276–283.

Hoffman DJ. 1994. Measurements of toxicity and critical stages of development. In: Kendall RJ, Lacher Jr TE, editors. Wildlife toxicology and population modeling: Integrated studies of agroecosystems. Boca Raton FL: Lewis. p. 47–67.

Hoffman DJ, Ohlendorf HM, Marn CM, Pendleton GW. 1998. Association of mercury and selenium with altered glutathione metabolism and oxidative stress in diving ducks from the San Francisco Bay region, USA. *Environ Toxicol Chem* 17:167–172.

Hothem RL, Roster DL, King KA, Keldsen TJ, Marois KC, Wainwright SE. 1995. Spatial and temporal trends of contaminants in eggs of wading birds from San Francisco Bay, California. *Environ Toxicol Chem* 14:1319–1331.

Huggett RJ, Kimerle RA, Mehrle Jr PM, Bergman HL. 1992. Biomarkers. Biochemical, physiological, and histological markers of anthropogenic stress. SETAC Special Publication Series. Boca Raton FL: Lewis. 347 p.

Jackson WB. 1952. Populations of the wood mouse (*Peromyscus leucopus*) subjected to the applications of DDT and parathion. *Ecol Monographs* 22:259–281.

Jett DA, Nichols JD, Hines JE. 1986. Effect of Orthene® on an unconfined population of the meadow vole (*Microtus pennsylvanicus*). *Can J Zool* 64:243–250.

Johnson DH. 1994. Population analysis. In: Bookhout TA editor. Research and management techniques for wildlife and habitats. Bethesda MD: The Wildlife Society. p. 419–444.

Kaiser KLE. 1998. Correlations of *Vibrio fischeri* bacteria test data with bioassay data for other organisms. *Environ Health Perspect* 106:583–591.

Karnofsky DA. 1965. Mechanisms of action of certain growth-inhibiting drugs. In: Wilson JG, Warkany J, editors. Teratology: Principles and techniques. Chicago IL: The University of Chicago Press. p 185–214.

Kenaga EE. 1981. Comparative toxicology of 131,596 chemicals to plant seeds. *Ecotoxicol Environ Safety* 5:469–475.

Kendall RJ, Bens CM, Cobb III GP, Dickerson RL, Dixon KR, Klaine SJ, Lacher Jr TE, LaPoint TW, McMurry ST, Noblet R, Smith EE. 1996. Aquatic and terrestrial ecotoxicology. In: Klaassen, CD, editor. Casarett and Doull's Toxicology: The basic science of poisons. New York: McGraw Hill. p. 883–905.

Kendall R, Dickerson R, Geisy J, Suk W. 1998. Principles and processes for evaluating endocrine disruption in wildlife. Pensacola FL: SETAC. 491 p.

King KG, Sanger GH. 1979. An oil vulnerability index for marine oriented birds. In: Bartinek JC, Nettleship DN, editors. Conservation of marine birds of North America. Washington DC: U.S. Fish and Wildlife Service Research Report 11. p. 227–239.

Kubiak TJ, Harris JH, Smith LM, Schwartz TR, Stalling DL, Trick JA, Sileo L, Docherty DE, Erdman TC. 1989. Microcontaminants and reproductive impairment of the Forster's tern on Green Bay, Lake Michigan – 1983. *Arch Environ Contam Toxicol* 18:706–726.

Kushlan JA. 1993. Colonial waterbirds as bioindicators of environmental change. *Colonial Waterbirds* 16:233–251.

Lancia RA, Nichols JD, Pollock KH. 1994. Estimating the number of animals in wildlife populations. In: Bookhout TA, editor. Research and management techniques for wildlife and habitats. Bethesda MD: The Wildlife Society. p. 215–253.

LeBlanc GA, Bain, LJ. 1997. Chronic toxicity of environmental contaminant: Sentinels and biomarkers. *Environ Health Perspect* 105:65–80.

Leighton FA. 1991. The toxicity of petroleum oils to birds: An overview. In: White J, Frink L, Williams TM, Davis RW, editors. The effects of oil on wildlife. Hanover PA: The Sheridan Press. p 43–57.

Lower WR, Kendall RJ. 1990. Sentinel species and sentinel bioassay. In: McCarthy JF, Shugart LR, editors. Biomarkers of environmental contamination. Boca Raton FL: Lewis. p. 306–331.

Luebke RW, Hodson PV, Faisal M, Ross PS, Grasman KA, Zelikoff J. 1997. Aquatic pollution-induced immunotoxicity in wildlife. *Fund Appl Toxicol* 37:1–15.

Ludke JL, Hill EF, Dieter MP. 1975. Cholinesterase (ChE) responses and related mortality among birds fed cholinesterase inhibitors. *Arch Environ Contam Toxicol* 2:1–21.

Lundholm CE. 1987. Thinning of eggshell in birds by DDE: Mode of action on the eggshell gland. *Comp Biochem Physiol* 88C:1–22.

Lundholm CE. 1997. DDE-induced eggshell thinning in birds: Effects of p,p'-DDE on the calcium and prostaglandin metabolism of the eggshell gland. *Comp Biochem Physiol* 118C:113–128.

Lundholm CE, Mathason K. 1986. Effects of some metal compounds on the Ca^{2+} binding and Ca^{2+}–Mg^{2+}–ATPase activity of the eggshell gland mucosa homogenate from the domestic fowl. *Arch Pharmacol Toxicol* 59:410–415.

Matter JM, Crain DA, Sills-McMurry C, Pickford DB, Rainwater TR, Reynolds KD, Rooney AA, Dickerson RL, Guillette Jr LJ. 1998. Effects of endocrine-disrupting contaminants in reptiles: Alligators. In: Kendall RJ, Dickerson RL, Geisy JP, Suk W, editors. Principles and processes for evaluating environmental endocrine disruption in wildlife. Pensacola FL: SETAC. p. 267–289.

Mayer FL, Versteeg DJ, McKee MJ, Folmar LC, Graney RL, McCume DC, Rattner BA. 1992. Physiological and nonspecific biomarkers. In: Huggett RJ, Kimerle RA, Mehrle Jr PM, Bergman HL, editors. Biomarkers. Biochemical, physiological, and histological markers of anthropogenic stress. SETAC Special Publication Series. Boca Raton FL: Lewis. p. 5–85.

McBee K, Bickham JW. 1990. Mammals as bioindicators of environmental toxicity. In: Genoways HH, editor. Current mammology, Vol.2. New York: Plenum Press. p 37–88.

McCarthy JF. 1990. Concluding remarks: Implementation of a biomarker-based environmental monitoring program. In: McCarthy JF, Shugart LR, editors. Biomarkers of environmental contamination. Boca Raton FL: Lewis. p 429–439.

McCarthy JF, Shugart LR. 1990. Biological markers of environmental contamination. In: McCarthy F, Shugart LR, editors. Biomarkers of environmental contamination. Boca Raton FL: Lewis. p 3–14.

Melancon MJ. 1996. Development of cytochromes P450 in avian species as a biomarker for environmental contaminant exposure and effect: Procedures and baseline values. In: Bengston DA, Henshel DS, editors. Environmental toxicology and risk assessment. Vol 5 ASTM STP 1306. Philadelphia PA: American Society of Testing and Materials. p. 95–108.

Mineau P, Collins BT, Baril A. 1996. On the use of scaling factors to improve interspecies extrapolation of acute toxicity in birds. *Regul Toxicol Pharmacol* 24:24–29.

Moriarty F, Bell AA, Hanson H. 1986. Does p,p'-DDE thin eggshells? *Environ Pollut* 40:257–286.

Newman JR, Schreiber RK. 1988. Air pollution and wildlife toxicology: An overlooked problem. *Environ Toxicol Chem* 7:381–390.

O'Brien RD. 1967. Insecticide—actions and metabolism. New York: Academic Press. 332 p.

Ohlendorf HM, Hothem RL. 1995. Agricultural drainwater effects on wildlife in central California. In: Hoffman DJ, Rattner BA, Burton Jr AG, Cairns Jr J, editors. Handbook of ecotoxicology. Boca Raton FL: Lewis. p. 577–595.

Ohlendorf HM, Hothem RL, Bunck CM, Aldrich TW, Moore JF. 1986. Relationships between selenium concentrations and avian reproduction. *Trans N Amer Wildl Nat Resour Conf* 51:330–342.

Ouellet M, Bonin J, Rodrigue J, DesGranges J-L, Lair S. 1997. Hindlimb deformities (ectromelia, ectrodactyly) in free-living anurans from agricultural habitats. *J Wildl Dis* 33:95–104.

Pain DJ. 1989. Haematological parameters as predictors of blood lead and indicators of lead poisoning in black duck (*Anas rubripes*). *Environ Pollut* 60:67-81.

Paine RT. 1966. Food web complexity and species diversity. *Amer Nat* 100:65–75.

Patnode KD, White DH. 1991. Effects of pesticides on songbird productivity in conjunction with pecan cultivation in southern Georgia: A multiple-exposure experimental design. *Environ Toxicol Chem* 10:1479–1486.

Peakall DB. 1992. Animals biomarkers as pollution indicators. New York: Chapman and Hall. 291 p.

Peakall DB. 1996. Dieldrin and other cyclodiene pesticides in wildlife. In: Beyer WN, Heinz GH, Redmon-Norwood AW, editors. Environmental contaminants in wildlife: Interpreting tissue concentration. SETAC Special Publication Series. Boca Raton FL: Lewis. p. 73–97.

Peakall DB, Fairbrother A. 1998. Biomarkers for monitoring and measuring effects. In: Douben PET, editor. Pollution risk assessment and management. New York: J. Wiley. p. 351–376.

Peakall DB, Walker CH. 1996. Comment. *Ecotoxicol* 5:227.

Pearce PA, Peakall DB, Erskine AJ. 1976. Impact on forest birds of the 1975 spruce budworm spray operation in New Brunswick. *Can Wildl Serv Progr Notes* 62:1–7.

Ratcliffe DA. 1967. Decrease in eggshell weight in certain birds of prey. *Nature* 215:208–210.

Rattner BA, Fairbrother A. 1991. In: Mineau P, editor. Cholinesterase-inhibiting insecticides: Their impact on wildlife and the environment. Vol 2 Chemicals in agriculture. New York: Elsevier. p. 89–107.

Rattner BA, Hoffman DJ, Marn CM. 1989. Use of mixed-function oxygenases to monitor contaminant exposure in wildlife. *Environ Toxicol Chem* 8:1093–1102.

Robel RJ, Stalling CD, Westfahl ME, Kadoum AM. 1972. Effects of insecticides on populations of rodents in Kansas: 1965–1969. *Pest J* 6:115–121.

Rogers JM, Kavlock RJ. 1996. Developmental toxicology. In: Klaassen, CD, editor. Casarett and Doull's Toxicology: The basic science of poisons. New York: McGraw Hill. p 301–331.

Roland DA, Sloan DR, Wilson HR, Harms RM. 1973. Influence of dietary calcium deficiency on yolk and serum calcium, yolk and organ weights and other selected production criteria of the pullet. *Poult Sci* 52:2220–2225.

Rowley MH, Christian JJ, Basu DK, Pawlikowski MA, Paigen B. 1983. Use of small mammals (voles) to assess a hazardous waste site at Love Canal, Niagara Falls, New York. *Arch Environ Contam Toxicol* 12:383–397.

Sheffield SR, Matter JM, Rattner BA, Guiney PD. 1998. Fish and wildlife species as sentinels of environmental endocrine disruptors. In: Kendall RJ, Dickerson RL, Geisy JP, Suk W, editors. Principles and processes for evaluating environmental endocrine disruption in wildlife. Pensacola FL: SETAC. p. 369–430.

Shugart L, Bickham J, Jackim G, McMahon G, Ridley W, Stein J, Steinert S. 1992. DNA alterations. In: Huggett RJ, Kimerle RA, Mehrle Jr PM, Bergman HL, editors. Biomarkers. Biochemical, physiological, and histological markers of anthropogenic stress. SETAC Special Publication Series. Boca Raton FL: Lewis. p. 125–153.

Smith AJ. 1974. Changes in the average weight and shell thickness of eggs produced by hens exposed to high environmental temperature: A review. *Trop Anim Health Prod* 6:237–244.

Sparling DW. 1995. Acid deposition: A review of biological effects. In: Hoffman DJ, Rattner BA, Burton Jr GA, Cairns Jr J, editors. Handbook of ecotoxicology. Boca Raton FL: Lewis. p. 301–329.

Spies RB, Rice SD, Wolfe DA, Wright BA. 1996. Effects of the *Exxon Valdez* oil spill on the Alaskan coastal environment. In: Rice SD, Spies RB, Wolfe DA, Wright BA, editors. Proceedings of the *Exxon Valdez* oil spill symposium. American Fisheries Society Symposium 18. Bethesda MD: American Fisheries Society. p. 1–16.

Stickel WH. 1981. Pesticides and eggshells: What can we believe? Washington DC. U.S. Fish and Wildlife Service, Research Information Bulletin Number 81–34. 3 p.

Stromborg KL, Grue CE, Nichols JD, Hepp GR, Hines JE, Bourne HC. 1988. Postfledging survival of European starlings exposed as nestlings to an organophosphorus insecticide. *Ecol* 69:590–601.

Suter II GW. 1990. Endpoints for regional ecological risk assessments. *Environ Manage* 14:19–23.

Talmage SS, Walton BT. 1991. Small mammals as monitors of environmental contaminants. *Rev Environ Contam Toxicol* 119:47–145.

Tank SL, Brewer LW, MJ Hooper, Cobb III GP, Kendall RJ. 1993. Survival and pesticide exposure of northern bobwhites (*Colinus virginianus*) and eastern cottontails (*Sylvilagus floridanus*) on agricultural fields treated with Counter® 15G. *Environ Contam Toxicol* 12:2113–2120.

Tew TE, MacDonald DW, Rands MRW. 1992. Herbicide application affects microhabitat use by arable wood mice (*Apodemus sylvanicus*). *J Appl Ecol* 29:532–539.

Thalken CE, Young AL. 1983. Long-term field studies of a rodent population continuously exposed to TCDD. In: Tucker RE, Young AL, Gray AP, editors. Human and environmental risks of chlorinated dioxins and related compounds. New York: Plenum Press. p. 357–372.

[USDOC] U.S. Department of Commerce. 1996. Natural resource damage assessments: Final rule. *Federal Register* 61:439–590.

[USDOI] U.S. Department of the Interior. 1986. Natural resource damage assessments: Final rule. *Federal Register* 51:27674–27753.

[USEPA] U.S. Environmental Protection Agency. 1992. Framework for ecological risk assessment. Washington DC. EPA/630/R-92/001. 41 p.

[USEPA] U.S. Environmental Protection Agency. 1993. Wildlife criteria portions of the proposed water quality guidance for the Great Lakes System. Washington DC. EPA/822/R-93/006. 67 p.

[USEPA] U.S. Environmental Protection Agency. 1994. Pesticide reregistration rejection rate analysis. Washington DC. EPA/738/R-94/035. 167 p.

[USEPA] U.S. Environmental Protection Agency. 1995. Final water quality guidance for the Great Lakes system: Final rule. *Federal Register* 60:15366–15425.

[USEPA] U.S. Environmental Protection Agency. 1998a. Guidelines for ecological risk assessment. Washington DC. EPA/630/R-95/002F. 171 p.

[USEPA] U.S. Environmental Protection Agency. 1998b. Report of the Endocrine Disruptor Screening and Testing Advisory Committee. Final Report. Volumes I and II. U.S. Environmental Protection Agency.

van Gestel CAM, van Brummelen TC. 1996. Incorporation of the biomarker concept in ecotoxicology calls for redefinition of terms. *Ecotoxicol* 5:2176–225.

Walker CH. 1998. Biomarker strategies to evaluate the environmental effects of chemicals. *Environ Health Perspect* 106:613–620.

Walker C, Kaiser K, Klein W, Lagadic L, Peakall D, Sheffield S, Soldan T, Yasuno M. 1998. 13th Meeting of the scientific group on methodologies for the safety evaluation of chemicals (SGOMSEC): Alternative testing methodologies for ecotoxicity. *Environ Health Perspect* 106:441–451.

Webb RE, Horsfall Jr F. 1967. Endrin resistance in the pine mouse. *Science* 156:1762.

Westlake GE, Martin AD, Stanley PI, Walker CH. 1983. Control enzyme levels in the plasma, brain and liver from wild birds and mammals in Britain. *Comp Biochem Physiol* 76C:15–24.

White DH, Hoffman DJ. 1995. Effects of polychlorinated dibenzo-*p*-dioxins and dibenzofurans on nesting wood ducks (*Aix sponsa*) at Bayou Metro, Arkansas. *Environ Health Perspect* 103:37–39.

White DH, Seginak JT, Simpson RC. 1990. Survival of northern bobwhites in Georgia: Cropland use and pesticides. *Bull Environ Contam Toxicol* 44:73–80.

Whitten, ML, Marden BT, Kendall RJ, Brewer LW. 1998. Use of radiotelemetry to investigate postfledging survival of European starlings following an agricultural spraying of methyl parathion. In: Brewer LW, Fagerstone KA, editors. Radiotelemetry applications for wildlife toxicology field studies. Pensacola FL: SETAC Press p. 93–101.

Wiemeyer SN. 1996. Other organochlorine pesticides in birds. In: Beyer WN, Heinz GH, Redmon-Norwood AW, editors. Environmental contaminants in wildlife: Interpreting tissue concentration. SETAC Special Publication Series. Boca Raton FL: Lewis. p. 99–115.

Wiemeyer SN, Porter RD. 1970. DDE thins eggshells of captive American kestrels. *Nature* 227:737-738.

Wiemeyer SN, Spitzer PR, Krantz WC, Lamont TG, Cromartie E. 1975. Effect of environmental pollutants on Connecticut and Maryland ospreys. *J Wildl Manage* 39:124–139.

Wiens JA, Crist TO, Day RH, Murphy SM, Hayward GD. 1996. Effects of the *Exxon Valdez* oil spill on marine bird communities in Prince William Sound, Alaska. *Ecol Appl* 6:828–841.

Wobeser GA. 1981. Diseases of wild waterfowl. New York: Plenum Press. 300 p.

Wren CD. 1991. Cause-effect linkages between chemicals and populations of mink (*Mustela vison*) and otter (*Lutra canadensis*) in the Great Lakes basin. *Environ Res* 33:549–585.

CHAPTER 4

Statistical Design of Wildlife Toxicology Studies

John R. Skalski

The choice of response variables in a wildlife toxicology study has a critical effect on the success of the investigation. The study goals may dictate the response to measure. However, more often than not, the choice is dependent upon the skills of the investigator in selecting a relevant endpoint that can be precisely and accurately estimated. Sample size and power calculations can help in identifying which responses can be adequately measured. Goals and logistics, on the other hand, will determine their relevance and feasibility.

The interconnection between goals, study design, data analysis, and inferences (Figure 4-1) cannot be forgotten. Only at the risk of an ineffectual study should these elements be considered separately. The study goals help identify the response variables to measure, which, in turn, determines the analytical approach to sample-size calculations and subsequent study design. Overlaid upon these considerations are the additional considerations of cost and logistical feasibility.

The purpose of this paper is to discuss basic design principles of well-crafted wildlife toxicology investigations. The discussion will then focus on the choices of response variables available to investigators and their relative merits in making inference to effects on wild populations. Throughout this discussion, the relationships among study goals, statistical design, analysis, and inferences will be emphasized.

Chapter Preview
Study design principles 96
Population-level response variables 100
Design recommendations for field experiments 101
Accident assessment 102
Study performance 104
Conclusions 105

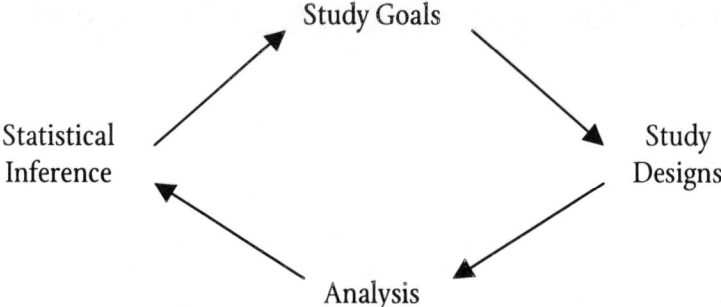

Figure 4-1 Schematic emphasizing the interrelationships between study design and analysis elements

Study Design Principles

The twin essential elements of the scientific method are experimentation and formulation of inferences (Das and Giri 1979). Although observational studies comprise the majority of today's wildlife investigations, manipulative experiments provide the strongest evidence of cause-and-effect relationships. The regulatory and litigative demands on toxicological studies drive investigators to the more rigorous requirements of true experimental studies.

A well-designed field experiment does not occur by accident, but rather by careful attention to basic design principles. Cox (1958) gave 5 criteria necessary for a good experiment:

- elimination of systematic errors that may bias or confound treatment comparisons,
- precise estimates of treatment contrasts,
- conclusions with a useful range of inference,
- simplicity in design, and
- ability to estimate error variances.

The best choice of a study design is an experimental design that provides the necessary precision to make valid inferences in a cost-effective manner. The simplest design that provides adequate precision and meaningful inferences is usually preferable.

Fisher (1947) recommended adherence to the 3 principles of replication, randomization, and error control in performing a successful study. Replication and randomization are quintessential elements that are necessary for any valid experiment. Error control is used to reduce the error variance that is necessary to precisely estimate treatment contrasts. Too often, wildlife experiments fail because they ignore 1 or more of these 3 design principles. Choice of response variable eliminates the necessity to satisfy these minimum design principles. The implications of Fisher's 3 principles of experimental design are discussed below in more detail.

Replication

Replication of experimental units is essential to demonstrate the reproducibility of the treatment effects and in estimating the magnitude of the experimental error variance. An experimental unit is defined as the smallest unit that independently received the treatment. For instance, the application of an herbicide on a field crop typically implies that the wildlife population at that site is a single experimental unit. Replication then requires multiple control and treatment sites distributed across the landscape for valid estimates of an error variance. Response measures for such an investigation could include the effects of environmental toxicants on population abundance or density. The individual animals on study plots contribute to the measurement process but do not provide independent assessments of toxicological effects.

For individual animals, breeding pairs, or egg clutches to serve as experimental units, the exposures to pollutants or toxicants need to be independently applied, measured, and assessed at these respective levels of investigation. In these cases, error variances are calculated from the variability in responses among these individuals or nesting groups. Inferences to effects are now at the individual level, expressing changes in survival or natality rates within exposed populations.

For any wildlife response (θ_i) estimated by $\hat{\theta}_i$ at each replicate site $(i = 1, \ldots, n)$, the variance of a sample mean $(\bar{\hat{\theta}})$ is estimated by

$$\hat{Var}(\bar{\hat{\theta}}) = \frac{s_{\hat{\theta}}^2}{n} \text{ or } \hat{SE}(\bar{\hat{\theta}}) = \frac{s_{\hat{\theta}}}{\sqrt{n}} \qquad \text{(Equation 4-1)}.$$

The variance among replicate estimates of $\hat{\theta}_i$ is composed of 2 error sources:

- natural variation in the response θ_i among replicate plots, denoted by σ_θ^2 and
- average measurement error associated with <u>estimating</u> rather than directly measuring wildlife responses, denoted by $Var(\hat{\theta}_i | \theta_i)$.

Hence, the standard error of the mean $\bar{\hat{\theta}}$ can be written as

$$SE(\bar{\hat{\theta}}) = \frac{\sqrt{\sigma_\theta^2 + \overline{Var(\hat{\theta}_i | \theta_i)}}}{\sqrt{n}} \qquad \text{(Equation 4-2)},$$

for any measured response $\hat{\theta}$. Careful attention to detail and greater within-plot sampling can often decrease measurement error, i.e., $Var(\hat{\theta}_i|\theta_i)$, but only replication can reduce the effect of natural variation in response (i.e., σ_θ^2). Because the measurement error in many wildlife tagging studies is proportional to the parameter being estimated

$$Var(\hat{\theta}_i|\theta_i) \propto \theta_i \quad \text{(Equation 4-3)},$$

or proportional to the square of the parameter where $Var(\hat{\theta}_i|\theta_i) \propto \theta_i^2$, considerable within-plot as well as between-plot sampling effort is needed to dampen the magnitude of Equation 4-2. What some investigators forget to appreciate is that elimination of the measurement error, i.e., achieving $Var(\hat{\theta}_i|\theta_i) = 0$, on 1 to few study plots does not eliminate the variance component associated with the natural variation (i.e., σ_θ^2) in Equation 4-2.

In establishing replicate study sites, investigators need to assure that population responses will be independent between sites. Interplot competition can occur if the treatment or wildlife investigation on 1 plot affects response measures on another site. For example, the deleterious effects of a pesticide on animal abundance can cause a "sink," reducing the animal abundance of nearby sites as well. Wide spacing between study plots, the use of enclosures, and the randomization of plots to treatments can minimize the effects of interplot competition on experimental results. Investigators should strive to find solutions to interplot competition that minimize disturbances to the demographics within plots while providing unbiased estimates of treatment means.

Randomization

Inherent differences between habitats and population characteristics at the study plots can confound treatment comparisons if randomization is not used. Randomization of treatments to study plots is an essential to assure that no known or unknown systematic differences are biasing treatment contrasts. Randomization of treatments to study areas eliminates the need to guess what biases might have occurred and to defending the belief that all sources of error have been controlled.

Under the null hypothesis that treatments have no effect on the wild population, treatment means are equal in expectation under randomization. This assumption of equality under the null hypothesis is central to the ability of experimental designs to establish cause-and-effect relationships. Hence, Fisher (1947) states that a study that fails to incorporate randomization and replication is more an experience than an experiment. In wildlife toxicology, the goal of the investigations should indeed be to provide defensible manipulative experiments, not simply to provide training for inexperienced investigators.

In experimentation, randomization should interdisperse the various treatments among replicate plots. However, in wildlife investigations, where the number of plots might be small, certain randomizations may cluster treatments together by the processes of chance alone. The possibility of excessively systematic or clumped arrangements of treatment conditions increases as the number of study locations decreases. For this reason, Hurlbert (1984) formally includes interspersion of study plots as another important design element of a well-crafted study design. Cox (1958) suggested 3 possible statistical solutions to the problem of poor plot interspersion as a result of chance randomization schemes. He suggests that an extremely aggregated layout should not be implemented simply because the design was generated randomly. The objective should be to perform individual experiments that are well-conducted and do not rely solely on long-term mathematical properties.

Error control and reduction

Equation 4-2 provides the essential guidance needed to reduce the magnitude of the experimental error variance. Cox (1958) suggests 4 ways to reduce the size of the experimental variance, including

- greater use of experimental controls (i.e., reduce σ_θ^2),
- greater replication (i.e., increase n),
- refined experimental techniques including greater sampling precision within experimental units, i.e., reduce $\overline{Var(\hat{\theta}_i | \theta_i)}$, and
- more-sophisticated experimental designs based on blocking and the use of covariates (i.e., reduce σ_θ^2).

Suggestions 2 and 3 have already been discussed above.

Greater use of control measures is always tempting in wildlife investigations. Caged animal studies and animal colonies increase control over environmental factors in order to reduce unwanted variation. However, these approaches sacrifice some realism, making inferences to wild populations more difficult or impractical. For example, radiotelemetry studies provide greater detection rates of animals than do passive tagging studies, thereby providing more precise estimates of survival for a given sample size of tagged animals. The increase in sampling precision must be tempered by possible effects of the transmitters on animal behavior that may invalidate inferences to wild populations.

The principle of experimental blocking is analogous to stratification in sample survey designs (Cochran 1977). The use of pairing or blocking of experimental plots attempts to assign as much of the heterogeneity between experimental units as possible to differences between blocks. In so doing, within-block comparisons are very precise and the overall magnitude of the experimental error reduced. Blocking or pairing of experimental units need not be perfect in order to benefit from this practice of error control. Rather, study plots within a block need only to be as similar as possible, and blocks as dissimilar as possible, in order to benefit from blocked designs. Proximity, topography, and habitat similarity are often the basis for pairing and assigning test plots to blocks.

Population-level Response Variables

Skalski and Smith (1998) provide recommendations on the design of avian toxicological studies whose purpose is to assess effects on survival. Their paper provides detailed recommendations on how radiotagging studies need to be coordinated both within and between sites. They also compare and contrast the design requirements for conducting a multi-plot manipulative experiment versus a toxicological study of individual-based exposures using an epidemiological approach to assessment. Anderson et al. (1995) discuss alternative approaches to risk assessment for wildlife populations. In this chapter, responses of populations or communities to environmental toxicants will be considered.

The distinction between population-level and individual-level investigations may not be as clear as first impressions might suggest. Certainly, investigations can focus directly on estimating animal abundance or density in manipulative experiments (Skalski and Robson 1992). Differences in mean abundance between control and treatment plots can provide unambiguous inferences to population effects. The greatest strength of these experiments is the ability to provide direct inferences to population effects based solely on the empirical evidence gathered in the field. Minimum assumptions are necessary and inferences are design-based rather than model-based in these population-level experiments (Edwards 1998).

However, conducting manipulative experiments in which replication is at the population level can be a daunting and expensive endeavor. In some circumstances, the responses of abundance or density measured at the study plots can be misleading or biased. For example, a toxic effect that kills resident populations can be masked or diminished by the immigration of animals from off-site. Mortality may have increased and natality may have decreased, but the net result on animal abundance may be negligible because of immigration from surrounding control areas. The likelihood of unwanted immigration increases as plot size decreases, thereby increasing edge effects.

The ability to sort out the opposing effects of the toxicant on survival and natality from immigration is dependent on separately examining the survival and birth processes of the exposed population. Hence, the population-level investigation may ultimately revert to measuring responses of survival and natality of individuals in order to properly interpret the higher-level responses at the level of the population.

Alternatively, population-level investigations may be designed at the onset to investigate toxic effects on survival and natality processes. The endpoint of the trials is then a model-based assessment of the toxicant or pollutant on demographic potential. In these cases, the field investigations should focus on providing age-specific estimates of fecundity and survival. These parameter estimates can then be combined to provide an estimate of the finite annual rate of population growth (i.e., λ) under control and treatment conditions. Leslie matrix models (Caswell 1989) provide a mathematical framework for the projection of the long-term effects of survival and fecundity rates on population growth and the estimation of λ.

Analytical approaches for projecting the variances of the survival and fecundity rates to estimates of $\hat{\lambda}$ are not generally available or accessible (Alvarez-Buylla and Slatkin 1994). Instead, Monte Carlo simulation techniques are often used to estimate the error variance for λ. Alternatively, the variance for the finite rate of growth could be calculated from the empirical variance among the estimates of $\hat{\lambda}_i$ calculated at each replicate study plot. In order to use the empirical variance, values of $\hat{\lambda}_i$ must be derived independently at each study site, adding to both the rigor and the cost of the investigation. Again, cost and logistical difficulties do not abrogate the need for replicated and randomized field trials.

As an alternative to deterministic matrix models, population viability analysis (PVA) provides a stochastic approach to assessing population health. The PVA incorporates temporal variability in survival and fecundity in assessing future population trends. Adding the stochastic element to the analysis often results in a more pessimistic perspective on population recovery. For a review of PVA, see Caughley and Gunn (1996).

The process-based approach to population responses to toxicants has the benefit of isolating and identifying the mechanisms that may affect the demographic potential of a population. However, these analyses are also model-based and have the inherent limitations of using any model, i.e., that it is a simplification of reality. Furthermore, inferences are to the long-term growth potential of the population that may or may not be reflected in short-term, small-scale population responses observed in field trials.

Design Recommendations for Field Experiments

Wildlife field investigators must contend with the logistics of performing replicated field plot investigations across the landscape and with the rigorous requirements of tag-recapture or radiotag procedures to estimate animal abundance, density, or survival. Neither the number of study plots nor the within-plot tagging techniques should dominate the study design. Both design elements need to be in balance to have an effective study design. Both elements need to be coordinated to generate comparable data for between-treatment comparisons. The assumptions and requirements of the various animal-tagging methods used in estimating the demographic parameters need to be adhered to as well. Readers are encouraged to review the literature on animal survey techniques (Seber 1982; White et al. 1982; Buckland et al. 1993; Thompson et al. 1998) and survival analysis (Lebreton et al. 1992;

Skalski et al. 1993; Smith et al. 1994) prior to designing and performing an investigation.

General recommendations on the design of multi-plot population investigations include the following suggestions:
- the study should be conducted during the season and at the location of desired inference,
- the study must have replicate study plots with treatment levels randomized to location,
- study plots need to be adequately dispersed to provide independence among sites,
- tagged animals should be a probabilistic sample of the wild populations of inference,
- survey periods should be synchronized and coincident across all study plots or plots within blocks,
- principles of pairing or blocking should be considered in order to reduce experimental variance,
- survey periods should be of equal duration over the course of the study, and
- application of treatments should be consistent and coincident among replicate plots in the study.

These basic design elements should be an essential part of any well-crafted field experiment. A final recommendation in the design of wildlife investigations is the use of precision or power calculations and design optimization techniques to assure an adequate and efficient study design. Field studies are too complex and too costly to base sample-size issues on convenience alone. Preliminary survey data can provide the necessary information to objectively perform the needed sample size calculations and determine an adequate study design.

Accident Assessment

To this point, the discussion has focused on the ability to design and conduct manipulative experiments to test the effects of pollutants or toxicants on wild populations. However, many situations arise in which a site-specific assessment of a chemical or oil spill may be necessary. These *post facto* investigations are deprived of the inferential strength afforded by the principles of replication and randomization. Very different strategies of investigation are therefore necessary in order to make inferences to the effects of pollutants on the demography of the exposed population.

Skalski (1995) reviewed various strategies for the design and analysis of accident assessment studies. Among the recommended approaches is to test for parallelism in the time profiles of unexposed and exposed populations (Skalski and Robson 1992). Under the null hypothesis of no effect, animal abundance among exposed and unexposed populations would fluctuate over time solely in response to regional influences of climate and weather (Figure 4-2a). However, if the chemical event did affect local populations, time profiles would fluctuate out of synchrony (i.e., nonparallel) until recovery had occurred (Figure 4-2b). Absence of parallelism immediately following the incident, with parallelism in succeeding years, would provide presumptive evidence of pollution effects. The study design requires a minimum of 3 years of post-accident monitoring with statistical power and resolution increasing with additional time. Multiple study sites within and outside the zone of potential influence are necessary to perform the multivariate profile analysis. The test of parallelism in the profile analysis is equivalent to a test of a time-by-treatment interaction. Analysis of the demographic data must, however, be conducted on the proper scale for valid inferences. Correctly formulating the response model is therefore an essential element of the statistical analysis. Wiens and Parker (1995) provide an additional review of strategies for analyzing the effects of environmental accidents.

A second approach to accident assessment of terrestrial populations is a demographic analysis of the exposed populations. Population analysis of wildlife populations consists of assessing the vitality of a population and predicting the status and trends of abundance over time (Eberhardt 1969; Caughley 1977; Downing 1980; Johnson 1996). Here again, survival and natality processes are characterized for the population at risk, and the population growth rate is calculated to assess long-term

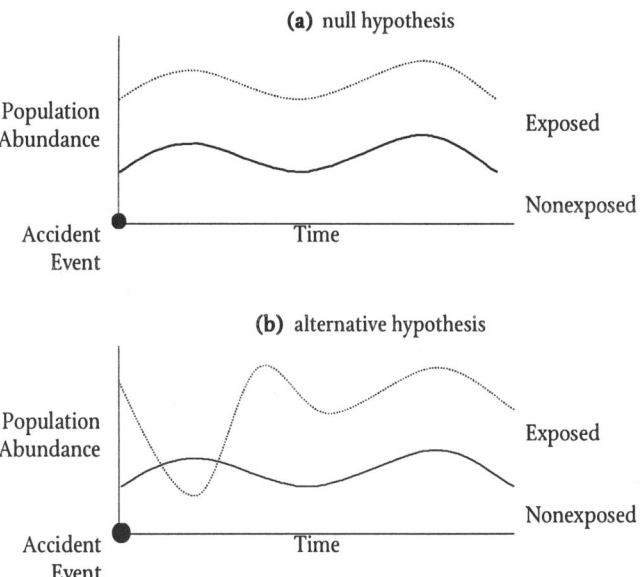

Figure 4-2 Schematics of time profiles for nonexposed and exposed populations under (a) null hypothesis and (b) alternative hypothesis of toxic effects on wild populations following a chemical accident at time 0

trends. If the chemical accident caused the death of individuals in the population, population growth might be projected to determine the number of years until the population recovered from such losses (Bowman et al. 1995). Alternatively, the finite population growth rate (λ) for the exposed population may be compared to reported rates for healthy populations to measure the extent of impact. Essential statistical elements of the demographic analysis are 1) an estimate of the uncertainty (i.e., variance) associated with the estimate of λ for the exposed population and 2) a range of reliable values of λ and associated variances for healthy, nonexposed populations. Nevertheless, the demographic analysis alone is incapable of ascribing a cause-and-effect relationship between a depressed value of λ and a pollution event. Inferences in the case of a demographic analysis will require more than a statistical basis to infer that estimated values of λ are a direct consequence of chemical exposure.

Study Performance

Preliminary surveys and sample-size calculations should be an integral component of study-design considerations. If parameter estimation is the goal of the study, then sample size calculations to achieve a prescribed level of precision should be performed. For example, if the desired level of precision can be described by the relative error in estimation, then formally

$$P\left(\left|\frac{\hat{\bar{\theta}} - \bar{\theta}}{\bar{\theta}}\right| < \varepsilon\right) = 1 - \alpha \qquad \text{(Equation 4-4)}$$

describes the objective function of the investigation. Hence, the relative error in estimation $\frac{\hat{\bar{\theta}} - \bar{\theta}}{\bar{\theta}}$ should be the less than ε, $(1 - \alpha)$ 100% of the time. In turn, the error in estimation is anticipated to be

$$\varepsilon \leq Z_{1-\frac{\alpha}{2}} \cdot \frac{\sqrt{Var(\hat{\bar{\theta}})}}{\bar{\theta}} \qquad \text{(Equation 4-5)},$$

and, where $Z_{1-\frac{\alpha}{2}}$ satisfies the inequality, $P\left(z > z_{1-\frac{\alpha}{2}}\right) = \frac{\alpha}{2}$, Equation 4-5 can further be characterized as

$$\varepsilon \leq Z_{1-\frac{\alpha}{2}} \cdot \frac{\sqrt{\frac{\sigma_\theta^2 + \overline{Var(\hat{\theta}_i | \theta_i)}}{n}}}{\bar{\theta}} \qquad \text{(Equation 4-6)}$$

by using Equation 4-3. Preliminary surveys should be used to obtain an estimate of σ_θ^2. Parametric models used in estimating abundance or survival can provide analytic expressions for the anticipated measurement error, e.g., $Var(\hat{\theta}_i|\theta_i)$, that will be a function of numbers of animals tagged and their recapture probabilities. Sample-size calculations would then consist of finding combinations of field replication (n) and within-plot tagging effort that satisfy the Equation 4-6. Sample-size calculations as found in Cochran (1977) or Seber (1982) do not consider partitioning the overall variance into a nonparametric component for natural variations, i.e., σ_θ^2, and a model-based component, i.e., $Var(\hat{\theta}_i|\theta_i)$, for the animal surveys. In practice, these considerations are central to proper sample size calculations for wild animal investigations.

This partitioning of the overall variance into elements of natural variability and measurement error from animal tagging studies is also essential if study designs are to be optimized (Skalski and Robson 1992). A necessary element of design optimization is the construction of a cost function describing within- and between-plot study costs as a function of effort (Skalski 1985). Optimization then consists of minimizing Equation 4-3 under the constraint of the cost function. Often, a field study will have sufficient performance only when field-sampling efforts are optimally allocated. Inefficient allocation of effort, conversely, can result in costly and ineffectual field studies.

Alternatively, if the objective of the investigation is to test hypotheses, power calculations should be performed to determine adequate sample sizes. Statistical power is the probability of rejecting the null hypothesis when the alternative hypothesis is true (e.g., the toxicant affected wildlife populations). Cohen (1988) provides a comprehensive description of power calculations for a variety of hypothesis tests. The exact nature of the power calculations is specific to the statistical test that will ultimately be used in analyzing the field study. An ancillary benefit of power calculations is the necessity to concurrently consider both the design and the analysis of the field investigation and consequently, to assure an integrated statistical approach to the investigation (Figure 4-1). Skalski and Robson (1992) present power calculations that can be used in the design of manipulative experiments, impact assessments, and accident assessments on the abundance of wild populations. They also present calculations for optimizing the power of tests based on field-study costs and cost functions. With sources of variation high, costs of investigation great, and the consequences grave, optimization of field-study effort should be considered an essential element of any study design.

Conclusions

Assessing the effects of toxicants at the population level of higher terrestrial vertebrates is quite feasible. However, these studies need to adhere to the principles of classical experimental design, including replication and randomization at the scale of the study plot, and to employ design measures that will control the size of experimental error. Sample-size calculations and design optimization should be essential features of any well-crafted field toxicology study. This is particularly true in wildlife toxicology, where plot replication is expensive, measurement error within a study plot can be large, and the costs of measuring population responses (i.e., abundance, density) can be great. Design optimization can help to properly allocate field-sampling efforts between and within study plots to obtain the best performance for available research funds.

References

Alvarez-Buylla ER, Slatkin M. 1994. Finding confidence limits on population growth rates: Three real examples revised. *Ecology* 7:255–260.

Anderson DR, White GC, Burnham KP. 1995. Some specialized risk assessment methodologies for vertebrate populations. *Environ Ecol Statistics* 2:91–115.

Bowman TD, Schempf PF, Bernatoruicz JA. 1995. Bald eagle survival and population dynamics in Alaska after the *Exxon Valdez* oil spill. *J Wildl Manage* 59:317–324.

Buckland ST, Anderson DR, Burnham KP, Laake JL. 1993. Distance sampling. Estimating abundance of biological populations. New York: Chapman and Hall. 446 p.

Caswell H. 1989. Matrix population models: Construction, analysis, and interpretation. Sunderland MS: Sinauer. 328 p.

Caughley G. 1977. Analysis of vertebrate animal populations. New York: Wiley. 234 p.

Caughley G, Gunn A. 1996. Conservation biology in theory and practice. Cambridge MA: Blackwell Science. 459 p.

Cochran WG. 1977. Sampling techniques. Third edition. New York: Wiley. 428 p.

Cohen J. 1988. Statistical power analysis for the behavioral sciences. Hillsdale NJ: Lawrence Erlbaum Associates. 567 p.

Cox DR. 1958. Planning of experiments. New York: Wiley. 308 p.

Das MN, Giri NC. 1979. Design and analysis of experiments. New Delhi India: Wiley Eastern Limited. 295 p.

Downing RL. 1980. Vital statistics of animal populations. In: Schmenitz SD, editor. Wildlife management techniques manual. Washington DC: The Wildlife Society. p 247–267.

Eberhardt LL. 1969. Population analysis. In: Giles RH, editor. Wildlife management techniques. Washington DC: The Wildlife Society. p 457–495.

Edwards D. 1998. Issues and themes for natural resources trend and change detection. *Ecol Appl* 8:323–325.

Fisher RA. 1947. The design of experiments. Fourth edition. New York: Hafner. 260 p.

Hurlbert SH. 1984. Pseudoreplication and the design of ecological field experiments. *Ecol Monogr* 54:187–211.

Johnson DH. 1996. Population analysis. In: Bookhout TA, editor. Research and management techniques for wildlife and habitats. Bethesda MD: The Wildlife Society. p 419–444.

Lebreton JD, Burnham KP, Clobert J, Anderson DR. 1992. Modeling survival and testing biological hypotheses using marked animals: Case study and recent advances. *Ecol Monogr* 62:67–118.

Seber GAF. 1982. Estimation of animal abundance and related parameters. New York NY: MacMillan. 654 p.

Skalski JR. 1985. Construction of costs functions for tag-recapture research. *Wildl Soc Bull* 13:273–283.

Skalski JR. 1995. Statistical considerations in the design and analysis of environmental damage studies. *J Environ Manage* 43:67–85.

Skalski JR, Robson DS. 1992. Techniques for wildlife investigations: Design and analysis of capture data. San Diego CA: Academic Press. 237 p.

Skalski JR, Smith SG. 1998. Design principles for radiotelemetry experiments to assess effects on avian survival. In: Brewer L, Fagerstone K, editors. Society of Environmental Toxicology and Chemistry (SETAC) Pellston Workshop on Avian Telemetry in Support of Pesticide Field Studies; 6–8 Jan 1993; Pacific Grove CA. Pensacola FL: SETAC Press. p 123–313.

Skalski JR, Hoffmann A, Smith SG. 1993. Testing the significance of individual- and cohort-level covariates in animal survival studies. In: Lebreton SD, North PM, editors. Marked individuals in the study of bird populations. Boston MA: Birkhauser. p 9–28.

Smith SG, Skalski JR, Schlechte JW, Hoffmann A, Cassen V. 1994. SURPH.1 Manual: Statistical survival analysis of fish and wildlife tagging studies. Portland OR: Bonneville Power Administration. 528 p.

Thompson WL, White GC, Gowan C. 1998. Monitoring vertebrate populations. San Diego CA: Academic Press. 365 p.

Wiens JW, Parker KR. 1995. Analyzing the effects of accidental environmental impacts: Approaches and assumptions. *Ecol Appl* 5:1069–1083.

White GC, Anderson DR, Burnham KP, Otis DL. 1982. Capture-recapture and removal methods for sampling closed populations. Los Alamos NM: Los Alamos National Laboratory. LA–8787–NERP.

CHAPTER 5

Approaches for Assessment of Terrestrial Vertebrate Responses to Contaminants: Moving Beyond Individual Organisms

Peter H. Albers, Gary H. Heinz, Russell J. Hall

A long-standing goal of biologists studying the effects of environmental contamination has been to understand how contaminants affect populations, communities, and ecosystems. Despite the keen interest, environmental pollution journals contain reports mostly of research focused below the population level and on direct (versus indirect) effects (Clements and Kiffney 1994). Similarly, ecologists express the desire to understand "higher-order interactions" and "indirect effects" while conducting research on simplified relations involving 1, or just a few, species (Kareiva 1994). The success of efforts to expand contaminants biology and toxicology into larger-scale and more-robust environmental assessments (e.g., ecotoxicology [Cairns and Pratt 1993] and landscape ecotoxicology [Cairns and Niederlehner 1996]) depends on improvements in our ability to directly measure or estimate the effects of contaminants in complex situations and at organizational levels above that of the individual.

An important use for knowledge of the effects of contaminants on wildlife is in formal evaluations of the potential for undesirable (from the human perspective) environmental responses to contaminant exposure, such as for ecological risk assessment (Suter 1993). "Predictive" assessments deal with possible future events involving hypothesized sources, estimated exposures, and potential effects of exposure. They are often performed by regulatory agencies. "Retrospective" assessments deal with past or ongoing events involving recognized contaminant situations and often involve research and management efforts by federal and state natural resources agencies. Retrospective assessments incorporate actual measures of the source, exposure, or effects and complementary laboratory investigations. In addition, they are site- or contaminant-specific and are amenable to evaluation with

CHAPTER PREVIEW

Definitions, perspectives, and regulatory considerations 110
Birds 114
Mammals 123
Reptiles and terrestrial amphibians 134
Conclusions 141

epidemiological criteria (Suter 1993; Gilbertson 1997). Qualitative or quantitative procedures (e.g., mathematical models) frequently are used to establish relations between "known and less-known" components of the assessment. The use of established cause-and-effect and exposure relations in different circumstances is usually referred to as "extrapolation," which is defined as "to infer or estimate by projecting or extending known information" (Severynse 1995). Extrapolation can occur among effects, taxonomic entities, and organizational levels and through time and space. Uncertainty, which grows as the magnitude of extrapolation increases, is countered with the incorporation of "safety factors" (Chapman et al. 1998).

The purpose of this chapter is to review published literature on the effects of environmental contaminants on all 4 classes of wild terrestrial vertebrates and to present examples of efforts to assess terrestrial vertebrate responses, evaluate attempts at information extrapolation, and make recommendations for improving our ability to determine effects of contaminants at population, community, and ecosystem levels of organization. The quantity and nature of relevant literature differed so much among classes that the reviews will be presented as 3 separate evaluations. The vast quantity of avian literature was reviewed selectively with an emphasis on examples of successful assessments of the effects of contaminants on populations. The lesser quantity of mammalian and herpetological literature was reviewed more broadly with an emphasis on an overall critique of published information and a few assessment examples. The literature review for mammals, amphibians, and reptiles was arbitrarily limited to the period from 1980 to 1997.

Evaluating the effects of environmental contaminants beyond the level of the individual poses problems of definition and perception. The terms "population," "community," and "ecosystem" have variable definitions and are perceived in similarly variable fashion by biologists. For purposes of literature review and assessment evaluation, we interpret these terms in the broadest possible sense. These terms are discussed in more detail in the next section. In a related matter of perception, the literature on mammals is reviewed by 1 author and the literature on amphibians and reptiles by another author, both of whom employ the same method of literature evaluation. Despite efforts by both reviewers to coordinate the mechanics of the process, differences of literature interpretation and perception are possible.

Definitions, Perspectives, and Regulatory Considerations

Because so many authors use the term "population" without actually defining what they mean, we will first discuss how the term can be defined. In a paper whose title, "Populations, metapopulations, and species populations: what are they and who should care?," suggests that some authors are sloppy in their use of terms, Wells and Richmond (1995) stress how important it is to spell out what is meant by the term "population" and how various terms have been defined by ecologists and others.

These authors include a table that lists 30 different terms for describing groups of individuals, including such terms as "local population" and "overall population," which sound vague compared to "island population" and "local colony," which are easier to visualize. The authors suggest that the term "population" be reserved for "cases in which groups of individuals are spatially, genetically, or demographically disjunct from other groups." They further suggest a set of 7 terms to describe groups of animals (Table 5-1). From among the 7 terms in Table 5-1, the term "local population" (a group of conspecifics within an area that is arbitrarily defined by the investigator) probably applies to most contaminant studies in which effects on populations are discussed. Wells and Richmond went on to recommend that biologists use the terminology they outlined or, at the very least, that authors carefully define what they mean when writing about populations. This last admonition, to clearly describe the population under study, is probably all that a wildlife toxicologist needs to remember. Ideally, wildlife toxicologists should communicate with one another using commonly accepted terms for different kinds of populations. However, most wildlife toxicologists either 1) will not be familiar with the definitions that population ecologists use or 2) will not be able to agree on these definitions, or their appropriateness, for specific contaminants studies. Therefore,

Table 5-1 Suggested terms to describe natural groupings of animals[1]

Term	Definition
Species population	All individuals of a species
Metapopulation	A set of spatially disjunct populations, among which there is some immigration
Population	A group of conspecific individuals that is demographically, genetically, or spatially disjunct from other groups of individuals
Aggregation	A spatially clustered group of individuals
Deme	A group of individuals more genetically similar to each other than to other individuals, usually with some degree of spatial isolation as well
Local population	A group of individuals within an investigator-delimited area smaller than the geographic range of the species and often within a population (as defined above). A local population could be a disjunct population as well
Subpopulation	An arbitrary spatially-delimited subset of individuals from within a population (as defined above)

[1]Source: Wells and Richmond (1995)

clearly defining what we mean when we claim an effect on a population may be the best that we can do.

With regard to the highest levels of biological organization, we need to give serious thought to what we mean when we refer to "effects on communities or ecosystems." As with the term "population," the lack of standardized definitions means that we should be sure to clearly describe the measured relation between contaminant exposure and organizational response.

Why is it so important to define terms clearly when we publish papers that address effects of contaminants on wildlife populations? It is important because, as all of those 7 terms reveal, there can be great differences among our various definitions of a population. At the extreme end, either few people would care that a contaminant affected a few local animals or the effect would be so geographically widespread that we would wonder why we allowed things to get so bad that a major reduction had occurred in the abundance of a species.

At the lowest extreme, it is true that the species population of, for example, American robins (*Turdus migratorius*), is reduced when a pesticide kills just 1 robin in a backyard. At that instant, the entire North American population of robins has been reduced. It does not make any difference that, 1 second later, somewhere else in North America, two baby robins hatch, thus resulting in a net increase of 1 robin. At the earlier instant, contaminants were responsible for a continent-wide population reduction. But, assuming other robins were not being killed by this pesticide in other backyards, most wildlife toxicologists would not consider the removal of a single individual a meaningful population decrease. The word "meaningful" is the key to describing the effects of contaminants on populations. Few wildlife toxicologists would call the 1 robin's death meaningful, but nearly all would call a 50% reduction in robins, in even 1 state, a serious problem if contaminants caused the reduction. It is between the 2 extremes that we might disagree.

The previous example of a statewide decline in robin abundance relates to what ecologists call "questions of scale." Scale refers to the size of the area suffering the population decline. There is no scientific consensus on how widespread a contaminant-induced population decline must be in order to be considered meaningful. Nevertheless, it is an important practical matter in wildlife toxicology because it influences the experimental design of field studies.

In addition to the spatial extent and magnitude of the population decline, the robin example also has important temporal aspects. How long does a reduction in the robin population have to persist before wildlife toxicologists consider it more than just a temporary loss? It is well known that game birds and mammals can absorb a large loss of individuals year after year as a result of hunting, and still rebound to normal numbers after the reproductive season. Suppose contaminants imposed a similar large loss, but the loss was replaced by reproductive gains each year. Is this a worrisome population effect? This is mostly a societal judgment. From a biological point of view, the affected population is tolerating that loss if, each year, it can rebound to its former numbers and the overall "fitness" of the affected population does not decline.

As will be discussed later, DDE-induced eggshell thinning, probably combined with heavy adult mortality from other organochlorine pesticides, caused precipitous population declines over large areas and for many years in some predatory birds. These declines were not just statewide or regional; they covered nearly the entire U.S. or North American continent. Because of the magnitude, spatial scale, and duration of these declines, few scientists would argue that they did not constitute a very serious population effect from a group of contaminants.

A final consideration given to population, as well as community or ecosystem, effects of contaminants is the need to place them in perspective with lesser levels of effect. From a biological point of view, preventing contaminant-induced population declines is probably the most important goal in wildlife toxicology. Many terrestrial vertebrate species are able to absorb heavy losses of adults or low reproductive success from natural causes and still maintain relatively stable, year-to-year populations. However, if a contaminant causes more than a temporary and spatially small decrease in the abundance of a species, the species is being subjected to an unusual and potentially serious stress. By the time a contaminant has reached a level in the environment sufficient to cause a large, extensive, and persistent reduction in the abundance of a bird, mammal, reptile, or amphibian species, a major catastrophe has already occurred. Experience with DDE and other organochlorines with birds has shown that such problems can take years or decades to reverse. Consequently, demonstrating effects on populations should not become a regulatory or common-sense requirement for controlling a contaminant. Too much damage will already have been allowed to occur. The role of looking for effects at population, community, or ecosystem levels is mostly a policing effort to see if contaminants either have slipped through regulatory cracks or have become too concentrated in the environment, and are causing what most people would agree is unacceptable harm.

Interestingly, the U.S. Environmental Protection Agency (USEPA), which is the main regulator of contaminants in the U.S., is under no requirement to demonstrate an effect on populations before taking regulatory action (Barton 1994). The USEPA looks for adverse ecological effects, but, at least as far as pesticides are concerned, the USEPA's criterion for adverse effects states that "the uses of the pesticide may result in residues in the environment of non-target organisms at levels which equal or exceed concentrations acutely or chronically toxic to such organisms, or at levels which produce adverse reproductive effects in such organisms." There is no mention of these adverse effects having to cause population declines, even though such a decline would be of interest to the USEPA. As Barton (1994) explained,

Population reduction is not mentioned because there is no way to measure directly population reduction from a specific pesticide use. Furthermore, in the opinion of Office of Pesticide Programs (OPP) scientists, there are no validated models that would allow us to estimate or predict the degree of such reduction from specific pesticide uses. Because population reduction is an endpoint of concern, models are needed that will be useful in pesticide risk assessment.

Birds

Examples of population effects

For birds there is an abundance of scientific literature in which authors have gone beyond measuring effects on individuals to assessing effects on "populations." Many times they have actually measured what they called "populations." Chapter 2 of this book has reviews of many cases in which contaminants were stated to have affected populations of birds. We will not review all of this literature but, instead, will present 3 examples of how scientists either extrapolated their findings from contaminant effects on individuals to effects on populations or presented a weight-of-evidence approach in claiming effects on populations. We proceed from an example of a population effect of large geographic scale and long duration to 1 of small geographic scale and short duration. As might be expected, there is much less literature on extrapolating from effects on individuals to effects on communities or ecosystems, but we will present 1 such example.

DDE-induced eggshell thinning in combination with organochlorine pesticide poisoning

The most impressive example of how wildlife toxicologists went beyond measurements of harm to individuals to measurements of harm to populations was with the combination of DDE-induced eggshell thinning and organochlorine pesticide-induced death of adults that affected populations of many predatory birds. There is no other example so scientifically compelling and biologically meaningful. This example does not represent a case where rigorous mathematical modeling was used to demonstrate population effects. Instead, it represents a situation in which so much evidence from the field and laboratory was collected that one did not need a mathematical model to say conclusively that a group of contaminants was causing widespread and long-lasting population declines.

Many species were affected, but the peregrine falcon (*Falco peregrinus*) is perhaps the best example of how biologists linked eggshell thinning and death of adults to declines of species populations in North America and Europe. As Ratcliffe (1980) related in his book about peregrines, the peregrine suffered a history of persecution from gamekeepers, egg collectors, falconers, and pigeon racers. Many people wanted either to kill peregrines because peregrines killed other birds valued by people, to rob their nests for private egg collections, or to steal young from nests for falconry purposes. However, the peregrine weathered these human attacks better than it did the attacks from modern pesticides that were introduced largely in the 1940s and 1950s.

In 1960, Ratcliffe was asked by the Nature Conservancy to undertake a study of the status of the peregrine falcon in Great Britain. The details of his investigations, the strength of his arguments, and the work of others he cites are all in his book; we only summarize what he painstakingly documented. Ratcliffe discovered quickly that a once-flourishing population in southern England had all but disappeared and that the populations in Wales and other regions were much reduced as well. For all of Britain, there was a 56% decrease in occupied breeding territories compared with the average during the period 1930 to 1939. In addition, few young were being produced.

In his search for the causes of the decline, Ratcliffe (1980) discovered 2 important things: many dead adults were found and eggs were broken in the nest. The use of pesticides such as aldrin, dieldrin, and heptachlor as seed dressings was found to be responsible for the deaths of seed-eating birds and, secondarily, for the deaths of raptors, e.g., peregrines, that fed on the dead and dying smaller birds. The presence of these pesticides at lethal levels in the dead peregrines confirmed their cause of death. The fact that declines of peregrine populations were greatest where land-use practices favored agriculture supported the belief that the death of adults was caused by their consumption of contaminated prey.

In response to Ratcliffe's investigations, the use of aldrin, dieldrin, and heptachlor as spring seed dressings was banned in England in 1962. However, the peregrine population continued to decline. After additional pesticide restrictions, peregrine numbers stabilized at a reduced level. Over the years, Ratcliffe and others had noticed the presence of broken eggs in peregrine nests, and sometimes adults were seen eating eggs. Broken and missing peregrine eggs became more prevalent during the time that England began using dichlorodiphenyltrichloroethane (DDT) (about 1948 to 1949); this was prior to the beginning of the use of aldrin, dieldrin, and heptachlor as seed dressings (about 1955 to 1956).

In 1966, friends of Ratcliffe suggested that he measure the weight or thickness of peregrine eggshells before and after the pesticide era began. The result of those suggestions was Ratcliffe's 1967 ground-breaking paper, "Decrease in eggshell weight in certain birds of prey." Eggshell thinning had been discovered. Peregrine eggshells collected after 1947 in England were, on average, almost 20% thinner than earlier eggshells. Ratcliffe deduced that, initially, peregrines got their DDT from eating homing pigeons that had been dusted with the pesticide to control ectoparasites. Later, DDT was used more for protecting agricultural crops, and exposure would have come from eating small birds that ingested it.

Subsequent to Ratcliffe's (1967) discovery of eggshell thinning, a great deal of laboratory work was done to demonstrate that p,p' DDT and p,p' DDE are the primary contaminants that cause eggshell thinning. As early as 1969, an experimental study demonstrated that DDT thinned eggshells of American kestrels (*Falco sparverius*), a member of the same genus as the peregrine falcon (Porter and Wiemeyer 1969). Heath et al. (1969) showed eggshell thinning and reproductive impairment in mallards (*Anas platyrhynchos*) that were fed DDE. Black ducks (*Anas rubripes*) also were shown to be vulnerable to DDE-induced shell thinning (Longcore et al. 1971). A great deal of additional experimental work and field research, summarized by Stickel (1975), demonstrated conclusively that DDE, the breakdown product of DDT, caused eggshell thinning in at least 54 species of 10 orders of birds.

The measurements that Ratcliffe and others made on individual peregrine falcons were death of adults and eggshell thinning. This same group of scientists also directly measured the percentage of historical peregrine nesting sites that were occupied each year, the incidence of broken eggs in nests, and the number of young falcons fledging from nests. Consequently, they did not merely extrapolate measurements of harm to individuals to an estimated reduction in the population; they actually measured the population decrease. In sum, they attributed the population effect to death of adults and eggshell thinning. The scientists felt confident doing this because a vast amount of published information from both the field and laboratory supported such a conclusion.

The final proof that DDE-induced eggshell thinning, coupled with the death of adults from organochlorine pesticide poisoning, had caused the peregrine falcon population decline in England came from the observations of population recovery after these pesticides were banned. Stickel (1981) characterized this phase of research as "a great natural experiment," in which scientists documented the recovery of predatory bird populations once DDT and other pesticide use stopped.

In England, several censuses of peregrine falcon populations were made under the direction of the British Trust for Ornithology (Ratcliffe 1980). In 1963, only 44% of historical peregrine breeding territories were occupied, and only 16% of the breeding pairs brought off young. By 1971, territory occupancy was up to 54%, and 25% of the breeding pairs were successful. By 1979, the peregrines occupied 75% of historical territories.

Ratcliffe's (1967) publication on eggshell thinning in peregrines stimulated Hickey and Anderson (1968) to look for the same phenomenon in peregrine eggshells from the U.S. They also found about 20% thinning, beginning between 1946 and 1947, and were the first researchers to use correlation analysis to demonstrate that eggshell thickness was inversely correlated with DDE concentrations in eggs. Hickey (1969), in his book on the peregrine falcon in North America, reviewed the severe population decline that took place at the same time both in this hemisphere and in Europe.

This early detective work on eggshell thinning in birds has been reviewed by Stickel (1973), Stickel (1975), Blus (1996), and Keith (1996). Blus (1996) concluded that if a population of birds experiences about 18% or more eggshell thinning over a period of years, a population decline is likely. Stickel (1975), after a thorough review of the eggshell thinning literature, concluded that "only DDE has caused serious thinning and great drops in reproduction at low, realistic dosages." Peakall and Lincer (1996) came to the same conclusion in a separate review.

What characterized the population declines from eggshell thinning, coupled with adult mortality, was not only the vast geographic extent of the problem but also the difficulty involved in discovering and proving the thinning effect. From the time it was understood that some of the organochlorine pesticides killed large numbers of birds, but that this mortality by itself could not explain why many predatory bird populations were in serious decline, it took nearly a quarter of a century to unravel the mystery of DDE-induced eggshell thinning. We are tempted to think, and hope, that a repeat of something as insidious as eggshell thinning could not escape our detection today. Pesticides today are not as persistent as the old organochlorines, and they do not accumulate as much in food chains. Wildlife toxicologists also measure many more things than they did decades ago. However, if a modern-day contaminant had a harmful effect on individual birds that was as difficult to detect as eggshell thinning was in its day, such an undiscovered effect conceivably could be causing otherwise unexplainable population declines in some birds. The lessons learned from eggshell thinning are that 1) we should look for all possible associations between contaminant use or distribution and declining populations of birds and 2) it likely will be difficult to make a convincing connection between cause and effect and doing so will require several lines of investigation.

Potts model for the grey partridge

Decades of painstaking research on the grey partridge (*Perdix perdix*) in Great Britain, resulting in a mathematical model of a pesticide-induced population decline, was summarized by Potts (1986). Like the population effects of DDE-induced eggshell thinning and organochlorine-induced death of adults in predatory birds, the declines in grey partridge numbers extended well beyond Great Britain. Populations crashed in all of Europe and Asia and also in Canada and the U.S., to which the species had been introduced. Potts (1986) stated that it was possible that grey partridge populations had declined in all of the 31 countries in which it was found. Evidence for these widespread declines came from 2 sources: a decrease in the number of partridges shot by hunters each year and a decrease in the number of breeding pairs in the spring. Prior to 1940, the mean number of partridges shot by hunters in Great Britain each year was about 2 million. By 1984, the kill had dropped to about 100,000 birds. Similar, or even greater, drops in the kill occurred during this same time period in nearly all of the countries in which the partridge was hunted.

In Great Britain, game keepers monitored the densities of breeding pairs of grey partridges on many farms; the decline in breeding pairs was from about 25 pairs per km^2 in 1952 to about 5 pairs in the mid-1980s. When Potts (1986) reviewed all of the detailed studies from Europe and North America, he estimated that the average number of breeding pairs per square kilometer prior to 1953 was about 11, compared with only 3 pairs in 1985. Worldwide, the population of grey partridges had declined by about 80%.

During the course of this decline in partridge numbers, Potts (1986) began a Partridge Survival Project and established a 62-km^2 area in West Sussex, England, as his main study area. From 1968 to 1985, Potts investigated the effects of hunting, mowing, predation, disease, habitat changes, weather, and pesticides on partridge numbers. His approach differed from that of Ratcliffe and others who studied peregrine falcons and other predatory bird declines in that he constructed a mathematical model, the Sussex Model, to try to identify the major causes of the partridge decline.

During 18 years of study in the Sussex site only 2 partridges were confirmed to have died from exposure to pesticides. In contrast, Potts estimated that during those 18 years at least 10,000 partridge chicks died from what he called "the ecological changes caused by the use of pesticides." The most important ecological change that Potts identified was the reduction in insects needed as food by partridge chicks. This reduction was caused by 2 things: 1) the widespread use of herbicides in fields of cereal crops that eliminated the weeds and thus the insects that lived on the weeds and 2) the direct mortality of insects from insecticide use (Potts 1977). Partridge chicks were starving to death. The model Potts used (Potts 1980, 1986) showed that the death of chicks was the primary effect that was responsible for the decline in partridge numbers. Potts was able to use mathematical simulations within his model to conclude that herbicides and insecticides were indirectly responsible for increases in chick mortality by way of reductions in insects.

Potts not only identified indirect effects of insecticides and herbicides as the major cause of the decline in the population of grey partridges, but also he recommended something that could be done to help reverse the decline. He estimated that by not spraying the margins of cereal grain fields, which would amount to only a 4% reduction in the use of pesticides, partridge numbers could be restored.

When overwhelming evidence of harm (e.g., the combined effects of death of adults and eggshell thinning on populations of predatory birds) is not available, mathematical modeling probably represents the most rigorous way to examine population effects of contaminants on birds. This is not a new idea. In 1990, a conference was held to discuss the mathematical modeling of pesticide effects on bird populations, with a book resulting (Kendall and Lacher 1994). One of the chapters in that book dealt with the use of individual-based toxicological data to model population effects of pesticides on animals, specifically, aquatic organisms, although the logic applies equally well to birds (Hallman and Lassiter 1994). This approach to modeling focuses on the effects of a contaminant on individual organisms within a population. The reason for using the individual as the focus for data collection is that death and reproductive problems act on individual organisms, and any effect at the population level is the summation of all the effects on individuals within that population. The model uses measurements of effects on individual organisms, combined with mathematical equations to estimate the effect on the population from the accumulated effects on individuals. The authors state that the drawbacks of this modeling approach are that it is very data-intensive and very complex in its computations. In general, any attempt to mathematically model the effects of contaminants on birds is only as good as the data that are used in the model. Often the data are not as complete or reliable as one might want.

One of the biggest criticisms of models, presented in another chapter in the Kendall and Lacher book, is that one cannot begin to model contaminant effects on bird populations until all of the other factors (e.g., weather, food, predation, disease) are also understood; in other words, non-contaminant effects must be modeled first (Heinz et al. 1994). Potts was able to address this problem by studying all of the major possible limiting factors on grey partridge numbers. The grey partridge model took decades of intensive field work to piece together and is, by far, the best example of modeling the effects of contaminants on a bird population.

Selenium at the Kesterson National Wildlife Refuge

As discussed earlier, populations affected by contaminants can be defined as very large, such as the North American and European declines in peregrine falcons, or small, such as the example that follows. In 1983, Ohlendorf et al. (1986) discovered severe reproductive problems in American coots (*Fulica americana*), eared grebes (*Podiceps nigricollis*), black-necked stilts (*Himantopus mexicanus*), American avocets (*Recurvirostra americana*), and various species of ducks at Kesterson Reservoir, which was part of the Kesterson National Wildlife Refuge in California. Death of embryos was often associated with deformities. Death of adults also was found in some of the species (Ohlendorf et al. 1988; Ohlendorf 1989) at Kesterson. Many years of additional field and laboratory work showed conclusively that selenium, found in irrigation drainwater from agricultural fields, was the contaminant responsible for the problem. Much of this research was summarized by Ohlendorf (1996), Ohlendorf and Hothem (1995), and Heinz (1996).

In an effort to estimate the total impact of selenium poisoning on bird numbers at Kesterson, Ohlendorf (1989) utilized information on 3 effects measured on individual birds. The first effect was the death of embryos. After reviewing the literature on hatching success of eggs at uncontaminated sites, Ohlendorf (1989) assumed that 90% of the bird eggs at Kesterson would have hatched were it not for the selenium poisoning. At Kesterson, only 78% of the eggs hatched. Based on the total number of nests and the eggs they contained, he estimated that about 540 embryos were lost to selenium poisoning. The second effect was death of hatchlings. By comparing survival of hatchlings at Kesterson with that of hatchlings in uncontaminated areas, Ohlendorf estimated that about 450 hatched chicks died of selenium poisoning at Kesterson. The third effect was excess death of adults caused by selenium poisoning. In total, Ohlendorf estimated that "at least 1000 migratory birds (adults, embryos, and chicks) died at Kesterson during 1983 to 1985" from selenium poisoning.

Although Ohlendorf was not able to measure the pre-selenium population and compare it directly to the post-selenium population, he was able to extrapolate his results from the 3 measures of harm to individuals to a population reduction of about 1000 birds. One thousand fewer birds was a large number for the 500-hectare Kesterson Reservoir. Other areas in California, and perhaps elsewhere in the western U.S., probably experienced selenium contamination to an extent that would have increased bird mortality over this greater geographic area (Olson and Welsh 1993), but Ohlendorf (1989) extrapolated his findings only to Kesterson.

The Kesterson example is a valid case that demonstrates that a contaminant had an effect on local populations of certain birds. This example differs greatly from the peregrine falcon example in 3 ways. First, the Kesterson population under study was much smaller than the peregrine populations of England and the U.S. Second, the peregrine population was shown to have been reduced for decades, whereas the Kesterson population of aquatic birds was studied only for a few years until the reservoir was filled with soil. Finally, Ohlendorf (1989) estimated the excess adult mortality and reproductive loss of birds at Kesterson, but was unable to place it in context with any earlier estimate of the normal population. The peregrine researchers, on the other hand, had access to extensive historical data on peregrines and, additionally, were able to track the recovery of peregrine numbers after the bans on organochlorine pesticides.

The Kesterson example also differs greatly from the Potts grey partridge example. Potts (1986) used a formal mathematical model and simulated the effects of various factors on partridge numbers, eventually concluding that a pesticide-induced reduction in insect abundance was the most influential variable in his model. This modeling was not done at the Kesterson site, nor were decades of data acquired. However, the Kesterson example illustrates that valid conclusions can be drawn using some simple mathematical calculations for small geographic areas.

Great Lakes fish-eating birds: An example of an extrapolation to communities and ecosystems

By any definition of the word "ecosystem," the Great Lakes is a large ecosystem, within which are communities of birds and other animals. There is little question that a complex group of organochlorine contaminants contributed to severe disruptions in this ecosystem. Declines of populations of raptors and fish-eating birds caused by organochlorine pesticides and perhaps by polychlorinated biphenyls (PCBs), dioxins, and other chlorinated pollutants are discussed in Chapter 2 of this book. If contaminant-induced reductions in the populations of several species of birds and fish within the same ecosystem qualify as an ecosystem-level effect, then the Great Lakes experienced such an effect.

Heinz (1998) reviewed the population declines, starting in the 1950s, of several species of fish-eating birds in the Great Lakes and concluded that organochlorine pesticides, possibly in conjunction with other contaminants, were responsible. As with peregrine falcons, the effects measured on individual birds were mostly reproductive failure attributable to DDE-induced eggshell thinning and death of adults as a result of dieldrin and other organochlorine pesticides. Polychlorinated biphenyls, dioxins, and other chlorinated pollutants also cannot be ruled out as a cause of death and deformities in embryos. As was the case with the peregrine falcon, many field biologists were actively monitoring populations of fish-eating birds in the Great Lakes over several decades. Once again, it was not a case of wildlife toxicologists having to rely on extrapolations from field and laboratory observations to some possible reduction in populations of wild fish-eating birds. They actually measured the effects on both individual birds and the Great Lakes populations of these birds. Their success in linking death and reproductive failure to the observed declines in populations was similar to that of the studies done with peregrines.

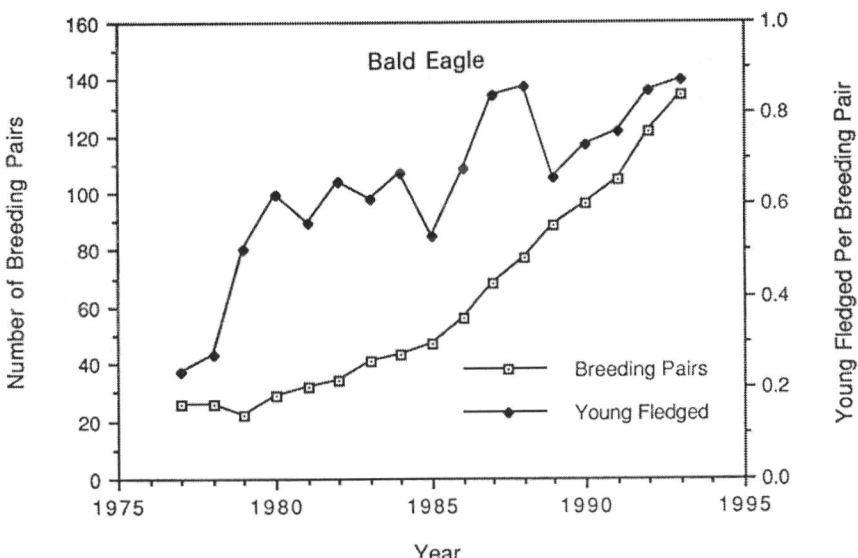

Figure 5-1 Changes in the number of breeding pairs of bald eagles in the Great Lakes (not including Lake Ontario) and along rivers supporting runs of anadromous fishes from the Great Lakes and changes in the number of young eagles fledged per breeding pair (from Bowerman 1993)

In the Great Lakes, and elsewhere in the U.S., regional declines of bald eagle (*Haliaeetus leucocephalus*) populations began in the 1950s and continued into the 1970s (Sprunt et al. 1973). Eggshell thinning and the death of adults were blamed (Cromartie et al. 1975; Wiemeyer et al. 1984). Figure 5-1, based on data collected by Bowerman (1993), shows the increase in bald eagle nesting pairs and reproductive success starting from the low point in the 1970s and extending into the 1990s, when most Great Lakes locations were free of the harmful effects of eggshell thinning and death of adults. As with the peregrine falcon, documentation of the recovery of the bald eagle population after the bans on organochlorine pesticides was just as important in showing the linkage between toxic effects on individuals and population declines as was the documentation of effects and population declines during the worst of the organochlorine era.

Other species of fish-eating birds that experienced population declines in the Great Lakes during the era of heavy use of organochlorine pesticides included the double-crested cormorant (*Phalacrocorax auritus*), herring gull (*Larus argentatus*), and various species of terns (Heinz 1998). The populations of these species also improved after the banning of many of the organochlorine pesticides.

Although this publication is about effects of contaminants on terrestrial vertebrates, it is important to mention briefly that ecosystem effects in the Great Lakes probably also included aquatic species. Concentrations of a complex mixture of chlorinated hydrocarbons were correlated with poor survival of the fry of some fish species. This evidence and other measurements on individual fish were reviewed by Giesy and Snyder (1998). Although the evidence may not be as easy to interpret as that for birds, the authors believed that synthetic halogenated hydrocarbons reduced populations of some species of fish in the Great Lakes.

Discussion

Effects of contaminants on bird populations

One of the great difficulties in extrapolating contaminant effects on individuals to higher levels of biological organization is to acquire enough information to know whether a population, community, or ecosystem has been affected. The case with DDE-induced eggshell thinning, coupled with death of adults from other organochlorine pesticides, is very unusual in that a vast amount of data was collected and it all pointed toward the same conclusion—many species populations had decreased because of the effects of these chemicals. Observations of large die-offs of adults and crushed, missing, and unhatched eggs made it easy for wildlife toxicologists to discern the ultimate consequences on large-scale populations. Many of the species (e.g., peregrine falcons, bald eagles, ospreys [*Pandion haliaetus*], and brown pelicans [*Pelecanus occidentalis*]) harmed by the organochlorine pesticides were highly visible and their populations had been monitored over many decades. Also, the discovery of eggshell thinning was facilitated by the fact that, for more than 100 years, egg collectors had saved extensive collections of eggshells from many of the affected species. One can conclude that the obvious and serious harm to birds was largely responsible for the great investigative effort directed at organochlorine pesticides. No similar effort has ever been mounted with any other type of contaminant.

With the great array of contaminants we are now responsible for studying, it will be almost impossible to marshal the resources to equal the data collection for the organochlorine pesticides. Even in the organochlorine pesticide era, other factors, such as habitat loss and human persecution of certain predatory birds interacted with the pesticides to cause population changes. If today's chemical threats are not as obvious as eggshell thinning and large numbers of dead adults, it will be even harder to sift out the contribution of contaminants to population declines. However, it is probably more important now than before to specify contaminant effects because wildlife, including birds, face many non-contaminant stresses from the increasing presence of humans.

Apart from death of adults and reproductive impairment, there has been no clear relation between other measurements of harm to individual birds and population declines. These other measurements include various behavioral and physiological responses to contaminants. It is first necessary to show that a given toxic response translates into death of adults or reproductive failure in nature before relating the response to a population change; this is not an easy translation.

Recommendations

Based on our review of the literature associated with the effects of contaminants on bird populations, we recommend the following:
- Develop case studies representative of classes of birds or contaminants that link survival, reproduction, and sublethal effects to measured effects on their local or regional populations. Representative studies are needed to improve the accuracy of extrapolation of effects on individuals to effects on populations and to reduce dependence on weight-of-evidence assessments.
- Develop methods to separate effects of contaminants from effects of environmental and other anthropogenic stressors.

Mammals

Review procedure

More than 1450 references generated by searches of electronic databases were reduced to 409 English-language references describing or reviewing research performed on wild or wild-strain mammals from 1980 to 1997. Eliminated from consideration were references on cetaceans, sirenians, livestock, laboratory animals, or veterinary medicine; theses and dissertations; specialized reports on physiological function or methods of development; and abstracts. Reports of baseline contaminant concentrations without associated biological interpretation also were excluded. Some of the retained references employed laboratory or domestic mammals and wild mammals in combination.

The remaining 409 references were scanned to identify the different types of study approaches:
- A = chemical concentrations,
- B = effects on individuals,
- C = effects on populations, and
- D = effects on communities or ecosystems.

In an effort to reduce the total number of references, subsets of references were arbitrarily chosen from those tentatively placed in the 4 (A through D) study categories. A smaller proportion of references was selected from the large categories (studies focused on tissue concentrations or toxic effects in individuals) than from the small categories (studies of populations or communities). This resulted in a working subset of 150 references for wild mammals. The percent of total references addressing specific contaminants in the sets of 150 and 409 is shown in the form "contaminant (% of 150; % of 409)":
- PCBs (27%; 26%),
- other organochlorines (24%; 23%),
- Hg (11%; 15%),
- other metals (22%; 25%),
- petroleum (24%; 17%),
- organophosphates or carbamates (15%; 9%),
- miscellaneous pesticides (8%; 6%), and
- others (12%; 12%).

The percent of total references addressing specific mammals in the sets of 150 and 409 is shown in the form "mammal (% of 150; % of 409)":
- small mammals (50%; 40%),
- mink or otter (23%; 22%),
- pinnipeds (15%; 21%),
- sea otter (10%; 9%), and
- others (20%; 26%).

Percentages do not total 100% because some references contained multiple categories of contaminants or vertebrates.

For all 150 references, attempts at information extrapolation within or among taxa, or across organizational heirarchies, were assigned to subcategories of extrapolation and evaluated according to the credibility of the effort to extrapolate findings to unmeasured biological responses, taxa, or levels of organization. References in Study Categories A through D that contained unrealized opportunities to extrapolate findings were placed in a new study category (E), and their unrealized extrapolation potential was assigned to subcategories. Some references were included in more than 1 category of study. The categories of study and subcategories of extrapolation were:

- Category A: Concentrations of contaminants in soil, food, or tissue are compared to effects attributed to the contaminant in other investigations of 1) individuals of the same species; 2) individuals of other species; 3) populations; 4) communities; or 5) ecosystems.
- Category B: Demonstrated effects on individuals of a species are evaluated for their relation to 1) other potential effects on individuals of the same species; 2) the same effects, or their consequences, on wild individuals of the same species; 3) the same or other potential effects on individuals of other species; 4) potential effects on populations of the same or other species; 5) potential effects on communities; or 6) potential effects on ecosystems.
- Category C: Demonstrated effects on populations are evaluated for their relation to 1) other potential effects on populations of the same species; 2) the same effects or other potential effects on populations of other species; 3) potential effects on communities; or 4) potential effects on ecosystems.
- Category D: Demonstrated effects on communities are evaluated for their relation to 1) potential effects on other communities or 2) potential effects on ecosystems.
- Category E: No effort is made to evaluate demonstrated effects on 1) individuals for their relation to other individual effects, other species, populations, communities, or ecosystems; 2) populations for their relation to other population effects, other species populations, communities, or ecosystems; or 3) communities for their relation to other communities or ecosystems.

The strength of each attempt at information extrapolation was rated as either "speculative," "limited," or "substantial." The term "speculative" characterized an attempt accompanied by little or no supporting information. The term "limited" characterized an attempt accompanied by supporting information on some aspects of the extrapolation. The term "substantial" characterized an attempt accompanied by supporting information on all, or nearly all, aspects of the extrapolation. The ratings within each category were converted to an index of "credibility" by multiplying the proportional frequencies of speculative, limited, and substantial ratings by factors of 1, 2, and 3, respectively, and converting to a 100-point scale. The result is an assessment of the believability of proposed links between measured and unmeasured effects that is weighted toward the presence of well-supported arguments.

Investigations into the effects of contaminants often have design characteristics or circumstances that affect data analysis and interpretation. Most of these "special conditions" can be classified into 3 categories. All references were scanned for incorporation of the following categories of conditions in the study design or discussion of results:

- Space: contaminant distribution is heterogeneous or characterized by spatial gradients; exposure is affected by seasonal movement of animals.
- Time: contaminant effects that change with time, seasonal cycles, multi-year population cycles, successional change in disturbed habitat, and successive generations.
- Interaction: combinations of contaminants; severe weather and geological events; logging; fire; agricultural practices; temperature; influence of age, sex, health, behavior, nutrition, or hibernation.

Results of review

Levels of biological organization according to their frequency of occurrence in the initial set of 409 references and in the subset of 150 references were Individual > Population > Community > Ecosystem. None of the 409 references presented data or discussion about the effects of environmental contaminants on mammals in ecosystems.

Organochlorines (including PCBs, dibenzodioxins, dibenzofurans, and pesticides) and metals were the most frequently studied contaminants in the subset of 150. Petroleum, cholinesterase-inhibiting pesticides, and herbicides were studied to a lesser degree (Table 5-2). Nearly all of the research involving tissue or soil concentrations with data extrapolation (Category A) was performed on organochlorines and metals. Research involving effects on individuals or populations with data extrapolation (Categories B and C) or effects on individuals, populations, or communities without data extrapolation (Category E) was dispersed over a greater array of contaminants.

Small mammals (insectivores and small rodents) were the most commonly studied classification of terrestrial mammal. The mink (*Mustela vison*) and river otter (*Lontra canadensis*) combination was the second most common classification studied. Pinnipeds and sea otters (*Enhydra lutris*) were studied to a lesser degree (Table 5-2). Small mammals were dominant in all categories of extrapolation except for the dominance of mink and otter in studies involving chemical concentrations (Category A).

The most commonly measured endpoints for effects on individuals were death, body weight and length, behavior, reproduction, and tissue abnormalities. Slightly less common were organ weight, blood and tissue chemistry, cholinesterase (ChE) activity, metabolic activity, blood characteristics, thermoregulation, genetic alterations, and immune function. Measures of growth, disease, endocrine function, and life span were used occasionally.

Space conditions that were addressed in the references were mostly contaminant gradients and, to a lesser extent, regional differences. Time conditions were mostly effects that changed with some measure of time and included lesser occurrences of seasonal cycles and plant succession. Interaction conditions were mostly sex and age, with lesser occurrences of animal health and ambient temperature.

Reports with information extrapolation involving concentrations of contaminants in tissue or soil (Subcategories A1 through A5; Table 5-3) were common in the selected references. None of the 44 extrapolations attempted to extend the interpretation of chemical concentrations beyond population consequences. Extrapolations restricted to individuals of the species for which information was collected (Subcategory A1) had the highest overall credibility rating. Extrapolations linking concentrations to population effects (Subcategory A3) had a credibility rating lower than the overall Category A rating but incorporated space, time, and interaction conditions more frequently than did extrapolations limited to individuals (Subcategories A1 and A2). Speculative extrapolation was rare for all of Category A.

Reports with information extrapolation involving effects of contaminants on individuals (Subcategories B1 through B6; Table 5-3) most often were limited to the organizational level of the individual (Subcategories B1 through B3). Extrapolations from individuals to populations (Subcategory B4) were slightly less common. The Subcategory B4 extrapolations had a credibility rating higher than the overall Category B rating and had the highest frequencies of occurrence, within Category B, for space, time, and interaction conditions. Only 1 reference contained information extrapolation from individuals to communities (Subcategory B5).

Few studies of the effects of contaminants on populations contained extrapolations of information to other population effects, other species populations, or communities (Subcategories C1 through C4; Table 5-3). The 5 examples of extrapolation for Subcategories C2 and C3 had high overall credibility. No studies of the effects of contaminants on communities contained extrapolation of information to other communities or to ecosystems (Subcategories D1 and D2).

Reports without extrapolations of information on the effects of contaminants (Subcategories E1 through E3; Table 5-3) had 2 notable patterns among special conditions. Reports about effects on individuals (Subcategory E1) rarely contained a spatial aspect, but, some aspect of time was included with increasing frequency as the level of organization increased from individual to population to community.

Discussion

Evaluation of references

References selected for this review dealt with classes of contaminants and mammals in approximately the same proportions as those occurring in the set of 409 references. Thus, with respect to contaminants and mammals, the selected references are thought to be representative of published literature on wild terrestrial mammals

Table 5-2 Contaminants and the most common mammals in the 150 selected references[1]

Classification	Study Category				
	A(44)	B(65)	C(5)	D(0)	E(94)
Contaminant					
PCBs[2]	61	39	0	-	16
Other organochlorines	55	20	20	-	13
Hg	36	17	0	-	3
Other metals	36	14	0	-	19
Petroleum	0	25	40	-	26
OPs, carbamates	0	18	0	-	17
Misc. pesticides[3]	0	6	20	-	15
Air pollutants[4]	0	2	0	-	4
Radionuclides	2	3	0	-	3
Others	9	8	20	-	4
Animals					
Small mammals	23	55	60	-	59
Mink, river otter	57	14	0	-	14
Pinnipeds	16	20	0	-	7
Sea otter	0	8	0	-	14
Bats	5	2	40	-	4
Muskrat	5	6	0	-	1
Weasels	0	8	40	-	2
Rabbits	2	5	0	-	5
Others	16	8	0	-	12

[1]Information is displayed according to Study Categories A (extrapolation of contaminant concentrations), B (extrapolation of effects on individuals), C (extrapolation of effects on populations), D (extrapolation of effects on communities), and E (no information extrapolation). The number in parenthesis after the category refers to the total number of extrapolations in Categories A-D and to opportunities for extrapolation in Category E; some references refer to more than one study category. Table numbers are percentages of total extrapolations or opportunities in that category. Cumulative percentages within each category do not equal 100% because some extrapolations involved multiple contaminants or mammals. [2]includes dibenzodioxins and dibenzofurans. [3]includes herbicides, economic poisons, and bacterial insecticides. [4]includes fluorides, ozone, and components of acid precipitation

during the years 1980 to 1997. The inclusion in the selected reference set of a majority of the references dealing with population or community effects permitted a better evaluation of these assessment methods than if all categories of study were subsampled proportionately. Consequently, the references in Subcategories A3–A5, B4–B5, C1–C4, D1–D2, and E2–E3 (Table 5-3) are numerically overrepresented compared with the other subcategories.

As expected, investigations involving effects of contaminants on individuals were far more common than were investigations of effects on populations or communities. The popularity of organochlorines and metals in studies with information extrapolation (Table 5-2) likely was due to a combination of the continuing environmental importance of these compounds, well-established analytical methods, and a large, detailed information landscape capable of supporting information extrapolation.

Table 5-3 Evaluation of attempts by authors to extrapolate measurements to unmeasured biological responses, taxa, or levels of organization and the incorporation of special conditions in the methods or discussion[1]

Subcategory	N	Extrapolation strength			C Index	Special conditions		
		Specul.	Limit.	Substan.		Space	Time	Interact.
A1	17	0	0.53	0.47	82	0.71	0.35	0.47
A2	11	0	1.00	0	67	0.45	0.27	0.36
A3	16	0.19	0.62	0.19	67	0.88	0.56	0.56
A4	0	--	--	--	--	--	--	--
A5	0	--	--	--	--	--	--	--
∑A1-5	44	0.07	0.68	0.25	73	0.70	0.41	0.48
B1	5	0.40	0.40	0.20	60	0	0.40	0.80
B2	13	0.46	0.31	0.23	59	0.23	0.23	0.69
B3	17	0.41	0.35	0.24	61	0.24	0.53	0.71
B4	29	0.21	0.41	0.38	72	0.38	0.76	0.93
B5	1	0	0	1.00	100	0	1.00	1.00
B6	0	--	--	--	--	--	--	--
∑B1-6	65	0.32	0.37	0.31	66	0.28	0.57	0.82
C1	0	--	--	--	--	--	--	--
C2	2	0	0.50	0.50	83	0.50	0.50	0.50
C3	3	0	0	1.00	100	0	1.00	0.33
C4	0	--	--	--	--	--	--	--
∑C1-4	5	0	0.20	0.80	93	0.20	0.80	0.40
D1	0	--	--	--	--	--	--	--
D2	0	--	--	--	--	--	--	--
E1	58	NA	NA	NA	NA	0.09	0.52	0.34
E2	29	NA	NA	NA	NA	0.38	0.79	0.59
E3	7	NA	NA	NA	NA	0.43	0.86	0.29
∑E1-3	94	NA	NA	NA	NA	0.20	0.63	0.41

[1] Refer to Review Procedure for descriptions of extrapolation subcategories. The column N contains the number of extrapolations in Subcategories A_x–D_x or opportunities in Subcategory E_x; some references were included in more than 1 subcategory. The numbers shown for classifications of extrapolation strength were frequencies of occurrence and were multiplied, from left to right, by factors of 1, 2, and 3 to produce an index of credibility (C Index) on a 100 point scale (($\sum_{1-3}$ (frequency × factor) ÷ maximum value of 3) × 100). The numbers shown for special conditions were also frequencies of occurrence. NA = not applicable

Similarly, the dominance of mink and river otter in studies involving chemical concentrations was the result of international concerns over mink and otter sensitivity to halogenated hydrocarbons, especially PCBs (Wren 1991; Kruuk and Conroy 1996). Small mammals were an overwhelming favorite of investigations documenting individual, population, and community effects, probably because of their ubiquitous distribution, small home range, high reproductive potential, and ease of capture and handling.

The credibility of the efforts of an investigator to extrapolate research results beyond the circumstances of collection is a function of the quality of the logical argument and the extent of supportive information from other sources. The large amount of published literature on chemical concentrations in soil and tissue (Category A) was responsible for minimal speculation on extrapolation of information and an overall credibility rating higher than that for extrapolations of information on effects on individuals (Category B) (Table 5-3). In general, the credibility of extrapolation was expected to be inversely related to its magnitude. This seemed to hold true for Category A, but information for effects on individuals extrapolated to populations (Subcategory B4) and the few examples of extrapolations of population information (Subcategories C2 and C3) had relatively good credibility. Further, efforts to extrapolate information on individuals to populations were numerous. It appears that during 1980 to 1997, investigators put forth a strong effort to link effects on individuals with effects on populations.

Environmental circumstances become an increasingly important determinant of contaminant effects as the level of organization increases because the number of individuals and species and the extent and variety of occupied habitat increases. Subcategories A3 (contaminant concentrations extrapolated to population effects) and B4 (extrapolation of effects from individuals to populations) had the highest frequencies of occurrence, within their categories, for special conditions (Table 5-3). Subcategory E2 (populations, no extrapolation) also had higher frequencies of occurrence for special conditions than did Subcategory E1 (individuals, no extrapolation). Spatial aspects of effects on individuals, populations, or communities (Categories B, C, E) were addressed less often than time and interaction aspects, whereas spatial aspects of contaminant concentrations (Category A) were addressed more often. The likely reason is the considerable cost and time required for examining effects of contaminants at multiple locations or on a regional scale, as compared to collecting tissue or soil samples.

Insufficient knowledge of the potential response of wild individuals to environmental contaminants is an obstacle for realistic design of experiments, interpretation of results of experiments, and accurate modeling of environmental effects. Of 35 extrapolations in Subcategories B1 through B3 (individual effects; extrapolation to same, wild, or other species), 43% were unconvincing (Table 5-3). Results of single-species laboratory toxicity tests are particularly troublesome to relate to wild species (Shore and Douben 1994). In addition, isolated laboratory or pen studies and short-term (partial life cycle or seasonal cycle) field studies are of limited usefulness in estimating the full range and complexity of natural exposure and subsequent response of wild individuals and populations. Although many of the references incorporated some characteristic of space, time, or interaction, most investigations were limited to contaminant gradients from a source, time sequence after exposure, and sex and age. Suggestions for improvement include 1) combination field and laboratory or pen studies (Westlake et al. 1982; Dickerson et al. 1994) that permit extrapolation of cause-and-effect relations to natural circumstances; 2) full-cycle or full-year (Rowley et al. 1983) and multiple-year studies (Duffy et al. 1994) that provide information on entire biological or seasonal cycles and the magnitude and trend of longer-term changes, some of which can be caused by environmental conditions; and 3) the inclusion of a manageable number of spatial and interaction conditions, in addition to time-dependent measures of response to contaminants (Olsson et al. 1994), for a realistic assessment of the potential effects of contaminant exposure. Contaminant combinations, weather, agricultural practices, behavior, and nutrition are examples of interactions that are in need of more attention.

The responses of local (Sullivan and Sullivan 1982; Gerell and Lundberg 1993) or regional (Garrott et al. 1993; Mason and Macdonald 1993) mammal populations or communities (Johnson et al. 1996; Sullivan et al. 1997) to environmental contaminants are less frequently measured than are those of individuals (Table 5-3). Measures at levels above the individual are valuable because they eliminate the uncertainty associated with extrapolation of information about individuals. Peckarsky et al. (1997), in a paper on benthic communities of streams, emphasized the difficulties of information extrapolation across organizational levels and the desirability of information collection at all organizational levels under consideration. Although the literature search did not find a published account of the effect of contaminants on mammals at the ecosystem level, whole-ecosystem impact assessments employing counts of "trophic groups" and species within groups are being used in Australia (Corbett 1997). Interestingly, the ecosystem assessment of a large uranium mine reported by Corbett (1997) does not include, or reference, collections of samples to confirm exposure to and distribution of radionuclides and other metals.

Mammals other than small mammals, mink, river otter, sea otter, and pinnipeds were poorly represented in the reference set (Table 5-2). Bats, muskrat (*Ondatra zibethica*), rabbits, and weasels were each studied in 4% of the references and 11 other species were each studied in $\leq$ 1.5% of the references. Increased information about the effects of contaminants on the many poorly studied species would permit better assessments of direct and indirect consequences to mammal communities and their habitat.

Examples of contaminants assessments involving mammals

Many of the reviewed references were associated with major assessments of specific contaminants or hazardous spill events. Some of the most notable:
- toxic effects of organochlorines on mink,
- small rodents and insectivores as indicators of metal contamination (smelters, hazardous waste sites, roads),
- effects of cholinesterase-inhibiting pesticides on small rodents and insectivores,
- habitat alteration from herbicides and its effect on small rodents and insectivores,
- effects of the 1989 *Exxon Valdez* oil spill on sea otters, and

- the interaction of organochlorines and disease in death and population declines of pinnipeds in the North Atlantic and Baltic Sea.

Summaries of 2 of these assessments are presented below.

Sea otters and the Exxon Valdez *oil spill*

On 24 March 1989, the tanker *Exxon Valdez* grounded in Prince William Sound, Alaska, and began discharging approximately 42 million liters of crude oil. A major effort was initiated by federal, state, and tribal trustees of Alaskan natural resources and the tanker owner (Exxon Oil Corporation) to determine the biotic and abiotic effects of the spilled oil in Prince William Sound and neighboring portions of the Gulf of Alaska. The studies were in response to requirements of the Natural Resource Damage Assessment (referenced in the Clean Water Act and Comprehensive Environmental Response, Compensation, and Liability Act). This example deals with efforts to determine short-term effects on the sea otter population of Prince William Sound and portions of the Gulf of Alaska, and to rescue and rehabilitate oiled otters. Counts of carcasses and live sea otters, population estimates based on population surveys and rates of carcass recovery, hydrocarbon analyses, and measures of toxicity to individuals were utilized.

Garrott et al. (1993) used data from a post-spill boat survey conducted in 1989 (Burn 1994) and the results of a 1984–1985 boat survey to estimate the loss of sea otters in Prince William Sound at 2650 (confidence limits of 500 to 5000). Using a carcass recovery estimate of 20% that was generated by a study conducted near Kodiak Island (DeGange et al. 1994), the authors estimated a total loss of 4600 for the entire spill. The work of Garrott et al. (1993) was criticized by Garshelis and Estes (1997) on the grounds that the boat-based information was inadequate for estimating otter losses. Garshelis (1997) used data on carcass collections and recovery estimates from sea otter and bird studies to estimate the loss in Prince William Sound at 750 (confidence limits of 600 to 1000). Johnson and Garshelis (1995) conducted boat surveys at 8 sites (3 oiled, 1 near an oiled area, 4 unoiled) in Prince William Sound in 1990 and 1991 using the same techniques used in boat surveys performed between 1976 and 1985. The results were compared to those of previous boat surveys and a helicopter survey of 1 island performed in 1973. The authors concluded that 1) by 1991 all oiled sites had as many otters as during the most recent (1985) pre-spill count, 2) reproduction appeared normal at all sites, 3) the number of carcasses on beaches was typical of pre-spill conditions, and 4) foraging behavior was unchanged. The absence of an observable population decrease (compared to the 1985 population) was attributed to immigration and a probable, but unmeasured, increase in the sea otter population between 1985 and 1989. The loss of sea otters in the vicinity of Kodiak Island was estimated at 494 with a model derived from information on the abundance and distribution of spilled oil and sea otters and the relation between oil abundance and otter deaths from Prince William Sound (Bodkin and Udevitz 1994). Investigators, in general, agreed that frequent population surveys (every 2 to 3 years), carcass recovery experiments, and survival estimates for sea otters at the time of the spill were needed in order to make accurate estimates of the effect on sea otter abundance. Investigators disagreed on which of these procedures were feasible considering the difficulty of conducting population surveys in remote locations and the crisis-environment characteristic of a major oil spill.

The rehabilitation effort for oiled sea otters produced much information on clinical symptoms and treatment (Wilson and McCormick et al. 1990; Wilson and Tuomi et al. 1990), macroscopic abnormalities (Groff et al. 1990), histological abnormalities (Lipscomb et al. 1993), and rehabilitation procedures and facilities (Davis 1990). Of 357 otters received by three rehabilitation centers, 197 were later released into Prince William Sound. Recovery of many of the captive sea otters, the gradual removal or degradation of spilled crude oil, and the presence of otters and their normal prey items in formerly oiled habitat (Williams 1990; Doroff and Bodkin 1994) implied that non-captive sea otters would have a reasonable chance of returning to normal function if they could survive the short-term effects of acute exposure.

Overall, the assessment of the effects of the *Exxon Valdez* oil spill on sea otters was more thorough than most evaluations of effects of major oil spills on terrestrial vertebrates. Questions remain about the long-term viability of sea otters in the affected areas, and these will probably be pursued with periodic surveys of local populations.

Effects of cholinesterase-inhibiting pesticides on small rodents and insectivores
Use of organophosphate (OP) and carbamate pesticides rapidly increased in the U.S. during the 1950s and replaced most of the banned or restricted organochlorine pesticides in the 1970s. By 1987, there were more than 100 OP and carbamate compounds registered for use in the U.S. (Smith 1987). These compounds were appealing because they had low environmental persistence and a low potential for bioaccumulation. High toxicity and a mode of action (ChE inhibition) that was not selective for invertebrate pests prompted investigations into the potential for harm to nontarget wildlife. In contrast to the sea otter example, this example is about a large group of functionally-similar compounds, an assemblage of vertebrates, and an assessment that spans decades and is not geographically limited. Investigators have employed laboratory bioassays, experiments with caged animals and animals in outdoor enclosures, and field experiments.

Cholinesterase inhibition has become the primary method of measuring exposure to OP and carbamate pesticides. Cholinesterase inhibition $\geq$ 50% is generally considered diagnostic of death in mammals and birds. Relations between ChE inhibition and other biological effects (survival, behavior, biochemistry) are complex, largely because of interspecies variability and interactions with circumstances of exposure (Grue et al. 1991; Shore and Douben 1994). Many sublethal effects of OP and carbamate poisoning (e.g., reduced tolerance to cold; impairment of endocrine function, immunocompetence, sensory perception, learning, and memory; changes in behavior; and reduced food and water intake) have been demonstrated, but few studies have attempted to relate them to survival or reproduction (Shore and Douben 1994).

Wild mammals have been killed by routine spraying of OP and carbamate pesticides, but intraspecies variation in ChE activity can make confirmation of the cause of death difficult (Greig-Smith 1991). Short-term reductions in abundance, survival, or recruitment have been reported for voles (Barrett 1988; Edge et al. 1996; Schauber et al. 1997) and mice (Schauber et al. 1997) in enclosure experiments. Short-term reductions in local populations of shrews (Innes and Bendell 1989) and mice (Johnson et al. 1991) and reduced survival of voles (Jett et al. 1986) have been reported in field studies. Enclosure and field studies have generated limited evidence for adverse effects on reproduction (e.g., Barrett 1988). Evidence for long-term or large-scale reductions in populations has not been reported.

Overall, the assessment of effects of ChE-inhibiting pesticides on small rodents and insectivores is incomplete. There is a need for combined laboratory and field studies to permit extrapolation of sublethal effects of experimental OP and carbamate poisoning to reproduction and survival (e.g., Dell'Omo and Shore 1996). Identification and determination of the importance of sublethal effects in wild species also is needed (Shore and Douben 1994).

Recommendations

Recommendations for an improved understanding of the effects of toxic contaminants on mammals at organizational levels above the individual are as follows:
- Acquire more information on the range and circumstances (i.e., interactions) of the response of wild individuals to contaminants.
- Conduct more studies involving measurements of the response of populations and communities to contaminant exposure.
- Determine the effects of contaminants on terrestrial mammals that have been poorly studied.

Reptiles and Terrestrial Amphibians

Review procedure

Literature surveys totaling 1441 citations were reviewed to evaluate the frequency with which studies attempted to extrapolate observed effects of contaminants on reptiles and terrestrial amphibians to populations, communities, and ecosystems. Published abstracts and unpublished theses and dissertations were eliminated. Some papers could not be obtained for examination. The majority of full papers included in these summaries had little bearing on the goals of the review and were also eliminated. Included in this category were numerous studies that examined amphibians as model systems used in a variety of physiological and pharmacological investigations unrelated to effects of contaminants on free-living animals. Examples of such papers were those in which preparations of frog skin were used to study normal mechanisms of transport across living membranes and the effects of various chemicals on these processes.

As the focus of the analysis was on terrestrial vertebrates, preliminary screening also sought to eliminate investigations that emphasized aquatic rather than terrestrial species. Certain general criteria were necessary to distinguish studies emphasizing terrestrial verses aquatic species, ecosystems, and routes of contamination. This exercise resulted in some arbitrary and imperfect distinctions. All studies addressing contamination of reptilian species in the environment were included in the analysis because the reptilian mode of reproduction does not involve an obligatory aquatic phase; reptiles are thus arguably terrestrial vertebrates, even though some species are closely tied to aquatic or marine environments. For amphibians, all studies in which transformed individuals were subjects of at least part of the investigation are included, with the exception of a few species (e.g., *Xenopus laevis*) having totally aquatic larvae and adults. Even highly terrestrial adult amphibians such as toads may be exposed to contaminants through aquatic environments (Hall and Swineford 1979), and many of the studies of reptiles examine species characteristic of aquatic or marine environments. One effect of restricting the focus of this analysis to terrestrial stages of amphibians is that it eliminates the great bulk of contaminant research performed on amphibians, most of which has used larval forms and has often applied the methods of aquatic toxicology. After screening, 95 studies were selected for analysis. All studies were categorized, evaluated for credibility of information extrapolation, and examined for incorporation of "special conditions" according to the procedure described for mammals.

Results

Studies based on animals exposed in the field were most frequent (73%), whereas controlled experimental investigations in the laboratory comprised the remaining 27%. Only 1 study (Baker 1985) had both field and controlled experimental components. Several of the field studies involved at least partial control, e.g., a study by Frisbie and Wyman (1992).

Despite the foregoing attempts to restrict the analysis to terrestrial systems, there is a strong bias in the literature herein regarded as "terrestrial" toward animals associated with aquatic and marine systems. Of 101 species or groups of species examined in the 95 studies chosen for analysis, 87 subjects of study were either turtles (37), frogs (30), crocodilians (8), or salamanders (12), most of which are strongly tied to aquatic or marine environments. Only 14 studies addressed lizards or snakes. While a few species of salamanders studied are totally terrestrial, some of

Table 5-4 Contaminants, amphibians, and reptiles in the 95 selected references.[1]

Classification	Study Category				
	A(41)	B(42)	C(7)	D(4)	E(7)
Contaminant					
PCBs[2]	57	17	14	0	0
Other organochlorines	41	12	28	0	42
Hg	17	12	14	0	0
Other metals	32	17	14	50	14
Petroleum	5	5	0	0	14
OPs, carbamates	7	17	14	50	0
Misc. pesticides[3]	0	14	0	0	0
Air pollutants[4]	2	14	28	0	14
Radionuclides	2	5	42	0	0
Animal					
Turtles	49	31	28	0	28
Frogs	20	40	28	50	28
Crocodilians	7	7	14	0	14
Lizards or snakes	17	9	14	25	14
Salamanders	5	14	28	25	14

[1] Information is displayed according to Study Categories A (extrapolation of contaminant concentrations), B (extrapolation of effects on individuals), C (extrapolation of effects on populations), D (extrapolation of effects on communities), and E (no information extrapolation). The number in parenthesis after the category refers to the total number of extrapolations in Categories A–D and to opportunities for extrapolation in Category E; some references were included in more than 1 study category. Table numbers are percentages of total extrapolations or opportunities in that category. Cumulative percentages within each category do not equal 100% because some extrapolations contained multiple contaminants or species. [2] includes dibenzodioxins and dibenzofurans. [3] includes herbicides, synthetic pyrethroids, piscicides, and economic poisons. [4] includes fluorides, ozone, and components of acid precipitation.

the snakes investigated have strong aquatic affinities, and the pattern of overall emphasis (85% to 90%) on aquatic or marine species remains.

A summary of the kinds of contaminants examined in the selected studies (Table 5-4) reveals a distribution similar to that of mammals except for a notable decrease in petroleum and an increase in air pollutants. The increase in air pollutants probably reflects concern over the effects of acid precipitation on amphibians.

The frequency with which reviewed studies attempted to extrapolate findings obtained from reptiles and amphibians to other species, or to higher levels of organization, was higher than that reported for mammals (Subcategories A2 through A5, B3 through B6; Tables 5-3 and 5-5). No attempt was made to extrapolate results in only 5% of the reviewed studies of reptiles and amphibians, whereas approximately 50% of reviewed studies of mammals had no attempt at extrapolation. Although objective criteria were developed for evaluating studies, some differences between the 2 reviewers as to what constitutes extrapolation were probably inevitable and could account for some of the differences between mammalian and herpetological references. Nevertheless, it seems likely that real differences exist. A qualitative examination of the selected studies indicated that a significant portion (at least 37%) of studies of reptiles and terrestrial amphibians were under-

Table 5-5 Evaluation of attempts by authors to extrapolate measurements to unmeasured biological responses, taxa, or levels of organization, and the incorporation of special conditions in the methods or discussion.[1]

Subcategory	N	Extrapolation Strength			C Index	Special Conditions		
		Specul.	Limit.	Substan.		Space	Time	Interact.
A1	9	0.56	0.33	0.11	52	0.33	0.11	0
A2	26	0.34	0.58	0.08	58	0.27	0.04	0
A3	5	0.80	0.20	0	40	0.20	0	0
A4	1	0	0	1.00	100	1.00	0	0
A5	0	--	--	--	--	--	--	--
$\sum$A1-5	41	0.44	0.46	0.07	52	0.29	0.05	0
B1	0	--	--	--	--	--	--	--
B2	5	0.40	0.40	0.20	60	0	0	0
B3	32	0.03	0.69	0.27	74	0.10	0.10	0
B4	5	0.40	0.40	0.20	60	0.20	0	0
B5	0	--	--	--	--	--	--	--
B6	0	--	--	--	--	--	--	--
$\sum$B1-6	42	0.12	0.62	0.26	71	0.10	0.07	0
C1	5	0	0.60	0.40	80	0.40	0	0
C2	0	--	--	--	--	--	--	--
C3	1	0	0	1.00	100	1.00	0	0
C4	1	0	1.00	0	67	0	0	0
$\sum$C1-4	7	0	0.57	0.43	81	0.43	0	0
D1	0	--	--	--	--	--	--	--
D2	4	0.50	0	0.50	67	0	0	0.50
$\sum$D1-2	4	0.50	0	0.50	67	0	0	0.50
E1	6	NA	NA	NA	NA	0.17	0.17	0.33
E2	0	NA	NA	NA	NA	--	--	--
E3	1	NA	NA	NA	NA	0	1.00	0
$\sum$E1-3	7	NA	NA	NA	NA	0.14	0.28	0.28

[1]Refer to Review Procedure for descriptions of extrapolation subcategories. The column N contains the number of extrapolations in Subcategories A_x–D_x or opportunities in Subcategory E_x; some references were included in more than 1 subcategory. The numbers shown for classifications of extrapolation strength were frequencies of occurrence and were multiplied, from left to right, by factors of 1, 2, and 3 to produce an index of credibility (C Index) on a 100 point scale (($\sum_{1-3}$ (frequency × factor) ÷ maximum value of 3) × 100). The numbers shown for special conditions were also frequencies of occurrence. NA = not applicable

taken because these animals were regarded as good indicators of general contamination in the environments in which they were found. These papers usually stated that assessment of contamination of the environment is the primary purpose of the study (e.g., Sabourin et al. 1984) or included this goal in the study title (e.g., Overmann and Krajicek 1995).

Reports with information extrapolations involving concentrations of contaminants (Category A; Table 5-5) were common, but only 1 interpretive attempt went beyond population consequences. Speculative extrapolation was common and the overall credibility ratings for studies of reptiles and amphibians were poor compared to those of mammals (Table 5-3). Reports with information extrapolations involving effects of contaminants on individuals (Category B; Table 5-5) had greatly improved credibility ratings, similar to those of mammals. No attempts were made to extrapolate results beyond populations. The small number of extrapolation attempts in studies of the effects of contaminants on populations (Category C; Table 5-5) or communities (Category D; Table 5-5) had good overall credibility. Special conditions were poorly represented in all of the categories.

Discussion

Evaluation of references

Because efforts to extrapolate effects observed in reptiles and amphibians to other species or to environmental media were more frequent than those based on studies of mammals, it seems useful to examine the way in which reptiles and amphibians have been viewed in general and by many investigators. Until recently, there was little general concern for the well-being of reptilian and amphibian populations (Hall and Henry 1992), and many investigators regarded their primary value as indicators of likely threats to more highly valued biological resources. Hence, in some cases there was a bias at the outset favoring the extrapolation of results to larger-scale systems, or at least to other species or groups of species. Likewise, special conditions tending to limit or modify findings were seldom recorded in studies of reptiles and amphibians because studies were focused not on the species or populations under study but rather on the general health of ecosystems or on threats to more valued components such as birds, mammals, or humans.

Many studies of reptiles reviewed in this analysis, particularly those focused on turtles (see Meyers-Schöne and Walton 1994), examined concentrations of contaminants in reptiles or amphibians as indicators of the overall levels of contamination in particular environments, with the apparent ultimate goal of assessing hazards to humans using those environments. Extrapolation of concentrations of contaminants to effects observed elsewhere or to populations, communities, or ecosystems is mostly limited to comparison with standards established for human consumption or are incidental and speculative. These studies technically fit the criteria for extrapolation in this exercise, and the approach may be useful in the few instances where reptiles or amphibians are sources of food for humans (e.g., Stone et al. 1980), but such extrapolation is of little use in understanding effects on reptile and amphibian populations.

In other instances (e.g., Ryan et al. 1986) the primary focus of investigation seemed to be to help understand the distribution of particular chemical contaminants. Reptiles or amphibians were viewed primarily as matrices useful for assessing sources, pathways, and sinks of those chemicals in the environment. While extrapolation based on such studies may not have been strong, the intent of investigations was directed toward ecosystems and they were scored as having engaged in extrapolation. In contrast, studies examining populations for which there is concern, e.g., reports on contaminants in sea turtles (e.g., Aguirre et al. 1994), usually attempted extension no further than to conspecifics.

Certain reptiles, and occasionally amphibians, have been touted as useful indicators of the overall levels of contamination within ecosystems because they are believed to be sinks for certain kinds of contaminants. The logic might proceed as follows:

- Many reptiles and amphibians occupy portions of aquatic environments, which are regarded as sinks for contaminants.
- They are more likely to be spatially restricted within their habitats than are birds and some mammals.
- Some, such as snakes and turtles, are predators or scavengers and reside high on food chains.
- As cold-blooded animals, their powers of catabolism and depuration are relatively slow compared with rates in endotherms.
- Some, particularly turtles, are regarded as relatively insensitive to a broad range of insults.
- Many reptiles have long life spans.
- This suite of characteristics makes them more likely to take up and retain environmental contaminants than are most other kinds of animals.

This chain of assumptions seems to have been widely accepted, despite lack of experimental proof that all or any of them are valid indications of the potential for reptiles and amphibians to accumulate environmental contaminants. Moreover, if all the assumptions are true and certain reptiles and amphibians are truly good indicators of contaminant accumulation, it could equally be argued that they are poor indicators of biological response within ecosystems.

Some studies (e.g., Yawetz et al. 1983) have tended to confirm that reptiles or amphibians have some of the qualities attributed to them, but others have shown negative (e.g., Heinz et al. 1980; Ford and Hill 1991) or inconsistent (e.g., Neithammer et al. 1985) results. Hall and Clark (1982) showed in 1 very limited comparison that lizards (*Anolis carolinensis*) closely resembled commonly tested birds in their responses to a range of OP insecticides, but there have been few other comparisons of this type. Some reptiles, including turtles have been shown to have temperature-influenced sex determination and to be unusually sensitive to hormone mimics or disruptors (Bergeron et al. 1994). Although they might be viewed as sensitive indicators of the presence of these chemicals, it would be more accurate to characterize their mode of sex determination as a special case with potential significance for reptile populations. Whether or not reptiles and amphibians can be regarded as reliable indicators of contamination, they should be recognized as free-living animals that exist as populations within biotic communities.

The value of the reviewed literature on reptiles and amphibians is reduced by the existence of few controlled experimental studies. Most studies dealt with mechanisms (e.g., Marty et al. 1995), a few (e.g., Roudebush 1988) dealt with the interpretation of field observations, and there was only 1 instance (Baker 1985) in which integrated field and laboratory investigations were applied to a single problem. The scarcity of this last approach, which has proven instrumental in evaluating effects on populations of birds, may have greatly hampered the effectiveness of extrapolation of documented effects on reptiles and amphibians beyond the immediate observations.

In an earlier report, (Hall 1980) reviewed the literature on effects of environmental contaminants on reptiles. Of 68 papers or reports examined, 27 recorded effects (death or no effect) on reptiles from operational applications of pesticides, 36 reported residues in the tissues of reptiles, and 6 reported body burdens of chemicals in animals believed to have been killed by pesticides. Only a few studies attempted to predict effects on populations of reptiles, usually in a speculative manner. While reports of death observed with pesticide applications usually were not technically sophisticated, they tended to report the scope and scale of effects on biota that was present, and, hence, they often gave a better indication of likely effects on populations, communities, and ecosystems than did carefully designed studies on 1 or a few species intended to establish statistical evidence of toxicity. This pattern holds true in the present analysis 2 decades later; those studies in which extrapolation was most likely were those that examined effects of contamination on a wide variety of vertebrate biota in the field and were often, on this basis, able to project likely effects on ecosystem structure and function.

Extrapolation of results from individuals to populations, communities, and ecosystems, or from any lower level of organization to higher levels, is certain to be less defensible scientifically than are the primary results of a study examined alone. The desire for scientific rigor, particularly in areas in which regulatory and legal strictures may apply, probably has prevented investigators from burdening their papers with material that is speculative, or at least less rigorous than the primary findings presented. Referees and journal editors have doubtless exacerbated these tendencies by counseling the removal of assertions not directly supported by data. Yet, asserting the application of the particular to the general is at the core of scientific induction, and what is seen by 1 scientist as engaging in speculation may be seen by another as developing testable hypotheses. Extrapolation of findings on reptiles and amphibians to higher levels of organization may become more acceptable and useful to the scientific community if such extrapolation is framed in terms of at least potentially testable hypotheses rather than as speculative assertions.

Examples of contaminants assessments involving reptiles or amphibians

Major assessments of specific contaminants or hazardous material incidents rarely utilized reptiles or amphibians as primary study species. A summary of probably the best example is presented below.

Acidic precipitation and amphibians

The most significant attempt at large-scale assessment of a particular contamination threat to amphibians or reptiles concerned the effects of acidic precipitation on amphibian populations. Most such work focused on aquatic life stages, and the hundreds of papers addressing this issue were not reviewed here. Relatively few studies have examined terrestrial stages, but some of these studies (e.g., Wyman and Hawsley-Lescault 1987; Wyman and Jancola 1992) have documented effects of soil pH on the local distribution of amphibian species or mechanisms capable of producing such distribution patterns. Despite growing concern for declining amphibian populations worldwide (Hall and Henry 1992) and much research in the laboratory and in the field, it has not been possible to relate these well-documented declines, including large-scale extirpations of amphibian populations, to acidic precipitation (e.g., Corn and Vertucci 1992).

Recommendations

Recommendations based on this evaluation of the literature on reptiles and amphibians are as follows:
- Conduct more field studies that broadly examine taxonomic groups within ecosystems, despite the fact that such investigations may be much less able to provide scientific proof than more closely focused studies.
- Support the interpretation of observations made in the field with controlled experimental investigations.
- Recognize that reptiles and amphibians may be no more or less able than other species of indicating the degree to which an ecosystem can be affected by certain kinds of contamination.
- Frame attempts to extrapolate findings to higher levels of organizations in terms of potentially testable hypotheses.

Conclusions

A need for a broader range of information on effects of contaminants on individuals exists among the 4 classes of terrestrial vertebrates, especially mammals, reptiles, and amphibians. Separation of contaminant effects from other effects and reduction of speculative extrapolation within and among species requires information that can be produced only by combined field and laboratory investigations that incorporate seasonal or annual cycles and important spatial and interaction conditions.

Assessments of contaminant effects at the population level and higher are frequently dependent on extrapolations from a lower organizational level. Actual measurements of the effects of contaminants on populations or communities, possibly in conjunction with case studies that establish relations between effects on individuals and effects on populations, are needed to reduce the uncertainty associated with these extrapolations. Associated with these assessment levels is the need for acceptable definitions of what we mean when we refer to a "meaningful population change" or an "effect on communities or ecosystems." At these higher levels of organization we are also confronted with the need for procedures useful for separating contaminant effects from effects caused by other environmental conditions.

Although the bulk of literature surveyed was of the focused cause-and-effect type that is necessary for proving relations between contaminants and wildlife, community or ecosystem field assessments, as sometimes performed with reptiles and amphibians, might be a useful alternative for estimating the potential of a contaminant to cause environmental harm. Assumptions about the special usefulness of reptiles and amphibians as environmental indicators ought to be tested with comparisons to mammals and birds.

Information on the effects of contaminants above the individual level is needed to generate accurate estimates of the potential consequences of anthropogenic pollution (e.g., ecological risk assessments). However, realized population, or higher, levels of effects should not be part of regulatory guidelines because the threshold of harm would be too high to be used as a catalyst for action. Measures of realized population or community effects could be used to evaluate the effectiveness of regulatory actions and assess chronic or difficult environmental problems.

Some of these information needs can be satisfied with modest effort and expense, but much of the suggested work that incorporates great complexity or long duration is likely to be difficult to accomplish. Cooperation among investigators with different specialties and a willingness by government, academia, and corporate organizations to support the most challenging work will be necessary. Because we are unlikely to have the financial resources to evaluate more than a small number of contaminants for effects at the levels of population, community, or ecosystem, we might need to thoroughly study a few contaminants and then extend the findings to functionally similar contaminants. If sufficient cooperation and organizational support does not materialize, the pursuit of estimation methods will overshadow the collection of actual information on relations between contaminants and wildlife.

References

Aguirre AA, Balazs GH, Zimmerman B, Galey FD. 1994. Organic contaminants and trace metals in the tissues of green turtles (*Chelonia mydas*) afflicted with fibropapillomas in the Hawaiian Islands. *Mar Pollut Bull* 28:109–114.

Baker KN. 1985. Laboratory and field experiments on the responses by two species of woodland salamanders to malathion–treated substrates. *Arch Environ Contam Toxicol* 14:685–691.

Barrett GW. 1988. Effects of Sevin on small-mammal populations in agricultural and oil-field ecosystems. *J Mamm* 69:731–739.

Barton AL. 1994. Ecological risk assessment in the Office of Pesticide Programs. In: Kendall RJ, Lacher TE, editors. Wildlife toxicology and population modeling: Integrated studies of agroecosystems. Boca Raton FL: Lewis. p 27–31.

Bergeron JM, Crews J, McLachlan JA. 1994. PCBs as environmental estrogens: Turtle sex determination as a biomarker of environmental contamination. *Environ Health Perspect* 102:780–781.

Blus LJ. 1996. DDT, DDD, and DDE in birds. In: Beyer WN, Heinz GH, Redmon-Norwood AW, editors. Environmental contaminants in wildlife: Interpreting tissue concentrations. Boca Raton FL: Lewis. p 49–71.

Bodkin JL, Udevitz MS. 1994. An intersection model for estimating sea otter mortality along the Kenai Peninsula. In: Loughlin TR, editor. Marine mammals and the *Exxon Valdez*. San Diego CA: Academic. p 81–95.

Bowerman WW. 1993. Regulation of bald eagle (*Haliaeetus leucocephalus*) productivity in the Great Lakes Basin: An ecological and toxicological approach [dissertation]. East Lansing MI: Michigan State University. 291 p.

Burn DM. 1994. Boat-based population surveys of sea otters in Prince William Sound. In: Loughlin TR, editor. Marine mammals and the *Exxon Valdez*. San Diego CA: Academic. p 61–80.

Cairns J Jr, Niederlehner BR. 1996. Developing a field of landscape ecotoxicology. *Ecol Appl* 6:790–796.

Cairns J Jr, Pratt JR. 1993. Trends in ecotoxicology. *Sci Total Environ* Suppl. 1993:7–22.

Chapman PM, Fairbrother, A, Brown D. 1998. A critical evaluation of safety (uncertainty) factors for ecological risk assessment. *Environ Toxicol Chem* 17:99–108.

Clements WH, Kiffney PM. 1994. Assessing contaminant effects at higher levels of biological organization. *Environ Toxicol Chem* 13:357–359.

Corbett L. 1997. Environmental monitoring at Ranger Mine: A whole-ecosystem approach. In: Demonstrating environmental excellence 97. Proceedings of the 22nd annual environmental workshop; 12–17 Oct 1997; Adelaide, South Australia. Dickson, Australian Capitol Territory: Minerals Council of Australia. p 240–254.

Corn PS, Vertucci FA. 1992. Descriptive risk assessment of the effects of acidic deposition on Rocky Mountain amphibians. *J Herpetol* 26:361–369.

Cromartie E, Reichel WL, Locke LN, Belisle AA, Kaiser TE, Lamont TG, Mulhern BM, Prouty RM, Swineford DM. 1975. Residues of organochlorine pesticides and polychlorinated biphenyls and autopsy data for bald eagles, 1971–1972. *Pestic Monit J* 9:11–14.

Davis RW. 1990. Advances in rehabilitating oiled sea otters: The Valdez experience. *Wildl J* 13:30–41.

DeGange AR, Doroff AM, Monson DH. 1994. Experimental recovery of sea otter carcasses at Kodiak Island, Alaska, following the Exxon Valdez oil spill. *Mar Mammal Sci* 10:492–496.

Dell'Omo G, Shore RF. 1996. Behavioral effects of acute sublethal exposure to dimethoate on wood mice, *Apodemus sylvaticus*: II — field studies on radio-tagged mice in a cereal ecosystem. *Arch Environ Contam Toxicol* 31:538–542.

Dickerson RL, Hooper MJ, Gard NW, Cobb GP, Kendall RJ. 1994. Toxicological foundations of ecological risk assessment: Biomarker development and interpretation based on laboratory and wildlife species. *Environ Health Perspect* 102 (Suppl. 12):65–69.

Doroff AM, Bodkin JL. 1994. Sea otter foraging behavior and hydrocarbon levels in prey. In: Loughlin TR, editor. Marine mammals and the *Exxon Valdez*. San Diego CA: Academic Press. p 193–208.

Duffy LK, Bowyer RT, Testa JW, Faro JB. 1994. Evidence for recovery of body mass and haptoglobin values of river otters following the Exxon Valdez oil spill. *J Wildl Dis* 30:421–425.

Edge WD, Carey RL, Wolff JO, Ganio LM, Manning T. 1996. Effects of Guthion 2S on *Microtus canicaudus*: A risk assessment validation. *J Appl Ecol* 33:269–278.

Ford WM, Hill EP. 1991. Organochlorine pesticides in soil sediments and aquatic animals in the Upper Steele Bayou watershed of Mississippi. *Arch Environ Contam Toxicol* 20:161–167.

Frisbie MP, Wyman RL. 1992. The effect of soil chemistry on sodium balance in the red-backed salamander; a comparison of two forest types. *J Herpetol* 26:434–442.

Garrott RA, Eberhardt LL, Burn DM. 1993. Mortality of sea otters in Prince William Sound following the *Exxon Valdez* oil spill. *Mar Mammal Sci* 9:343–359.

Garshelis DL. 1997. Sea otter mortality estimated from carcasses collected after the *Exxon Valdez* oil spill. *Conserv Biol* 11:905–916.

Garshelis DL, Estes JA. 1997. Sea otter mortality from the *Exxon Valdez* oil spill: Evaluation of an estimate from boat-based surveys. *Mar Mammal Sci* 13:341–351.

Gerell R, Lundberg KG. 1993. Decline of a bat *Pipistrellus pipistrellus* population in an industrialized area in south Sweden. *Biol Conserv* 65:153–157.

Giesy JP, Snyder EM. 1998. Xenobiotic modulation of endocrine function in fishes. In: Kendall RJ, Dickerson RL, Giesy JP, Suk WP, editors. Principles and processes for evaluating endocrine disruption in wildlife. Pensacola FL: SETAC. p 155–237.

Gilbertson M. 1997. Advances in forensic toxicology for establishing causality between Great Lakes epizootics and specific persistent toxic chemicals. *Environ Toxicol Chem* 16:1771–1778.

Greig-Smith PW. 1991. Use of ChE measurements in surveillance of wildlife poisoning in farmland. In: Mineau P, editor. Cholinesterase–inhibiting insecticides, their impact on wildlife and the environment. Chemicals in agriculture, volume 2. New York NY: Elsevier. p 127–150.

Groff JM, Blake JE, Rideout B, Basaraba R, Wilson D. 1990. Necropsy observations in Alaskan sea otters (*Enhydra lutis*) from Prince William Sound affected by the Exxon Valdez oil spill. *IAAAM Proc* 21:31–32.

Grue CE, Hart ADM, Mineau P. 1991. Biological consequences of depressed brain cholinesterase activity in wildlife. In: Mineau P, editor. Cholinesterase–inhibiting insecticides, their impact on wildlife and the environment. Chemicals in agriculture, Volume 2. New York: Elsevier. p 151–209.

Hall RJ. 1980. Effects of environmental contaminants on reptiles: A review. Washington DC: U.S. Fish and Wildlife Service. *Spec Sci Rep Wildl* 228:1–12.

Hall RJ, Clark, Jr., DR. 1982. Responses of the iguanid lizard Anolis carolinensis to four organophosphorus pesticides. *Environ Pollut (Ser. A)* 28:45–52.

Hall RJ, Henry PFP. 1992. Assessing effects of pesticides on amphibians and reptiles: Status and needs. *Herpetological J* 2:65–71.

Hall RJ, Swineford D. 1979. Uptake of methoxychlor from food and water by the American toad (*Bufo americanus*). *Bull Environ Contam Toxicol* 23:335–337.

Hallman TG, Lassiter RR. 1994. Individual–based mathematical modeling approaches in ecotoxicology: A promising direction for aquatic population and community ecological risk assessment. In: Kendall RJ, Lacher, Jr. TE, editors. Wildlife toxicology and population modeling: Integrated studies of agroecosystems. Boca Raton FL: Lewis. p 531–542.

Heath RG, Spann JW, Kreitzer JF. 1969. Marked DDE impairment of mallard reproduction in controlled studies. *Nature* 224:47–48.

Heinz, GH. 1996. Selenium in birds. In: Beyer WN, Heinz GH, Redmon-Norwood AW, editors. Environmental contaminants in wildlife: Interpreting tissue concentrations. Boca Raton FL: Lewis. p 447–458.

Heinz GH. 1998. Contaminant effects on Great Lakes fish-eating birds: A population perspective. In: Kendall RJ, Dickerson RL, Giesy JP, Suk WP, editors. Principles and processes for evaluating endocrine disruption in wildlife. Pensacola FL: SETAC. p 141–154.

Heinz GH, Brody M, Blus LJ. 1994. How valuable are the results of models and laboratory studies when extended to field situations? In: Kendall RJ, Lacher, Jr. TE, editors. Wildlife toxicology and population modeling: Integrated studies of agroecosystems. Boca Raton FL: Lewis. p 551–555.

Heinz GH, Haseltine SD, Hall RJ, Krinitsky AJ. 1980. Organochlorine and mercury residues in snakes from Pilot and Spider Islands, Lake Michigan—1978. *Bull Environ Contam Toxicol* 25:738–743.

Hickey JJ, editor. 1969. Peregrine falcon populations: Their biology and decline. Madison WI: University of Wisconsin Press.

Hickey JJ, Anderson DW. 1968. Chlorinated hydrocarbons and eggshell changes in raptorial and fish-eating birds. *Science* 162:271–273.

Innes DGL, Bendell JF. 1989. The effects on small-mammal populations of aerial applications of *Bacillus thuringiensis*, fenitrothion, and Matacil used against jack pine budworm in Ontario. *Can J Zool* 67:1318–1323.

Jett DA, Nichols JD, Hines JE. 1986. Effect of Orthene on an unconfined population of the meadow vole (*Microtus pennsylvanicus*). *Can J Zool* 64:243–250.

Johnson CB, Garshelis DL. 1995. Sea otter abundance, distribution, and pup production in Prince William Sound following the *Exxon Valdez* oil spill. In: Wells PG, Butler JN, Hughes JS, editors. Exxon Valdez oil spill: Fate and effects in Alaskan waters. ASTM STP 1219. Philadelphia PA: American Society for Testing and Materials. p 894–929.

Johnson IP, Flowerdew JR, Hare R. 1991. Effects of broadcasting and of drilling methiocarb molluscicide pellets on field populations of wood mice, *Apodemus sylvaticus*. *Bull Environ Contam Toxicol* 46:84–91.

Johnson KH, Olson RA, Whitson TD. 1996. Composition and diversity of plant and small mammal communities in Tebuthiuron–treated big sagebrush (*Artemisia tridentata*). *Weed Technol* 10:404–416.

Kareiva P. 1994. Editorial introduction to: Special feature. Higher order interactions as a foil to reductionist ecology. *Ecology* 75:1527–1528.

Kendall RJ, Lacher, Jr. TE, editors. 1994. Wildlife toxicology and population modeling: Integrated studies of agroecosystems. Boca Raton FL: Lewis.

Keith JO. 1996. Residue analyses: How they were used to assess the hazards of contaminants to wildlife. In: Beyer WN, Heinz GH, Redmon-Norwood AW, editors. Environmental contaminants in wildlife: Interpreting tissue concentrations. Boca Raton FL: Lewis. p 1–47.

Kruuk H, Conroy JWH. 1996. Concentrations of some organochlorines in otters (*Lutra lutra* L.) in Scotland: Implications for populations. *Environ Pollut* 92:165–171.

Lipscomb TP, Harris RK, Moeller RB, Pletcher JM, Haebler RJ, Ballachey BE. 1993. Histopathologic lesions in sea otters exposed to crude oil. *Vet Pathol* 30:1–11.

Longcore JR, Samson FB, Whittendale, TW Jr. 1971. DDE thins eggshells and lowers reproductive success of captive black ducks. *Bull Environ Contam Toxicol* 6:485–490.

Marty J., Michel XR, Narbonne JF, Ferrier V. 1995. In vivo metabolism of benzo(a)pyrene in a lower vertebrate, the newt *Pleurodeles waltl*. *Ecotoxicol Environ Safety* 32:51–57.

Mason CF, Macdonald SM. 1993. Impact of organochlorine pesticide residues and PCBs on otters (*Lutra lutra*): A study from western Britain. *Sci Total Environ* 138:127–145.

Meyers-Schöne L, Walton BT. 1994. Turtles as monitors of chemical contaminants in the environment. *Rev Environ Contam Toxicol* 135:93–153.

Neithammer KR, Atkinson RD, Baskett TS, Samson FB. 1985. Metals in riparian wildlife of the lead mining district of southeastern Missouri. *Arch Environ Contam Toxicol* 14:213–223.

Ohlendorf HM. 1989. Bioaccumulation and effects of selenium in wildlife. In: Selenium in agriculture and the environment. SSSA Special Publ. No. 23. Madison WI: American Society of Agronomy and Soil Science Society of America. p 133–177.

Ohlendorf HM. 1996. Selenium. In: Fairbrother A, Locke LN, Hoff, GL, editors. Noninfectious diseases of wildlife. Second edition. Ames IO: Iowa State University Press. p 128–140.

Ohlendorf HM, Hothem RL. 1995. Agricultural drainwater effects on wildlife in Central California. In: Hoffman DJ, Rattner BA, Burton Jr GA, Cairns Jr J, editors. Handbook of ecotoxicology. Boca Raton FL: Lewis. p 577–595.

Ohlendorf HM, Hoffman DJ, Saiki MK, Aldrich TW. 1986. Embryonic mortality and abnormalities of aquatic birds: Apparent impacts of selenium from irrigation drainwater. *Sci Total Environ* 52: 49–63.

Ohlendorf HM, Kilness AW, Simmons JL, Stroud RK, Hoffman DJ, Moore JF. 1988. Selenium toxicosis in wild aquatic birds. *J Toxicol Environ Health* 24:67–92.

Olson MM, Welsh D. 1993. Selenium and eared grebe embryos from Stewart Lake National Wildlife Refuge, North Dakota . *Prairie Naturalist* 25:119–126.

Olsson M, Karlsson B, Ahnland E. 1994. Diseases and environmental contaminants in seals from the Baltic and the Swedish west coast. *Sci Total Environ* 154:217–227.

Overmann SR, Krajicek JJ. 1995. Snapping turtles (*Chelydra serpentina*) as biomonitors of lead contamination of the Big River in Missouri's Old Lead Belt. *Environ Toxicol Chem* 14:689–695.

Peakall DB, Lincer JL. 1996. Do PCBs cause eggshell thinning? *Environ Pollut* 91:127–129.

Peckarsky BL, Cooper SD, McIntosh AR. 1997. Extrapolating from individual behavior to populations and communities in streams. *J N Am Benthol Soc* 16:375–390.

Porter RD, Wiemeyer SN. 1969. Dieldrin and DDT: Effects on sparrow hawk eggshells and reproduction. *Science* 165:199–200.

Potts GR. 1977. Population dynamics of the grey partridge: Overall effects of herbicides and insecticides on chick survival rates. *Proc Internat Congr Game Biol* 13:203–211.

Potts GR. 1980. The effects of modern agriculture, nest predation and game management on the population ecology of partridges *Perdix perdix* and *Alectoris rufa*. *Adv Ecol Res* 11:2–79.

Potts GR. 1986. The partridge: Pesticides, predation, and conservation. London: Collins.

Ratcliffe DA. 1967. Decrease in eggshell weight in certain birds of prey. *Nature* 215:208–210.

Ratcliffe D. 1980. The peregrine falcon. Vermillion SD: Buteo Books.

Roudebush RE. 1988. A behavioral assay for acid sensitivity in two desmognathine species of salamanders. *Hepetologica* 44:392–395.

Rowley MH, Christian JJ, Basu DK, Pawlikowski MA, Paigen B. 1983. Use of small mammals (voles) to assess a hazardous waste site at Love Canal, Niagara Fall, New York. *Arch Environ Contam Toxicol* 12:383–397.

Ryan JJ, Lau BP, Hardy JA, Stone WB, O'Keefe P, Gierthy JF. 1986. 2,3,7,8–tetrachlorodibenzo-p-dioxin and related dioxins and furans in snapping turtle (*Chelydra serpentina*) tissue from the upper St. Lawrence River. *Chemosphere* 15:537–548.

Sabourin TD, Stickle WB, Michot TC, Villars CE, Carton DW, Mushinsky HR. 1984. Organochlorine residue levels in Mississippi River water snakes in southern Louisiana. *Bull Environ Contam Toxicol* 32:460–468.

Schauber EM, Edge WD, Wolff JO. 1997. Insecticide effects on small mammals: Influence of vegetation structure and diet. *Ecol Appl* 7:143–157.

Severynse M, editor. 1995. Webster's II new college dictionary. New York: Houghton Mifflin.

Shore RF, Douben PET. 1994. Predicting ecotoxicological impacts of environmental contaminants on terrestrial small mammals. *Rev Environ Contam Toxicol* 134:49–89.

Smith GJ. 1987. Pesticide use and toxicology in relation to wildlife: Organophosphorus and carbamate compounds. Washington DC: U.S. Fish and Wildlife Service. Resource Publ. 170.

Sprunt, IV A, Robertson, Jr. WB, Postupalsky S, Hensel RJ, Knoder CE, Ligas FJ. 1973. Comparative productivity of six bald eagle populations. *Trans N Amer Nat Resour Conf* 38:96–106.

Stickel LF. 1973. Pesticide residues in birds and mammals. In: Edwards CA, editor. Environmental pollution by pesticides. London: Plenum. p 254–312.

Stickel WH. 1975. Some effects of pollutants in terrestrial ecosystems. In: McIntyre AD, Mills, CF, editors. Ecological toxicology research. New York: Plenum. p 25–74.

Stickel WH. 1981. Pesticide and eggshells: What can we believe? U.S. Fish and Wildlife Service, Research Information Bulletin No. 81–34.

Stone WB, Kiviat E, Butkas SA. 1980. Toxicants in snapping turtles. *New York Fish Game J* 27:39–50.

Sullivan TP, Sullivan DS. 1982. Responses of small-mammal populations to a forest herbicide application in a 20–year–old conifer plantation. *J Appl Ecol* 19:95–106.

Sullivan TP, Sullivan DS, Lautenschlager RA, Wagner RG. 1997. Long-term influence of glyphosate herbicide on demography and diversity of small mammal communities in coastal coniferous forest. *Northwest Sci* 71:6–17.

Suter II GW. 1993. Ecological risk assessment. Boca Raton FL: Lewis. 364 pp.

Wells JV, Richmond ME. 1995. Populations, metapopulations, and species populations: What are they and who should care? *Wildl Soc Bull* 23:458–462.

Westlake GE, Bunyan PJ, Johnson JA, Martin AD, Stanley PI. 1982. Biochemical effects in mice following exposure to wheat treated with Chlorfenvinphos and Carbophenothion under laboratory and field conditions. *Pest Biochem Physiol* 18:49–56.

Wiemeyer SN, Lamont TJ, Bunck CM, Sindelar CM, Gramlich FJ, Fraser JD, Byrd MA. 1984. Organochlorine pesticide, polychlorobiphenyl, and mercury residues in bald eagle eggs—1969–1979—and their relationships to shell thinning and reproduction. *Arch Environ Contam Toxicol* 13:529–549.

Williams TM. 1990. Evaluating the long term effects of crude oil exposure in sea otters: Laboratory and field observations. *Wildl J* 13:42–48.

Wilson RK, McCormick CR, Williams TD, Tuomi PA. 1990a. Clinical treatment and rehabilitation of sea otters. In: Bayha K, Kormendy J, editors. Sea otter symposium. Proceedings of a symposium to evaluate the response effort on behalf of sea otters after the T/V *Exxon Valdez* oil spill into Prince William Sound; 17–19 Apr 1990; Anchorage, Alaska. Biological Report 90(12). Washington DC: U.S. Fish and Wildlife Service. p 326–334.

Wilson RK, Tuomi P, Schroeder JP, Williams T. 1990b. Clinical treatment and rehabilitation of oiled sea otters. In: Williams TM, Davis RW, editors. Sea otter rehabilitation program: 1989 Exxon Valdez oil spill. International Wildlife Research. p 101–117.

Wren CD. 1991. Cause-effect linkages between chemicals and populations of mink (*Mustela vison*) and otter (*Lutra canadensis*) in the Great Lakes Basin. *J Toxicol Environ Health* 33:549–585.

Wyman RL, Hawsley-Lescault DS. 1987. Soil acidity affects distribution, behavior, and physiology of the salamander *Plethodon cinereus*. *Ecology* 68:1819–1827.

Wyman RL, Jancola J. 1992. Degree and scale of terrestrial acidification and amphibian community structure. *J Herpetol* 26:392–401.

Yawetz A, Sidis I, Gasith A. 1983. Metabolism of parathion and brain cholinesterase inhibition in Aroclor 1254–treated and untreated Caspian terrapin (Mauremys rivulata, Emydidae, Chelonia) in comparison with two species of wild birds. *Comp Biochem Physiol* 75C:377–382.

CHAPTER 6

Using Single-Species Measurements to Anticipate Community Consequences of Environmental Contaminants

E. E. Holmes and P. M. Kareiva

Conservation and environmental biologists ultimately seek to understand how humans damage ecosystems and how this damage can be mitigated or prevented. Unfortunately, it is very difficult to identify the symptoms of a "sick" or "damaged" community or ecosystem. As a result, we use indicators, simple model systems, and focal species to measure damage impact. But while practicality forces us to focus on simple assays, modern ecological theory raises doubts about the validity of such an approach. Communities and ecosystems are complicated webs of interacting species, with the effects of perturbations often yielding surprises because of the rich web of connections among species and the nonlinear nature of population dynamics (Yodzis 1988). What does this "system" or community view mean for ecotoxicology, which, like so much of environmental and conservation biology, is forced to design pragmatic studies of single species?

Lessons from Food-web Theory and Experiments

There has been a tremendous amount of theoretical and experimental work on food-web dynamics (reviewed in Pimm 1982; Lawton 1989; Pimm et al. 1993; Hall and Raffaelli 1993; Morin and Lawler 1995). The majority of this work has been in three areas: 1) predicting and explaining patterns found in real food webs, 2) understanding the relationship between community dynamics (especially stability) and the structure of food webs, and 3) predicting and describing the effects of perturbations on communities. Work on the latter 2 areas provides some direct insight into the possible impacts of contaminants on natural communities. Theories and studies on the effect of community structure on community dynamics (Area 2)

Chapter Preview

Lessons from food-web theory and experiments   149
The utility of focal species and lethality measures   152
Lethal versus sublethal effects   159
Discussion   168
Appendix   171

have a long history in the field of ecology. Up to the early 1970s, ecologists argued that complex systems were more stable and, by implication, less affected by small perturbations (e.g., Elton 1927; MacArthur 1955; Elton 1958). In 1972, May used dynamic food-web models to argue against the traditional view that complexity equals stability. In food-web models, a complex system is more likely to be unstable (and unable to support all species) than is a simple system. Ecologists quickly pointed out that May's approach was flawed (DeAngelis 1975; Lawler 1978). A randomly chosen complex food web may be less stable than a simple one, but real systems are not random. This spawned an enormous amount of research that continues today asking, "What makes real complex webs stable?" (given that they are not inherently stable) and "What structures or configurations tend to make them unstable?" In theory, answers to the latter question may help in predicting what types of communities will be more susceptible to perturbations from contaminants.

May's original analysis (1972) suggested that there is a relationship between the stability of a community and the product of the number of species (S) in the community and the number of links between species (C). Specifically, a community should be stable if

$$\alpha \, (S \, C)^{1/2} < 1 \qquad \text{(Equation 6-1)},$$

where α is the average interaction strength between species. This suggests that in real communities we should see a negative relationship between size of the community and connectance (C) (Pimm 1982). However, the field evidence has been equivocal. Some studies have found a strong negative and even the predicted hyperbolic relationship (Briand 1983; Schoenly et al. 1991), but other studies have found no relationship or even a positive relationship (Winemiller 1989; Warren 1990). However, even if the field evidence were not equivocal, Equation 6-1 has little value as a practical metric for ascertaining whether a given community will be susceptible to perturbations from contaminants. First, many authors have pointed out that there exist enormous practical difficulties in cataloguing all the species and connections within a large community and that aggregating species (e.g., lumping all the detritivores together) gives one a spurious (SC) value (Paine 1988; Warren 1990; Martinez 1991). Second, it would take enormous effort and years of experiments to estimate α, the mean interaction strength, for even a medium-sized community.

Another potential source of guidance is work on the types of food-web structures that tend to be more stable (Pimm and Lawton 1977, 1978). This work suggested that model food webs, even when parameterized for real communities, tend to be unstable if they have high omnivory. In addition, for the subset of stable food webs, omnivory tends to reduce the return time to equilibrium after a community is perturbed. Overall, it was argued that omnivory should be rare in natural systems because it is destabilizing (Pimm and Lawton 1978; Pimm 1982; Pimm et al. 1993).

This argument stimulated much research on whether omnivory is indeed rare. Some studies support this supposition and others contradict it (reviewed by Hall and Raffaelli 1993), but more pertinent to our discussion is whether omnivory is indeed destabilizing. With alternate conceptual frameworks for food webs, omnivory is stabilizing (Polis and Strong 1996). Furthermore, an experimental test of the effect of omnivory on a community's response to perturbation (Fagan 1997) found that omnivory was in fact a stabilizing factor. The consistent pattern seems to be that even if some structures (such as omnivory) tend to be more unstable than others, all food-web configurations can be stable depending on the particular strengths and functions of interactions within a community. It is hard to make general theoretical predictions as to what web structures will be stable, and experimental work is too limited to provide much evaluation of the theory.

Despite the limitations of food-web theory in terms of giving us specific metrics, food-web models can give enormous insight into the general responses of communities to perturbations. Two papers that address responses of communities to specific perturbations suggest that community response is non-intuitive. Yodzis (1988) used dynamic food-web models to study the effect of increased mortality of a single species on its abundance or the abundances of its predators, prey, or competitors. Yodzis's theoretical study has direct implications for ecotoxicology because it addresses the question of whether a contaminant that is lethal to a single species would actually cause a change in the abundance of that species or a species with which it directly interacts. It may seem logical that an increase in the death rate of a species will cause a decline in its abundance. Indeed, Yodzis found that when the death rate of a target species was increased, the abundance of the target species decreased as expected in as much as 73% of the cases. Although the expected change occurred more often than not, no change in the abundance of the targeted species was observed in many (27%) of the perturbations. In addition, for the prey, predators, and competitors of the directly impacted species, Yodzis found that one is most likely to observe no effect or an effect opposite to that expected. For example, in 50% of the cases no change was seen in prey abundances, and in 11% of the cases the prey actually decreased even though one would expect a prey species to increase if its predator suffered a higher death rate. Yodzis found similar counterintuitive community responses in a study of the effect of seal harvesting on a commercial fishery in the Benguela ecosystem (Yodzis 1998).

The causes of these counterintuitive behaviors reside in the complex web of interacting species in which indirect effects are ubiquitous (Wootton 1994a). Such dynamic systems do not behave in simple fashions. A change in the death rate of 1 species propagates to all the other species in the community, and those changes lead to feedback that produces more change, perhaps even opposite to the initial set of population responses. The ultimate change in population numbers after a complex community readjusts and reforms may be quite contrary to what one would expect from considering the interactions of 1, 2, or 3 species of the community in isolation.

The main message for ecotoxicology is that when a species is within a natural community, as opposed to being experimentally isolated, knowledge of the lethality of a contaminant to that species does not necessarily translate into a predictable population change for that species. In particular, the indirect responses of the prey, predators, or competitors of the targeted species can be contrary to expectation.

The Utility of Focal Species and Lethality Measures

Laboratory and field measurements of a contaminant's impact on the survival of selected focal species (e.g., daphnia, honey bees, or aphids) often are used as indicators of how detrimental a particular contaminant will be to a whole community. In recent years, this emphasis on LD50s and mortality rates has been criticized and better use of population and behavioral ecology has been advocated, (e.g., measurements of rates of population growth as opposed to simply mortality, subtle changes in fecundity, and sublethal changes in behavior) (Cohn and Macphail 1996; Clements 1997). Although population and behavioral ecology clearly offers a broader vision to ecotoxicology than has been traditional, even this vision is limited in its view of assessing effects of contaminants because the premise of much of environmental toxicology remains that impacts can be captured by quantifying changes in 1 or 2 species. This single-species approach is weakened by the food-web research (reviewed above) indicating that small changes in species can cascade through networks of interacting species to produce major community-wide changes and that impacts on a single species in a community often lead to counterintuitive changes in abundances.

Yodzis (1988) explored whether an increase in the mortality rate of a species is reflected in a change in the abundance of the directly impacted species or of the species with which it directly interacts. In this paper, we use similar multi-species community models to examine how well can we expect monitoring of mortality effects on a single species to capture impacts measured at the community level. Our intention is to examine to what extent the dynamic nature of multi-species interactions can thwart attempts to make predictions concerning community responses based on single-species information. We examine first whether lethality is expected to be a good predictor of the overall changes in abundances in the entire community. Second, we examine the relative sensitivity of communities to changes in death rates versus other changes that a contaminant might cause such as changes in prey-capture rates, changes in susceptibility to predation or disease, and changes in producer fecundity.

Are changes in mortality rates correlated with community impact?

There are many ways to pose this question and these will differ depending on how a contaminant affects the milieu of species in the community of concern. In this analysis, we explore 1 specific scenario. We suppose that the actual food web or the strengths of interactions between species in our community of concern is not known. However, we do know that there is only one highly sensitive species in the community, and we have been lucky enough to choose this species as our focal species. We examine only changes in the death rate of the focal species and assume that there are not direct impacts on other species or on the focal species' predation or consumption rates. Although contaminants can certainly affect multiple species, we assume here that it affects only 1 species, that we know which species is affected, and that we know to what degree the mortality is increased. In this setting, is a measurement of a contaminant's lethality a reasonable metric for the cascade of changes in species abundance that will occur when the death rate of the target species changes? This would be the case if, in our analysis, high lethality causes large community impacts and low lethality causes low impacts regardless of the identity of the focal species (the species whose death rate changes) or the community in which the species resides.

The communities

In order to contrast the single-species versus community perspective on toxicological impacts, we constructed a series of Lotka-Volterra community models (May 1973; Pimm 1982) for a variety of communities with 3 to 16 component species. In this framework, the interactions between species are rates that are a linear function of density. For example, the rate of predation of a herbivore on a producer is $a_{p,h}$ h. With this framework, a 3-species community model is (See Figure 6-1a)

$$dc/dt = c\,(-d_c + a_{c,h}\,h)$$
$$dh/dt = h\,(-d_h + a_{h,p}\,p - a_{h,c}\,c) \qquad \text{(Equation 6-2)},$$
$$dp/dt = p\,(b_p - a_{p,p}\,p - a_{p,h}\,h)$$

where
    dc/dt is the rate of change of the density of the top carnivore,
    dh/dt is the rate of change of the herbivore, and
    dp/dt is the rate of change of the producer.

The "a" parameters are measures of the interaction rates between species. The "d" parameters are the death rates of consumers, and the "b" parameter is the fecundity of the producer. In this framework, deaths of producers are due exclusively to consumption. Thus, their death rate is encapsulated in the $a_{i,j}$ Si Sj terms (S=species). Similarly, the reproduction rate of consumers is governed by their consumption of prey. The fecundity of consumers also is encapsulated in the $a_{i,j}$ Si Sj terms. Notice that species on the bottom trophic level (the producers) experience a self-damping intraspecific competitive effect (through the $a_{p,p}$ p term) but that all other species are controlled by what they eat or what eats them.

This type of model is a first-order approximation of real community dynamics. In real communities, the interactions between species are nonlinear and higher order interactions may be present (Wootton 1994b). It is well known that nonlinear effects can dramatically change the response of a community to perturbations (McCann et al. 1998). For example, in a linear model, there is 1 globally stable equilibrium (May 1973), whereas a nonlinear model can display multiple stable equilibria. When multiple stable equilibria exist, the community may jump to a completely new equilibrium in response to a relatively small perturbation (e.g., Spencer 1997). With a highly nonlinear model, we could expect much more dramatic community changes that are out of proportion to small changes in mortality. Thus, our linear modeling approach is likely to underestimate the actual changes in abundances.

We compared 2 archetypal food webs (Figures 6-1a and 6-1b) and 3 community food webs drawn from actual ecosystems (Figures 6-2, 6-3, and 6-4). These food webs provide a small sample of the webs of interactions that might be found between "types of organisms" in a real community. In the Mono Lake, California, salt marsh and Shoals food webs, the species in the community are lumped into functionally related species that may include a guild of species or simply a single species. In our discussion, we will refer to these functionally similar groups (the nodes in the food-web diagrams) simply with the term "species."

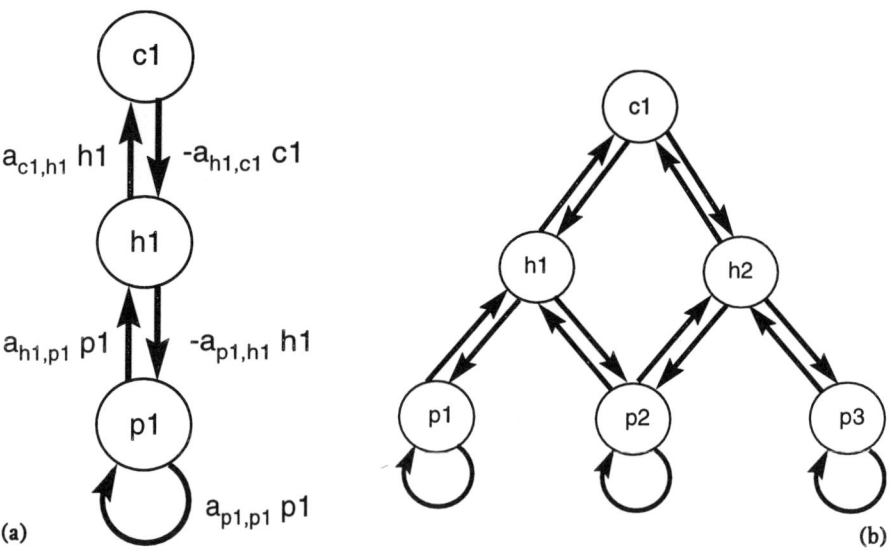

Figure 6-1 (a) Simple 3-species food web. The interaction rates between species (shown by the arrows) are a linear function of density. (b) 6-species food web. The web has 3 plant producers, 2 herbivores and 1 top predator.

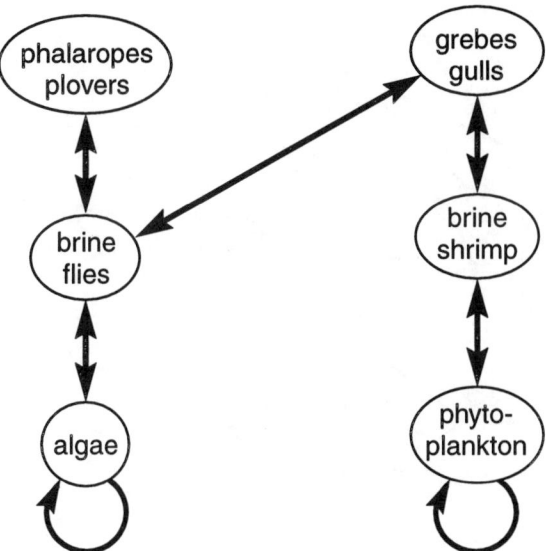

Figure 6-2 Food web of the Mono Lake ecosystem

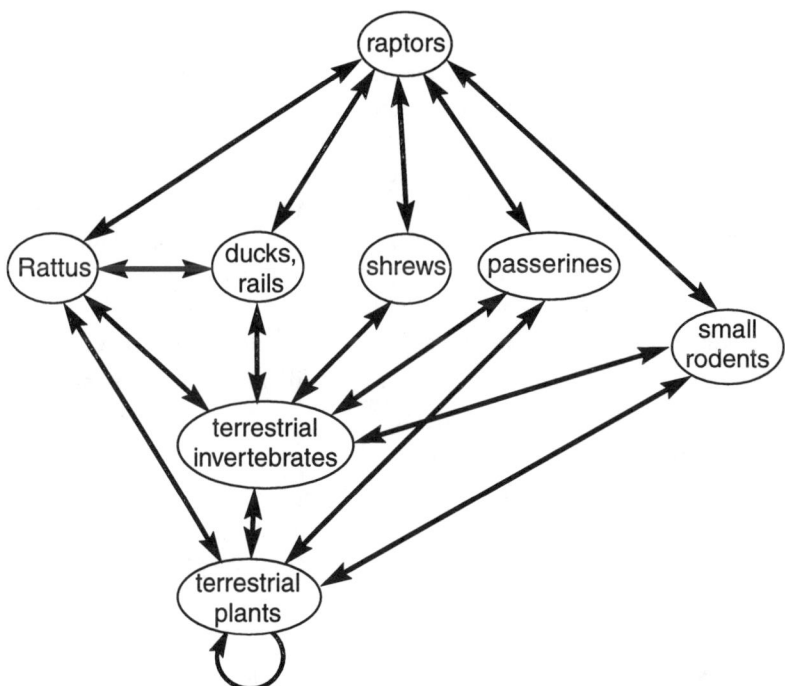

Figure 6-3 Food web of a California salt marsh (from Briand 1983). The small rodents are *Microtus, Reithrodontomys,* and *Mus*. The raptors are *Circus* and *Asia*.

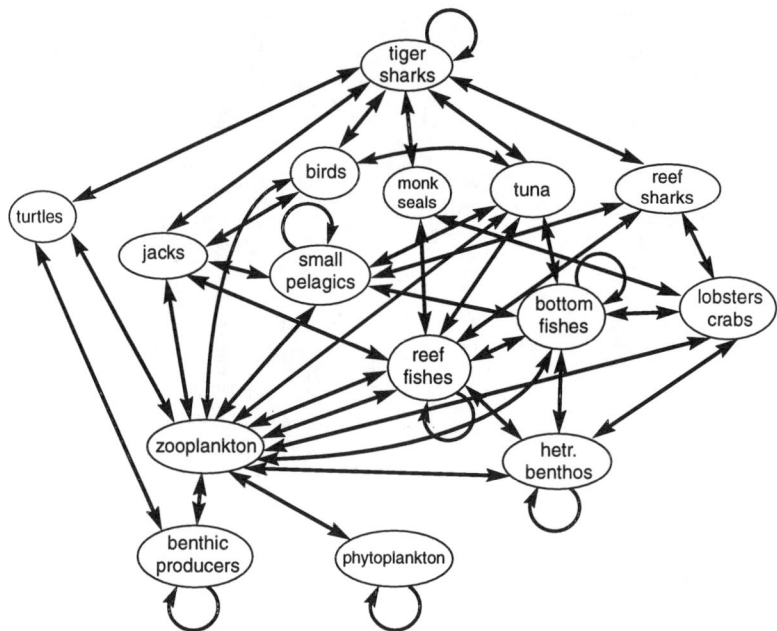

Figure 6-4 Food web of the French Frigate Shoals (from Polovina 1984)

The relative values of the interaction parameters (i.e., the $a_{i,j}$) allow us to describe different types of communities with the same basic food-web structure. For example, in a plant–grazing mammal interaction, individual herbivores typically have a heavy impact on individual plants, while individual plants have a low impact on individual mammals. The reverse is typical in a plant–phytophagus-insect interaction (individual insects have a low per capita impact on individual plants). For the simple 3-species food web (Figure 6-1a), we used 6 different sets of $a_{i,j}$ parameters representing 4 different types of communities:

plant–mammal herbivore–mammal predator (2 different sets of parameters);
plant–phytophagus insect–insect predator (2 different sets of parameters);
plant–phytophagus insect–parasitoid,
plant–phytophagus insect–disease.

For the 6-species food web (Figure 6-1b) with 3 plant species, 2 herbivores, and 1 carnivore, we selected different $a_{i,j}$s to represent a plant–mammal herbivore–mammal carnivore and a plant–phytophagus insect–insect predator community.

The real-world food webs, shown in Figures 6-2, 6-3, and 6-4, were based on the Mono Lake community (Wiens et al. 1993), the terrestrial community of a California salt marsh (Briand 1983), and the French Frigate Shoals (Polovina 1984). For the Mono Lake and salt marsh food webs, we had no estimates for the parameters, and thus we selected relative sizes of parameters that reflected the actual species in the community (Appendix). For the Shoals food web, Polovina (1984) estimated parameters from data on the component species using the ECOPATH framework (see also Christensen and Pauly 1992). Following the method suggested by Walters et al. (1997), we prepared a Lotka-Volterra community model using Polovina's parameters.

Community impact following changes in death rate

Communities with linear interaction rates have a single stable equilibrium with a particular density for each species of the community (May 1973). If the community is perturbed (e.g., by release of a contaminant that changes death rates), the equilibrium shifts, the densities of the species change, and some species may be lost altogether. To quantify the level of community impact, we look at the change in the equilibrium abundances of the species in the community caused by the increase in the death rate of the focal species. We define the community impact as the average percent change in the equilibrium species densities. For example, in a community with 3 species, if Species 1 experiences an 80% decline, Species 2 experiences a 20% increase, and Species 3 experiences a 200% increase, then the community impact would be 80 + 20 + 200)/3 = 100. A community impact greater than 100 is quite severe because this means that the average species either disappeared or more than doubled in density.

We took each species above the lowest trophic level and simulated chronic increases in its mortality rate. For a community with 6 species, this translated into 6 simulations in which a different species was treated as the focal and therefore experienced an increase in its mortality rate. We varied mortality enhancement from 0% to 300% in increments of 0.1%. It is worth noting that a 300% increase in instantaneous death rates (d_i in our model) corresponds to a 75% reduction in the life expectancy, and a 100% increase in instantaneous death rate reduces the life expectancy by half. Thus the range of 0% to 300% increase in death rate corresponds to a reasonable range of a 0% to 75% decrease in life expectancy.

Each line in Figure 6-5 represents the community impact observed with different focal species. We did not simulate enhanced mortality of primary producers. We also did not include the Shoals food web for this analysis. Two points are immediately apparent from Figure 6-5. First, death rate alone was not a good predictor of the change in the model communities that was due to increases in mortality rates. In some communities, a large change in a species death rate led to a large change in the densities of the component species of the community, whereas in other communities, a large change in the death rate led to little change. In general, but not always, changes in the death rate of top predators caused a greater community impact than did changes in the death rate of species at lower trophic levels. Note in Figure 6-5 that some of the dashed lines (for species at the second trophic level) are very high. The second related point is that the food-web structure of the community and the

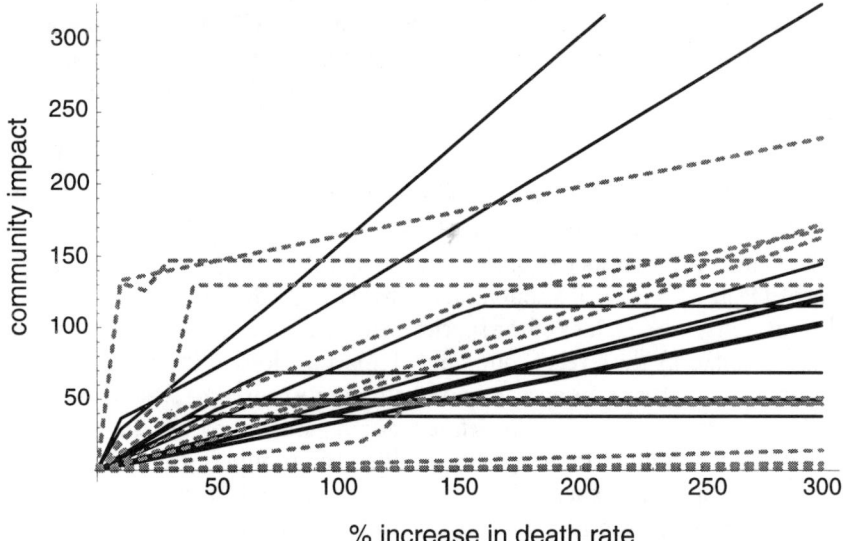

Figure 6-5 Impact of an increase in the death rate of 1 species on species abundances in the community as a whole. Community impact is defined as the percent change of the average species in the community. Black lines indicate species at the top of the food web. Thick gray dashed lines indicate species at lower trophic levels; the lines nearest the X-axis represent multiple lines on top of each other. These results include the 6 3-species webs giving 6 top-predator lines and 6 lower-trophic-level lines, the 2 6-species webs giving 2 top-predator lines and 4 lower-trophic-level lines, the Mono Lake web giving 2 top-predator lines and 2 lower-trophic-level lines, and the California salt marsh web giving 2 top-predator lines and 5 lower-trophic-level lines.

strengths of interactions between species must be known in order to estimate the impact of the change in death rate on the community.

In real-world assessments, the focal species must be chosen carefully and should not be chosen randomly. However, in many cases, we must give estimates of the potential impacts of a contaminant while having little information on the community or communities in which it will be released. In addition, in many cases a detailed food web and estimates of the strengths of interaction between species will not be known. With this in mind, it is worth asking whether certain categorizations of species might help us to predict when mortality changes will lead to large community impacts. For example, does enhanced mortality of generalists tend to cascade through webs of interacting species with greater impact than does the same amount of extra mortality imposed on specialists? In order to ask these sorts of questions, we divided the species in our food webs according to the following characteristics: abundance relative to other non-producers in the community, number of links to other species, and trophic level.

In Figure 6-6, we look at the community impact versus relative abundance. We quantify relative abundance for Species i as the ratio of Species i's abundance divided by the mean abundance of all non-producer species. The relative abundance seems to have low predictive value. Species whose relative abundance varies 10-fold show no clear pattern of community impact. In Figure 6-7, we decompose the relative abundance results into separate trend lines for the 10 different communities. For some communities, abundance and impact are positively related; in others they are negatively related. The trophic level and the number of links (Figure 6-8 and 6-9) show a more consistent trend with community impact, and, in both cases, a higher trophic level or higher number of links is associated with a higher impact. However, there is wide variance, and the best assessment is that a community is more likely to be sensitive to mortality increases that affect species with many links. But this is by no means certain.

Furthermore, when we examined the influence of trophic level and number of links within individual communities, we found no predictable patterns. For example, in the 3-species communities, high trophic level was associated with reduced impacts, and a greater number of links was associated with higher impacts. However, the 6-species communities showed the exact opposite trend (i.e., increased impacts with high trophic level and reduced impact with more links). The 2 other communities showed other combinations of effects. In the California salt marsh community, mortality changes applied to higher trophic levels yielded higher impacts, but there was no close association between number of links and impacts. In the Mono Lake community, mortality changes applied to higher trophic levels yielded reduced impacts, while a higher number of links correlated with higher impacts. Thus, for these model communities, knowledge of a single species was not sufficient to give even a qualitative estimate of the impact on the community caused by contaminant exposure. Instead the specific food web and strengths of interactions within the community had to be known.

Lethal Versus Sublethal Effects

Our analyses thus far have assumed that the perturbation from the contaminant alters only mortality rates. However, a great deal of recent research indicates that contaminants can have many sublethal effects that alter fecundity, behavior, and immune-system function. For example, the widely publicized decline in human sperm count, characterized by enormous geographic variation, is often attributed to combinations of environmental pollutants; estimated annual rates of decline are 1.5% in the U.S. and 3% in Europe (Swan et al. 1997). Animal foraging rates often are reduced by chemical contaminants (e.g., Peakall 1982; Donkin et al. 1995; Gopal and Ram 1995; Roper et al. 1995), with possible cascading effects through communities (Schmitz et al. 1997). Intraspecific interactions also are likely to change with exposure to certain compounds. For example, deer mice exposed to extremely low concentrations of pesticide combinations demonstrate dramatically altered levels of aggression, with implications for spacing behavior and population dynamics (Porter

160   *Environmental Contaminants in Terrestrial Vertebrates*

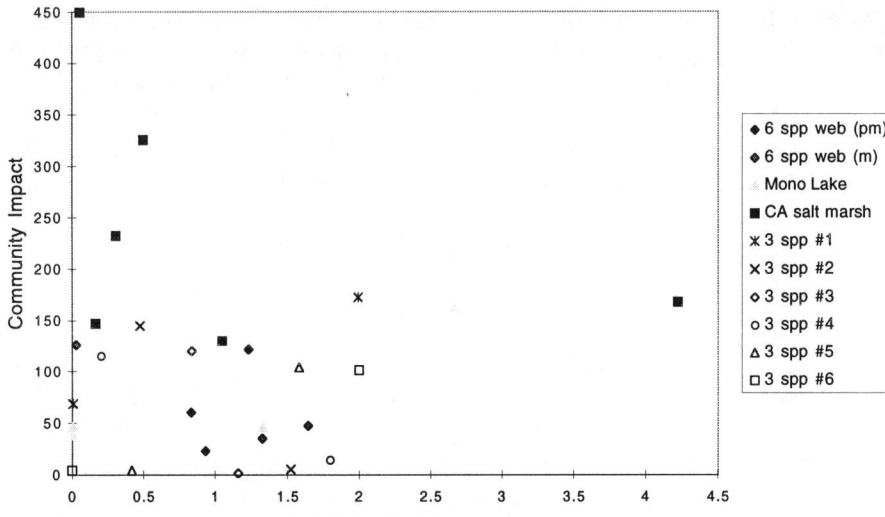

Figure 6-6 The relative abundance of a species in a community versus the community impact resulting from a 300% increase in that species' death rate (which corresponds to a 75% decrease in mean life span). The relative abundance is abundance divided by the mean abundance of all non-producers in the community. Results are shown for the following webs, Mono Lake, CA salt marsh, 2 6-species webs (m: plant–mammal–mammal and pm: plant–phytophagus insect–mammal) and 6 3-species webs (1 = plant–mammal–mammal A, 2 = plant–mammal–mammal B, 3 = plant–insect–insect A, 4 = plant–insect–insect B, 5 = plant–insect–parasitoid, 6 = plant–insect–disease). See Appendix for parameters.

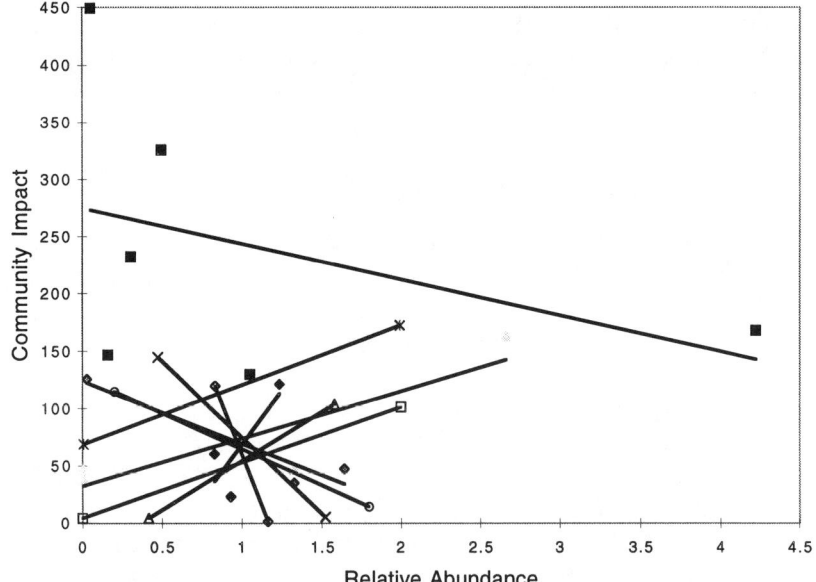

Figure 6-7 Abundance versus community impact. This is the same data that is in Figure 6-6 with trend lines for each of the 10 communities. For 5 communities, impact increases with abundance, and for the other 5 communities, impact decreases with abundance.

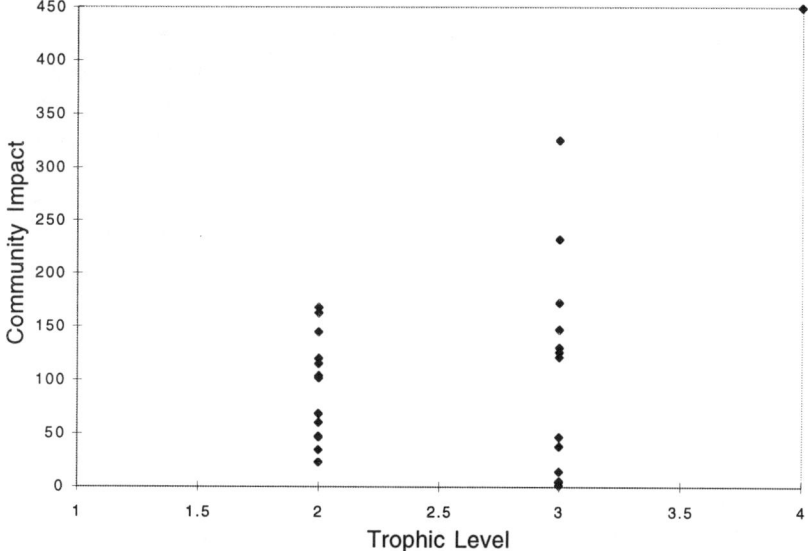

Figure 6-8 Trophic level of a species versus the community impact resulting from a 300% increase in that species' death rate. Data for all species in all 10 communities are lumped together. At this level, there is a positive trend of trophic level with community impact. However, as discussed in the text, there is no consistent trend when the data are examined at the level of individual communities.

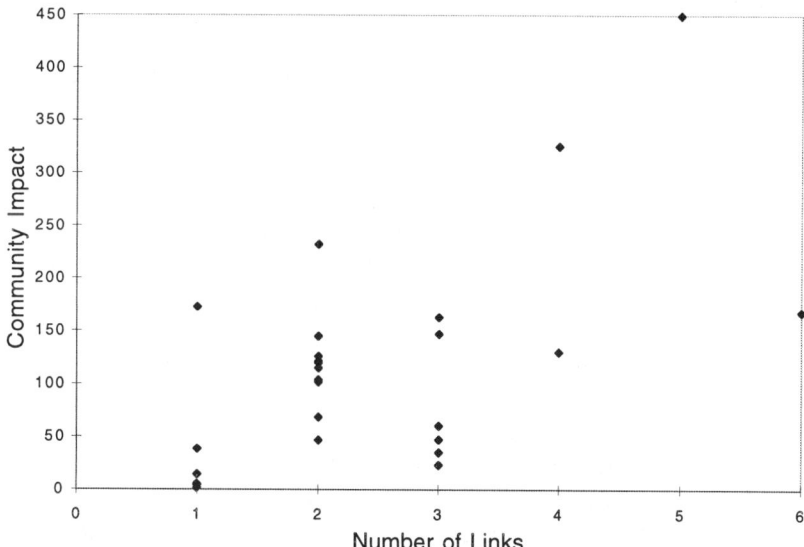

Figure 6-9 Number of links versus the community impact resulting from a 300% increase in death rate. Data for all species in all 10 communities are lumped together. In the lumped data, there is a positive trend for the number of links a species has with other species and the impact of an increase in the death rate of that species. However, as discussed in the text, there is no consistent trend when the data are examined at the level of individual communities. Half of the communities show a positive trend and half shows a negative trend.

et al. 1998). Probably the best documented evidence of toxicant-induced changes in species interactions comes from studies of chemical impacts on immune systems. It has been shown repeatedly that low dosages of toxicants can suppress immune systems in mice and rats (Porter et al. 1984; Olson et al. 1987; Porter et al. 1998). Such changes have obvious implications for the interactions between diseases and their hosts. In short, there is compelling evidence of changes in predator-prey, host–pathogen, and within-species interaction coefficients as a result of chemical exposure. There is also evidence of changes in fecundity that are due to chemical exposures. In terms of human health and risk analysis, many of these subtle effects are clearly important. However, when it comes to community dynamics, it is important to ask how the impact of these sublethal effects compares to the impact of direct alterations of death rates. Perhaps sublethal effects are dwarfed in importance by mortality changes, which would imply that ecotoxicologists should continue to focus on death rates as the key parameter that defines ecosystem impact.

Modeling the relative importance of sublethal effects

At first glance, it might seem that any toxicant's impact on mortality would always have a greater community-level impact than a proportionate change in predator–prey, competition, or fecundity rates, as death is much more final than a modest change in prey consumption. To address this formally, we used our community models to contrast small perturbations of fecundity (b_i in equations) or interaction coefficients ($a_{i,j}$) versus death rates (d_i). In this case, the interaction coefficients encompass rates that predators (including herbivores) capture prey, rates that prey (including plants) are captured by predators, and the rate of within-species competition. Specifically, we changed each parameter (Figure 6-10) by 0.1% and recorded the resultant community impact. The impact was then normalized by dividing by the highest community impact observed. Note that the actual amount that we changed the parameters was not critical and that the same results would have occurred if we had perturbed the parameters by 0.05% or 0.001%. This is because 1) we were looking at the response of the community to changes in mortality relative to changes in predation, competition, or fecundity and 2) these models behave linearly to small perturbations from equilibrium. It is important to note, however, that we are comparing the effects of equal percent changes in rates of mortality, interaction, and fecundity (i.e., mean lifespan, mean number of prey eaten in some time period, and mean number of offspring produced). There has been abundant work documenting decreased prey capture rates in response to low levels of contaminants. However, information on actual relative impacts of different levels of contaminants on life span versus number of prey captured per time period, number of individuals captured as prey per time period, or number of offspring produced is not available.

Figures 6-10, 6-11, 6-12, 6-13, and 6-14, demonstrate the relative impacts caused by the small changes in individual parameters in each of the 5 communities. On the

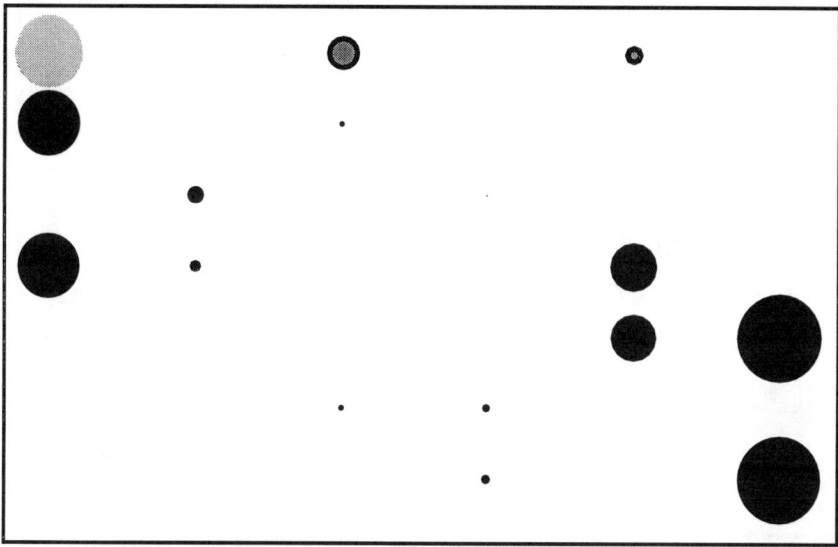

Figure 6-10 Sensitivity of the 6-species food web (Figure 6-1b) to changes in death, fecundity, and interaction (predation) rates (using plant–mammal–mammal parameters). Top: The diameter of each circle represents the relative community impact that is due to a small (0.1%) change in the parameter at the corresponding position on the right. Relative impacts that are due to changes in death rates are shown with dark gray circles with black perimeter; in fecundities, with light gray circles; and in predator–prey rates, with black solid circles. Predator–prey rates denote both the rate that predators capture prey and the rate that prey (including plants) are eaten by predators or herbivores. Bottom: Parameters corresponding to the circles on the top. On the top lines, the d_i and b_j are the death and fecundity rates of species i and j. On the lower lines, the i,j terms denote the $a_{i,j}$ interaction parameters; i,j represents the impact of species j on i. The i,j terms correspond to the species in Figure 6-1b: 1 = p1, 2 = p2, 3 = p3, 4 = h1, 5 = h2, 6 = c1.

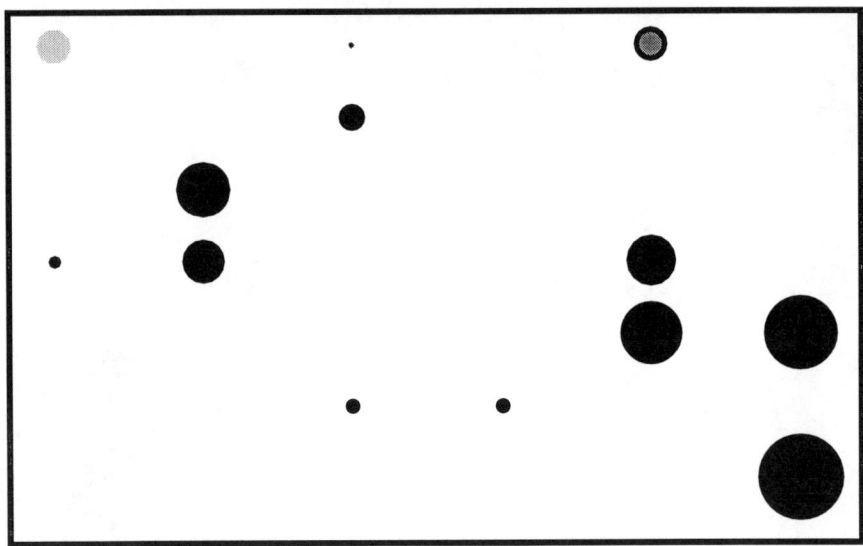

Figure 6-11 Sensitivity of the 6-species food web (Figure 6-1b) to changes in death, fecundity, and interaction (predation) rates (using plant–phytophagus-insect–mammal parameters). See Figure 6-10 for an explanation of the figure. Bottom: Parameters corresponding to the circles on the top. The i,j term represents the impact of species j on i. The i,j terms correspond to the species in Figure 6-1b: 1 = p1, 2 = p2, 3 = p3, 4 = h1, 5 = h2, 6 = c1.

6: Using single-species measurements to anticipate community consequences of environmental contaminants 165

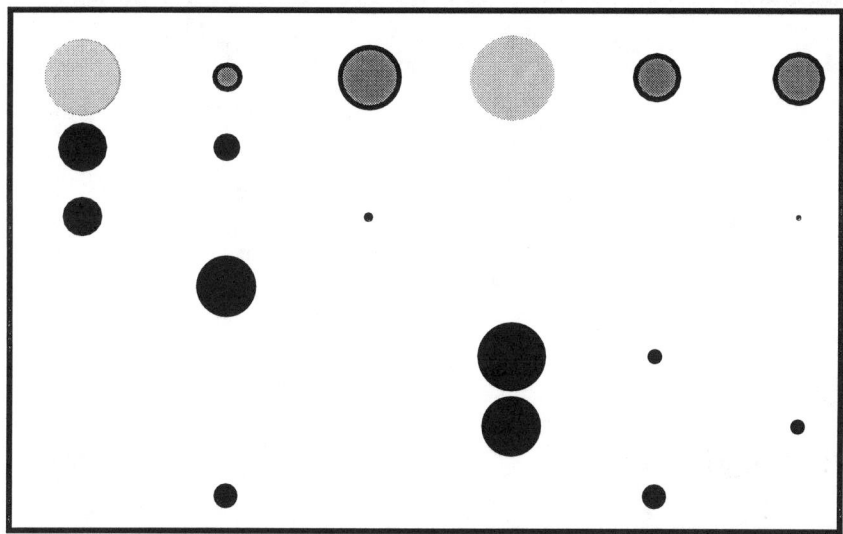

b1	d2	d3	b4	d5	d6
1 , 1	1 , 2	.	.	.	.
2 , 1	.	2 , 3	.	.	2 , 6
.	3 , 2	.	.	.	.
.	.	.	4 , 4	4 , 5	.
.	.	.	5 , 4	.	5 , 6
.	6 , 2	.	.	6 , 5	.

Figure 6-12 Sensitivity of the Mono Lake food-web (Figure 6-2) to changes in death, fecundity, and interaction (predation) rates. See Figure 6-10 for an explanation of the figure. Bottom: Parameters corresponding to the circles on the top. The i,j term represents the impact of species j on i. The i,j terms correspond to the species in Figure 6-2: 1 = algae, 2 = brine flies, 3 = plovers, 4 = phytoplankton, 5 = brine shrimp, 6 = gulls.

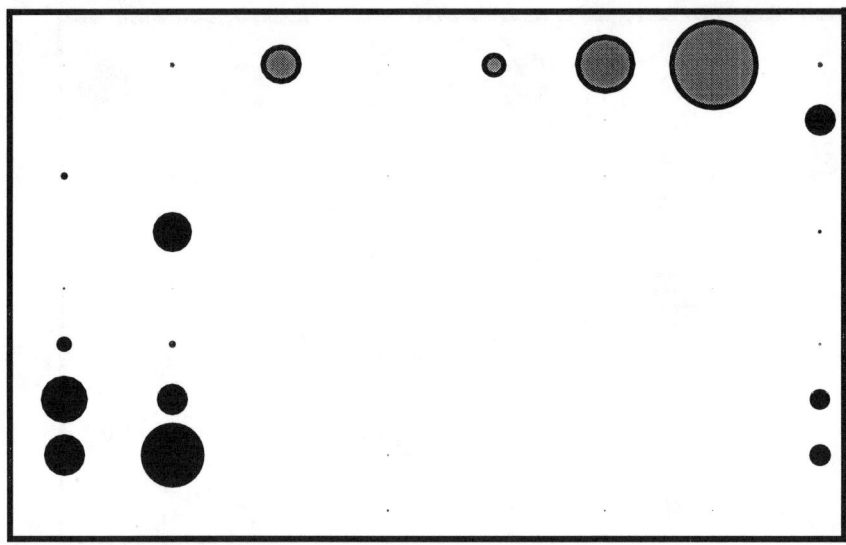

Figure 6-13 Sensitivity of the California salt marsh food web (Figure 6-3) to changes in death, fecundity, and interaction (predation) rates. See Figure 6-10 for an explanation of the figure. Bottom: Parameters corresponding to the circles on the top. The i,j term represents the impact of species j on i. The i,j terms correspond to the species in Figure 6-3: 1 = terrestrial plants, 3 = terrestrial invertebrates, 6 = shrews, 8 = rails, 10 = passerines, 11 = small rodents, 12 = rats, 13 = raptors.

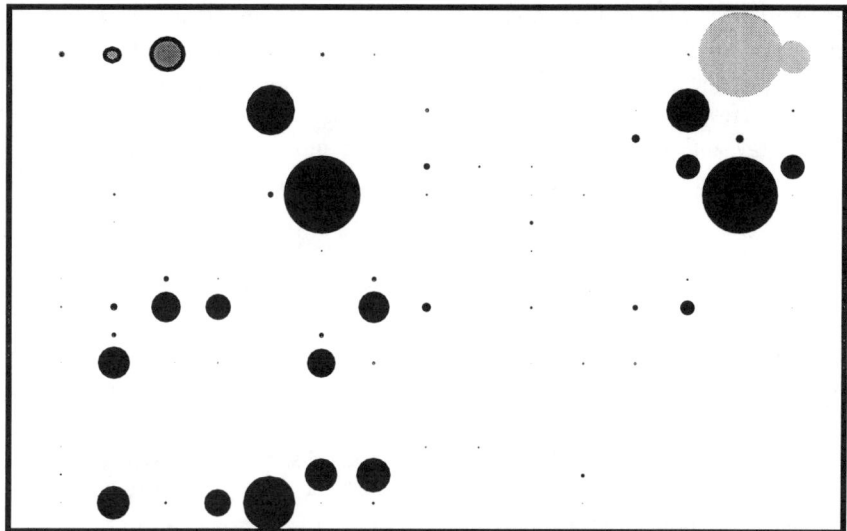

	d1	d2	d3	d4	d5	d6	d7	d8	d9	d10	d11	d12	d13	f14	f15
	.	.	.	.	.	.	.	.	.	.	.	.	.	.	.
	.	.	.	.	5,15	.	.	8,15	.	.	.	12,15	13,15	.	15,15
	.	.	.	.	.	.	.	.	.	.	.	12,14	.	14,14	.
	.	.	.	.	.	.	.	8,13	9,13	10,13	.	.	13,13	.	15,13
	.	2,12	.	.	5,12	6,12	.	8,12	9,12	10,12	11,12	.	.	14,12	15,12
	1,11	2,11	.	.	.	6,11	.	8,11	.	10,11	.	12,11	.	.	.
	.	.	.	.	.	6,10	.	8,10	9,10	10,10	11,10	12,10	13,10	.	.
	1,9	.	3,9	4,9	.	.	7,9	.	.	10,9	.	12,9	13,9	.	.
	1,8	2,8	3,8	4,8	.	.	7,8	8,8	.	10,8	11,8	12,8	13,8	.	15,8
	1,7	2,7	.	.	.	6,7	.	8,7	9,7	.	.	.	.	.	.
	1,6	2,6	.	4,6	.	6,6	7,6	.	.	10,6	11,6	12,6	.	.	.
	1,5	.	.	.	.	.	.	.	.	.	.	12,5	.	.	15,5
	1,4	.	.	.	.	6,4	.	8,4	9,4	.	.	.	.	.	.
	1,3	.	.	.	.	.	.	8,3	9,3	.	.	.	.	.	.
	1,2	.	.	.	.	6,2	7,2	8,2	.	.	11,2	12,2	.	.	.
	1,1	2,1	3,1	4,1	5,1	6,1	7,1	8,1	9,1	.	11,1	.	.	.	.

Figure 6-14 Sensitivity of the reef food web (Figure 6-4) to changes in death, fecundity, and interaction (predation) rates. See Figure 6-10 for an explanation of the figure. Bottom: Parameters corresponding the circles on the top. The i,j term represents the impact of species j on i. The i,j terms correspond to species in Figure 6-4: 1 = tiger sharks, 2 = birds, 3 = monk seals, 4 = reef sharks, 5 = turtles, 6 = small pelagics, 7 = jacks, 8 = reef fishes, 9 = lobsters, 10 = bottom fishes, 11 = tuna, 12 = zooplankton, 13 = benthos, 14 = phytoplankton, 15 = benthic producers.

bottom are the parameters of the model. Note that $a_{i,j}$ is the impact of species j on species i. On the top, the size of the circle indicates relative sensitivity of the community to each parameter. The locations of the circles on the top and the parameters on the bottom correspond. The color of the circles (dark gray circles with black perimeters, light gray, and solid black) indicate the relative impacts that are due to changes in death rates, fecundities, and interaction rates, respectively. Note that each producer has only a fecundity circle because its death rate is encapsulated in the $a_{i,j}$ terms, in which "i" is the producer and "j" is a consumer, and that each consumer has only a death-rate circle because its fecundity is encapsulated in the $a_{i,j}$ terms in which "i" is the consumer and "j" is a producer.

In general, the grey circles bordered in black (sensitivity to death rates) are not appreciably larger than the black circles (sensitivity to interaction rates). This indicates that, overall, the communities were equally sensitive to changes in the interaction rates and changes in the death rates. In the 6-species food webs (Figures 6-10 and 6-11) and the Shoals food web (Figure 6-14), the community was actually more sensitive to changes in the interaction rates than to changes in death rates. Clearly in these model food webs, a proportional change in interaction rates (e.g., prey captured per hour) had as much of an impact on the community as equal proportional changes in death rates did (Figure 6-15). In our comparison of 5 food webs, we discerned no simple pattern in terms of which interaction terms were important. Neither predation, intraspecific competition, nor fecundity terms of a specific trophic level were uniformly important in all webs. Instead, in each community a different set of interaction terms was important.

Discussion

In this paper, we used dynamic food-web models to explore whether a measurement of the lethality of a contaminant to a target species is a good predictor of the total change of species abundances in the community. If the lethality is a good predictor, then one would expect a tight relationship between lethality and community impact, i.e., high lethality leads to large impacts on communities, and low lethality leads to low impacts on communities. Instead, our analysis suggests that it is impossible to predict whether the contaminant will cause a large or small change in the community without knowing the organization of the affected community. A contaminant that causes only minor mortality in a target species might cause large impacts in one community but small impacts in another. Similarly, a highly lethal contaminant might cause small or large impacts depending on the community into which it is released. Therefore, detailed information is needed on the species in the community and the strengths of interactions between them; unfortunately, measuring these strengths is not trivial. However, there has been recent progress on practical methods for measuring the strengths of species interactions directly by

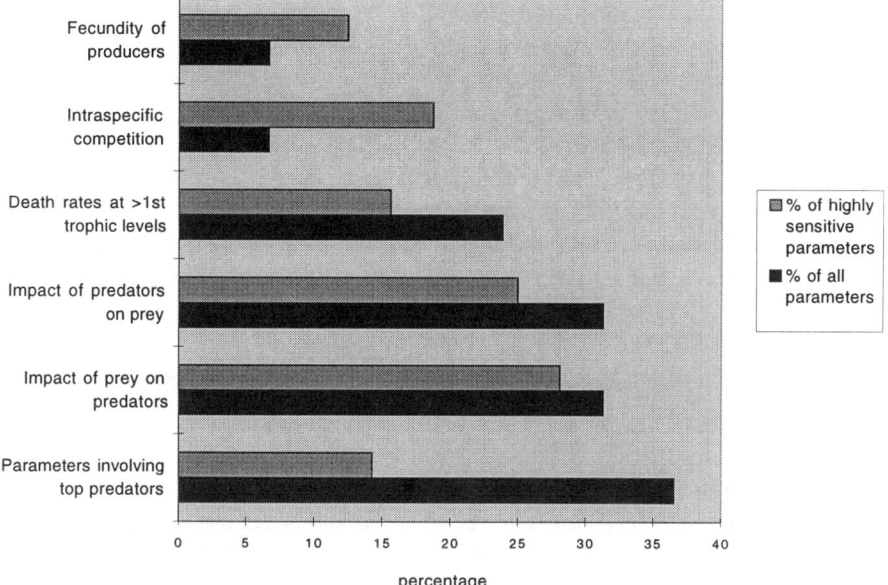

Figure 6-15 Summary of the parameters causing the greatest changes in community abundances. The gray bars show the percentage of all sensitive parameters (large circles) that were of a particular type. This is contrasted to the black bars, which show the percentage of all parameters (among all the models) that were of that type.

using experiments (Wootton 1994a) and indirectly by using population-level measures (Yodzis 1998).

Our results on the link (or lack thereof) between mortality of focal species and community-wide changes echo many of the general results found in studies of trophic cascades in which changes in 1 species cascade through the community (see reviews in Fretwell 1987; Power 1992; Wootton 1994a). This body of work has been concerned with the propagation of impacts down from top-level predators or up from producers. Our analysis differed in that we considered all species as potential target species; thus, we were not concentrating specifically on changes that propagate from upper or lower levels of a food web. Despite numerous famous examples of trophic cascades in natural communities, e.g., the shift in the Lake Victoria community with the introduction of an exotic cichlid, the general importance of trophic cascades in large complex communities has been a matter of intense debate. Many factors (e.g., omnivory, prey resistance, and predator competition and interference) tend to limit trophic cascades within communities, and enormous variability is seen in the extent and magnitude of these cascades (see reviews listed above and Morin and Lawler 1995). Thus, predicting whether a trophic cascade will occur and how large it will be depends on whether we know the community composition and the strengths of interactions between species.

Given the need to make predictions concerning contaminants in the presence of less-than-ideal information on the community, we explored characteristics of target species that might suggest that they will be better predictors of community impact. In our analyses simple characteristics of the target species (e.g., whether it is a generalist or specialist, is at a high or low trophic level, or is abundant or rare) did not increase the predictive value of the target species. Based on dynamic food-web models, the measurement of lethality of a contaminant—even with knowledge of the trophic level, abundance, or linkages of the target species—cannot be assumed by itself to indicate its potential impact on a community. The same weak link between impact and species characteristics has been found previously in analyses of keystone species (i.e., species that have a disproportionately high impact on a community). There may be general trends, e.g., species at higher trophic levels are more likely to be keystone species, but there are as many exceptions to this rule as there are adherents to it (Power et al. 1996).

In addition, we studied the relative impacts of mortality changes versus sublethal effects, namely, sublethal changes in fecundity and interactions (predator–prey and intraspecific competition rates). We found that these sublethal changes can be equally important or more important than changes in mortality. In 2 out of the 4 model communities, the community was more sensitive to changes in fecundity or interaction rates than to changes in mortality rates. Sensitivities were similar in the other communities. In our analyses, we compared proportionately equal changes in mortality and sublethal rates. For example, we compared a 0.1% decrease in mean life span with a 0.1% decrease in prey capture rates. The sensitivity of these model communities to proportionately equal changes in mortality and sublethal rates suggests that we need to pay attention to the magnitude of changes in sublethal rates (especially prey capture and fecundity rates) relative to changes in mortality rates.

Where do we go from here?

At first glance, our results appear discouraging and suggest that simple assays may not anticipate community-level consequences of toxicants. However, absence of simple answers does not mean it is hopeless. We believe that further exploration of community models could yield hypotheses about assays of impact that are of mid-level complexity. For example, it is possible that although the responses of single species yield little predictive power, studies that focus on 3 to 4 interacting species are effective assays. Thus, empirical studies of microcosms or extracts of communities might be good indicators of community impacts. This possibility could be tested with models that could be used to identify what type of microcosms offer the greatest benefit. Also, it is possible that field studies, which directly measure several community features, might capture the impact of contaminants. We will need to consult ecological theory for possible ways of summarizing the response of a community to a pollutant.

Finally, our analyses suggest that it is important to consider relative changes in fecundity, predator-prey, and/or feeding rates concurrently with changes in mean life spans. In simple models, such sublethal, short-term behaviors translate easily into population-level parameters; however, this translation is less clear in the natural environment. If these changes in behavior produce meaningful population changes in predation dynamics, fecundity, or competition, they can have an effect on the community that is equal to or greater than equivalent changes in death rates.

Overall, our analysis of dynamic food-web models shows that environmental assessments of contaminants that are based largely on results derived from simple lab systems are not likely to work. Environmental toxicology needs to be as sophisticated as nature in order to understand and manage the risks associated with environmental contamination.

APPENDIX TO CHAPTER 6

Models and Parameters

Three-Species Food Web

The subscripts on the parameters correspond to the species (S) in Figure 6-1a: 1 = p1, 2 = h1, 3 = c1. For the parameters, $a_{i,j}$ is the impact of species j on species i.

$d\,S1/dt = S1\,(b_1 - a_{1,1}\,S1 - a_{1,2}\,S2)$

$d\,S2/dt = S2\,(-d_2 + a_{2,1}\,S1 - a_{2,3}\,S3)$

$d\,S3/dt = S3\,(-d_3 + a_{3,2}\,S2)$

Plant–Mammal–Mammal A: $b_1 = 7$, $a_{1,1} = 0.7$, $a_{1,2} = 100$, $a_{2,3} = 3000$, $a_{2,1} = 0.1$, $a_{3,2} = 8$, $d_2 = 0.4$, $d_3 = 0.2$

Plant–Mammal–Mammal B: $b_1 = 7$, $a_{1,1} = 0.7$, $a_{1,2} = 0.05$, $a_{2,3} = 10$, $a_{2,1} = 10$, $a_{3,2} = 0.8$, $d_2 = 4$, $d_3 = 2$

Plant–Insect–Insect A: $b_1 = 7$, $a_{1,1} = 0.7$, $a_{1,2} = 0.05$, $a_{2,3} = 10$, $a_{2,1} = 10$, $a_{3,2} = 0.08$, $d_2 = 4$, $d_3 = 2$

Plant–Insect–Insect B (univoltine): $b_1 = 7$, $a_{1,1} = 0.7$, $a_{1,2} = 0.05$, $a_{2,3} = 10$, $a_{2,1} = 10$, $a_{3,2} = 0.08$, $d_2 = 1$, $d_3 = 1$

Plant–Insect–Parasitoid: $b_1 = 7$, $a_{1,1} = 0.7$, $a_{1,2} = 0.05$, $a_{2,3} = 10$, $a_{2,1} = 10$, $a_{3,2} = 0.8$, $d_2 = 4$, $d_3 = 2$

Plant–Insect–Disease: $b_1 = 7$, $a_{1,1} = 0.7$, $a_{1,2} = 0.05$, $a_{2,3} = 0.01$, $a_{2,1} = 10$, $a_{3,2} = 10$, $d_2 = 4$, $d_3 = 10$

Six-Species Food Web

The subscripts on the parameters correspond to the species (S) in Figure 6-1b: 1 = p1, 2 = p2, 3 = p3, 4 = h1, 5 = h2, 6 = c1.

$dS1/dt = S1\,(b_1 - a_{1,1}\,S1 - a_{1,3}\,S3)$

$dS2/dt = S2\,(b_2 - a_{2,2}\,S2 - a_{2,3}\,S3 - a_{2,4}\,S4)$

$dS3/dt = S3\,(-d_3 + a_{3,1}\,S1 + a_{3,2}\,S2 - a_{3,5}\,S5)$

$dS4/dt = S4\,(-d_4 + a_{4,2}\,S2 + a_{4,6}\,S6 - a_{4,5}\,S5)$

$dS5/dt = S5\,(-d_5 + a_{5,4}\,S4 + a_{5,3}\,S3)$

$dS6/dt = S6\,(b_6 - a_{6,6}\,S6 - a_{6,4}\,S4)$

Plant–Mammal–Mammal: $b_1 = 7$, $a_{1,1} = 0.7$, $a_{1,3} = 100$, $a_{1,4} = 6$, $b_2 = 7.1$, $a_{2,2} = 0.77$, $a_{2,3} = 7$, $a_{2,4} = 101$, $a_{3,5} = 3000$, $b_6 = 2$, $a_{6,6} = 1.4$, $a_{6,3} = 20$, $a_{6,4} = 30$, $a_{3,6} = 0.02$, $a_{4,6} = 0.5$, $a_{4,5} = 3000$, $a_{3,1} = 0.05$, $a_{4,1} = 0.01$, $a_{3,2} = 0.01$, $a_{4,2} = 0.15$, $a_{5,3} = 8$, $a_{5,4} = 9$, $d_3 = 0.2$, $d_4 = 0.3$, $d_5 = 0.1$

Plant–Insect–Insect: $b_1 = 7$, $a_{1,1} = 0.4$, $a_{1,3} = 100$, $a_{1,4} = 6$, $b_2 = 4.1$, $a_{2,2} = 2.1$, $a_{2,3} = 0.07$, $a_{2,4} = 0.01$, $a_{3,5} = 300$, $b_6 = 2$, $a_{6,6} = 1.4$, $a_{6,3} = 0.06$, $a_{6,4} = 0.08$, $a_{3,6} = 10$, $a_{4,6} = 15$, $a_{4,5} = 200$, $a_{3,1} = 20$, $a_{4,1} = 22$, $a_{3,2} = 12$, $a_{4,2} = 13$, $a_{5,3} = 8$, $a_{5,4} = 9$, $d_3 = 3$, $d_4 = 4$, $d_5 = 1.1$

Mono Lake

The parameter subscripts correspond to the species (S) in Figure 6-2: 1 = algae, 2 = brine flies, 3 = plovers, 4 = phytoplankton, 5 = brine shrimp, 6 = gulls.

$dS1/dt = S1\,(b_1 - a_{1,1}\,S1 - a_{1,2}\,S2)$

$dS2/dt = S2\,(-d_2 + a_{2,1}\,S1 - a_{2,3}\,S3 - a_{2,6}\,S6)$

$dS3/dt = S3\,(-d_3 + a_{3,2}\,S2)$

$dS4/dt = S4\,(b_4 - a_{4,4}\,S4 - a_{4,5}\,S5)$

$dS5/dt = S5\,(-d_5 + a_{5,4}\,S4 - a_{5,6}\,S6)$

$dS6/dt = S6\,(-d_6 + a_{6,2}\,S2 + a_{6,5}\,S5)$

$b_1 = 7$, $a_{1,1} = 0.7$, $a_{1,2} = 100$, $a_{2,3} = 3000$, $a_{2,1} = 0.1$, $a_{3,2} = 8$, $d_2 = 0.4$, $d_3 = 0.2$, $b_4 = 7$, $a_{4,4} = 0.7$, $a_{4,5} = 100$, $d_5 = 0.4$, $d_6 = 0.2$, $a_{5,4} = 0.065$, $a_{5,6} = 3000$, $a_{6,2} = 4$, $a_{6,5} = 8$, $a_{2,6} = 2000$

California Salt Marsh

The parameter subscripts correspond to the species (S) in Figure 6-3: 1 = terrestrial plants, 3 = terrestrial invertebrates, 6 = shrews, 8 = rails, 10 = passerines, 11 = small rodents, 12 = rats, 13 = raptors.

$d\, S1/dt = S1\, (b_1 - a_{1,1}\, S1 - a_{1,10}\, S10 - a_{1,11}\, S11 - a_{1,12}\, S12 - a_{1,3}\, S3 - a_{1,8}\, S8)$

$d\, S3/dt = S3\, (-d_3 + a_{3,1}\, S1 - a_{3,10}\, S10 - a_{3,11}\, S11 - a_{3,12}\, S12 - a_{3,6}\, S6 - a_{3,8}\, S8)$

$d\, S6/dt = S6\, (-d_6 - a_{6,13}\, S13 + a_{6,3}\, S3)$

$d\, S8/dt = S8\, (-d_8 + a_{8,1}\, S1 - a_{8,12}\, S12 - a_{8,13}\, S13 + a_{8,3}\, S3)$

$d\, S10/dt = S10\, (-d_{10} + a_{10,1}\, S1 - a_{10,13}\, S13 + a_{10,3}\, S3)$

$d\, S11/dt = S11\, (-d_{11} + a_{11,1}\, S1 - a_{11,13}\, S13 + a_{11,3}\, S3)$

$d\, S12/dt = S12\, (-d_{12} + a_{12,1}\, S1 - a_{12,13}\, S13 + a_{12,3}\, S3 + a_{12,8}\, S8)$

$d\, S13/dt = S13\, (-d_{13} + a_{13,10}\, S10 + a_{13,11}\, S11 + a_{13,12}\, S12 + a_{13,6}\, S6 + a_{13,8}\, S8)$

$b_1 = 7$, $d_3 = 13$, $d_6 = 1.5$, $d_8 = 0.7$, $d_{10} = 0.75$, $d_{11} = 1.2$, $d_{12} = 0.85$, $d_{13} = 0.2$, $a_{1,1} = 0.10$, $a_{1,3} = 0.2$, $a_{1,8} = 5.1$, $a_{1,10} = 24$, $a_{1,11} = 37$, $a_{1,12} = 5$, $a_{3,1} = 10.12$, $a_{3,6} = 14$, $a_{3,8} = 13$, $a_{3,10} = 25$, $a_{3,11} = 13.3$, $a_{3,12} = 14.4$, $a_{6,3} = 2.2$, $a_{6,13} = 14.55$, $a_{8,1} = 0.4$, $a_{8,3} = 0.33$, $a_{8,12} = 2.4$, $a_{8,13} = 12.2$, $a_{10,1} = 0.3$, $a_{10,3} = 0.35$, $a_{10,13} = 10.2$, $a_{11,1} = 0.5$, $a_{11,3} = 0.87$, $a_{11,13} = 44.5$, $a_{12,1} = 0.21$, $a_{12,3} = 0.86$, $a_{12,8} = 0.07$, $a_{12,13} = 22.4$, $a_{13,6} = 0.3$, $a_{13,8} = 0.45$, $a_{13,10} = 0.27$, $a_{13,11} = 0.42$, $a_{13,12} = 0.46$

French Frigate Shoals

The model for the French Frigate Shoals (Figure 6-4) was a series of 15 ordinary differential equations for each species (S) in the web. Each equation for each species was of the form

$d\, Si/dt = Si\, (d_i + a_{i,1}\, S1 + a_{i,2}\, S2 + a_{i,3}\, S3 + a_{i,4}\, S4 + a_{i,5}\, S5 + a_{i,6}\, S6 + a_{i,7}\, S7 + a_{i,8}\, S8 + a_{i,9}\, S9 + a_{i,10}\, S10 + a_{i,11}\, S11 + a_{i,12}\, S12 + a_{i,13}\, S13 + a_{i,14}\, S14 + a_{i,15}\, S15)$

with $a_{i,j}$ parameters as below. Missing $a_{i,j}$ parameters equal zero. Parameter values are from Polovina (1984). The parameter subscripts correspond to species in Figure 6-4: 1 = tiger sharks, 2 = birds, 3 = monk seals, 4 = reef sharks, 5 = turtles, 6 = small pelagics, 7 = jacks, 8 = reef fishes, 9 = lobsters, 10 = bottom fishes, 11 = tuna, 12 = zooplankton, 13 = benthos, 14 = phytoplankton, 15 = benthic producers.

$a_{1,1} = -0.95$, $a_{1,2} = 10$, $a_{1,3} = 0.63$, $a_{1,4} = 0.44$, $a_{1,5} = 0.38$, $a_{1,6} = 0.021$, $a_{1,7} = 0.064$, $a_{1,8} = 0.009$, $a_{1,9} = 0.058$, $a_{1,11} = 0.23$, $d_1 = -0.455$, $a_{2,1} = -90$, $a_{2,6} = 1.97$, $a_{2,7} = 1.39$, $a_{2,8} = 0.054$, $a_{2,11} = 2.45$, $a_{2,12} = 0.27$, $d_2 = -1.62$, $a_{3,1} = -5.71$, $a_{3,8} = 0.17$, $a_{3,9} = 0.37$, $d_3 = -2.76$, $a_{4,1} = -3.97$, $a_{4,6} = 0.0047$, $a_{4,8} = 0.011$, $a_{4,9} = 0.007$, $d_4 = -0.008$, $a_{5,1} -3.46$, $a_{5,12} = 0.015$, $a_{5,15} = 0.002$, $d_5 = -0.005$, $a_{6,1} = -0.19$, $a_{6,2} = -29.18$, $a_{6,4} = -0.102$, $a_{6,6} = -0.21$, $a_{6,7} = -0.16$, $a_{6,10} = -0.24$, $a_{6,11} = -1.36$, $a_{6,12} = 1.04$, $d_6 = -0.056$, $a_{7,1} = -0.58$, $a_{7,2} =$

-20.56, $a_{7,6} = 0.015$, $a_{7,8} = 0.018$, $a_{7,9} = 0.035$, $d_7 = -0.017$, $a_{8,1} = -0.084$, $a_{8,2} = -0.80$, $a_{8,3} = -2.27$, $a_{8,4} = -0.23$, $a_{8,7} = -0.20$, $a_{8,8} = -0.066$, $a_{8,10} = -0.11$, $a_{8,11} = -0.028$, $a_{8,12} = 0.257$, $a_{8,13} = 0.009$, $a_{8,15} = 0.005$, $d_8 = -0.075$, $a_{9,1} = -0.52$, $a_{9,3} = -4.99$, $a_{9,4} = -0.16$, $a_{9,7} = -0.38$, $a_{9,10} = -0.054$, $a_{9,12} = 0.011$, $a_{9,13} = 0.006$, $d_9 = -0.026$, $a_{10,6} = 0.021$, $a_{10,8} = 0.01$, $a_{10,9} = 0.005$, $a_{10,10} = -0.96$, $a_{10,11} = -4.76$, $a_{10,12} = 0.033$, $a_{10,13} = 0.001$, $d_{10} = -0.017$, $a_{11,1} = -2.045$, $a_{11,2} = -36.36$, $a_{11,6} = 0.17$, $a_{11,8} = 0.003$, $a_{11,10} = 0.59$, $a_{11,12} = 0.24$, $d_{11} = -0.029$, $a_{12,2} = -4.028$, $a_{12,5} = -0.35$, $a_{12,6} = -7.010$, $a_{12,8} = -1.63$, $a_{12,9} = -0.17$, $a_{12,10} = -0.38$, $a_{12,11} = -1.92$, $a_{12,14} = 9.57$, $a_{12,15} = 0.049$, $d_{12} = -2.009$, $a_{13,8} = -0.057$, $a_{13,9} = -0.104$, $a_{13,10} = -0.012$, $a_{13,13} = -0.018$, $a_{13,15} = 0.034$, $d_{13} = -0.15$, $a_{14,12} = -66.96$, $a_{14,14} = -1.84$, $b_{14} = 73.49$, $a_{15,5} = -0.042$, $a_{15,8} = -0.032$, $a_{15,12} = -0.34$, $a_{15,13} = -0.14$, $a_{15,15} = -0.017$, $b_{15} = 13.12$

References

Briand F. 1983. Environmental control of food web structure. *Ecology* 64:253.

Christensen V, Pauly D. 1992. ECOPATH II: A software for balancing steady-state models and calculating network characteristics. *Ecol Model* 61:169–185.

Clements WH. 1997. Effects of contaminants at higher levels of biological organization in aquatic ecosystems. *Rev Toxicol* 1:107–146.

Cohn J, Macphail RC. 1996. Ethological and experimental approaches to behavior analysis: Implications for ecotoxicology. *Environ Health Perspect* 104:299–305.

DeAngelis DL. 1975. Stability and connectance in food web models. *Ecology* 56:238–243.

Donkin P, Widdows J, Evans SV, Staff FJ, Yan T. 1995. Effect of neurotoxic pesticides on the feeding rate of marine mussels (Mytilus edulis). *Pest Sci* 49:196–209.

Elton C. 1927. Animal ecology. London UK: Sidgwick & Jackson.

Elton C. 1958. The ecology of invasions by animals and plants. London UK: Chapman & Hall.

Fagan WF. 1997. Omnivory as a stabilizing feature of natural communities. *Am Nat* 150:554–567.

Fretwell SD. 1987. Food chain dynamics: The central theory of ecology? *Oikos* 50:291–301.

Hall SJ, Raffaelli DG. 1993. Food webs: Theory and reality. *Adv Ecol Res* 24:187–239.

Gopal K, Ram M. 1995. Alteration in the neurotransmitter levels in the brain of the freshwater snakehead fish (*Channa punctatus*) exposed to carbofuran. *Ecotoxicology* 4:1–4.

Lawler LE. 1978. A comment on randomly constructed model ecosystems. *Am Nat* 112:445–447.

Lawton JH. 1989. Food webs. In: Cherrett JM, editor. Ecological concepts. Oxford UK: Blackwell Scientific. p 43–78.

May RM. 1972. Will a large complex system be stable? *Nature* 238:413–141.

May RM. 1973. Stability and complexity in model ecosystems. Princeton NJ: Princeton Monographs.

MacArthur RH. 1955. Fluctuations of animal populations and a measure of community stability. *Ecology* 36:533–536.

Martinez ND. 1991. Artifacts or attributes? Effects of resolution on the Little Rock Lake food web. *Ecol Monogr* 61:367–392.

McCann K, Hastings A, Huxel GR. 1998. Weak trophic interactions and the balance of nature. *Nature* 395:794–798.

Morin PJ, Lawler SP. 1995. Food web architecture and population dynamics: Theory and empirical evidence. *Ann Rev Ecol Systematics* 26:505–529.

Olson LJ, Erickson BJ, Hinsdill RD, Wyman JA, Porter WP, Binning LK, Bidgood RC, Nordheim EV. 1987. Aldicarb immunomodulation in mice: An inverse dose–response to parts-per-billion levels in drinking water. *Arch Environ Contam Toxicol* 16:433–439.

Paine RT. 1988. On food webs: Road maps of interactions of the grist for theoretical development? *Ecology* 69:1648–1654.

Peakall DB. 1982. Disrupted patterns of behavior in natural populations as an index of ecotoxicity. *R Neurobehav Toxicol* 104:331–335.

Pimm SL. 1982. Food webs. New York NY: Chapman and Hall. 219 p.

Pimm SL, Lawton JH. 1977. The number of trophic levels in ecological communities. *Nature* 268:329–331.

Pimm SL, Lawton JH. 1978. On feeding at more than one trophic level. *Nature* 275:542–544.

Pimm SL, Lawton JH, Cohen JE. 1993. Food web patterns and their consequences. *Nature* 350:669–674.

Polis GA, Strong DR. 1996. Food web complexity and community dynamics. *Am Nat* 147:813–846.

Polovina JJ. 1984. Model of a coral reef ecosystem. I. The ECOPATH model and its application to French Frigate Shoals. *Coral Reefs* 3:1–11.

Porter W, Hinsdill R, Fairbrother A, Olson LJ, Jaeger J, Yuill T, Bisgaard S, Hunter WG, Nolan K. 1984. Toxicant–disease–environment interactions associated with suppression of immune system, growth, and reproduction. *Science* 224:1014–1017.

Porter W, Jaeger J, Carlson I. 1999. Endocrine, immune and behavioral effects of aldicarb, atrazine and nitrate mixtures at groundwater concentrations. *Toxicol Ind Health* 15:133-50.

Power ME. 1992. Top–down and bottom–up forces in food webs: Do plants have primacy? *Ecology* 73:733–746.

Power ME, Tilman D, Estes JA, Menge BA, Bond WJ, Mills S, Daily G, Castilla JC, Lubchenco J, Paine RT. 1996. Challenges in the quest for keystones. *Bioscience* 46:609–620.

Roper DS, Nipper MG, Hickey CW, Martin ML, Weatherhead MA. 1995. Burial, crawling and drifting behavior of the bivalve Macoma Lillian in response to common sediment contaminants. *Mar Pollut Bull* 31:471–478.

Schmitz OJ, Beckerman AP, O'Brien KM. 1997. Behaviorally mediated trophic cascades: Effects of predation risk on food web interactions. *Ecology* 78:1388–1399.

Schoenly K, Beaver RA, Heumier TA. 1991. On the trophic relations of insects: A food web approach. *Am Nat* 137:597–632.

Spencer PD. 1997. Optimal harvesting of fish populations with nonlinear rates of predation and autocorrelated environmental variability. *Can J Fish Aquat Sci* 54:59–74.

Swan SH, Elkin EP, Fenster L. 1997. Have sperm densities declined? A reanalysis of global trend data. *Environ Health Perspect* 105:1228–1232.

Walters C, Christensen V, Pauly D. 1997. Structuring dynamic models of exploited ecosystems from trophic mass–balance assessments. *Rev Fish Biol Fish* 7:139–172.

Warren PH. 1990. Variation in food-web structure: The determinants of connectance. *Am Nat* 136:689–700.

Wiens JA, Patten DT, Botkin DB. 1993. Assessing ecological impact assessment: Lessons from Mono Lake, California. *Ecol Appl* 3:595–609.

Winemiller KO. 1989. Must connectance decline with species richness? *Am Nat* 134:960–968.

Wootton T. 1994a. Predicting direct and indirect effects: An integrated approach using experiments and path analysis. *Ecology* 75:151–165.

Wootton T. 1994b. The nature and consequences of indirect effects in ecological communities. *Ann Rev Ecol Systematics* 25:443–466.

Yodzis P. 1988. The indeterminacy of ecological interactions as perceived through perturbation experiments. *Ecology* 69:508–515.

Yodzis P. 1998. Local trophodynamics and the interaction of marine mammals and fisheries in the Benguela ecosystem. *J Animal Ecol* 67:635–658.

CHAPTER 7

Modeling Toxic Effects on Populations: Experience from Aquatic Studies

Glenn W. Suter II and Lawrence W. Barnthouse

There are a number of potential advantages to estimating the effects of contaminants on populations. The first is that effects on populations are generally considered more important than effects on individuals. Most ecologists, decision-makers, and members of the public are likely to agree that it is important to protect ecosystems over component populations and populations over component individuals because the utility of species as resources and their importance to ecosystem function are associated with population properties such as abundance, productivity, and age-class structure. Therefore, an assessment that predicts effects on populations should be more likely to compel action than one that predicts effects on individuals. Second, population analysis provides a consistent basis for interpreting toxicological information. That is, we can use population responses to integrate organismal responses, e.g., mortality in various life stages, reproductive decrements, and growth decrements, into a single response, i.e., population production. Finally, population responses potentially provide a more useful basis for decision-making. Population responses are easy to explain to decision-makers and stakeholders and are more readily incorporated into decision tools such as cost–benefit analysis.

Despite these advantages, there is a need to model toxic effects on populations only if there is a desire by decision makers to regulate toxic effects at that level of ecological organization. The U.S. has had no consistent policy concerning the protection of nonhuman organisms at the individual, population, or ecosystem levels. Individuals of wildlife and fish populations have been harvested with the intent of sustaining populations while exploiting their production. Similarly, the loss of fish in the cooling system of power plants and in hydroelectric turbines has been assumed to be acceptable if the effects on populations are not unacceptable. On the other hand, environmental protection from pesticides and other chemicals has been based largely on organism-level toxicology and ecoepidemiology (e.g., fish kills have long

CHAPTER PREVIEW
Toxicology data problems and solutions   178
Modeling problems and solutions   182
Results   183
Recommendations   184

been considered grounds for regulating aqueous effluents). The courts endorsed this approach when they ruled that the USEPA was correct in restricting the use of diazinon on grasses because of incidents of goose mortality. The U.S. Fifth Circuit Court of Appeals ruled that, although geese are harvested and the USEPA had not demonstrated an effect on goose populations, the mortality incidents were sufficient grounds for restricting diazinon use. The USEPA has been struggling with this issue of what to protect but has not yet developed a policy (Troyer and Brody 1994; Barton et al. 1997; USEPA 1998).

During the 1980s, the Environmental Sciences Division of ORNL was commissioned by the USEPA to develop methods for ecological risk assessment. Those activities included the development of a series of mathematical models of aquatic population and the toxicological datasets and statistical models needed to parameterize them. The following sections discuss the methods used, results obtained, and lessons learned that may be applicable to risk assessments for wildlife.

Toxicology Data Problems and Solutions

Most risk assessments must use existing toxicological data because assessors do not have the time or resources to perform tests ad hoc. However, toxicity tests are not designed to provide information for risk assessments, particularly not for population modeling. The principal limitations of existing ecotoxicity data are that exposure–response relationships are not provided for chronic responses and that test results are not available for species, life stages, and responses of interest. For example, effects on fecundity are seldom available for fishes.

The most fundamental problem is that standard chronic toxicity tests are designed as research projects; they test the null hypothesis that there are no toxic effects rather than determine the exposure–response relationships. The fact that a test did or did not find a statistically significant difference from controls tells the assessor nothing about the biological significance of the response and does not help the assessor to adjust the parameters of a population model to account for toxicity. Therefore, if chronic effects are to be modeled, one must first obtain the raw test results and then apply an exposure–response model. This was done at ORNL in the 1980s, resulting in a database of concentration–response data for freshwater fish and invertebrates and fitted parameters of logistic concentration–response models and variance parameters for those models. The ability to complete this process is limited by the test designs. If there are few exposure levels, or if they were not selected to cover a range of response levels, it is not possible to fit a function to the data. While this reduces the amount of data available for risk assessment, one must question whether a study that does not demonstrate that effects increase with exposure should be used for regulatory or management purposes.

Because test results typically are not available for the species, life stages, and responses of interest, it is necessary to develop extrapolation models to estimate the unreported responses from available data. For aquatic interspecies extrapolations, we used a dataset from the Columbia National Fisheries Research Laboratory that contained acute test results for a large number of species and chemicals (Mayer and Ellersieck 1986). We found that the best predictor of the relative susceptibility of species was the taxonomic relationship between the test species and the endpoint species (Suter et al. 1983; Barnthouse et al. 1990). The more closely related the species were, the more similar their response was. Therefore, we developed a set of regression models relating the response of 1 taxon to another for species within genera, genera within families, and families within orders (Figure 7-1). Wildlife risk assessors have used allometric models for inter-taxa extrapolation. Allometric factors are important; however, ever since it was found that the responses of standard test species were not good predictors of the effects of dichlorodiphenyl-trichloroethane (DDT) on falconiforms, it has been clear that there are taxonomic differences in susceptibility as well. More recently, a principal components analysis of LD50 data for 8 avian species found distinct clusters for the Phasianidae and Icteridae (Baril et al. 1994).

Extrapolation among life stages and responses is a more complex problem. One must develop a set of responses that are reported in toxicity tests and that are relevant to population responses. For fishes, we used egg survival, larval survival, juvenile and adult survival, fecundity, and juvenile or adult growth. We developed regression models to relate each of these responses to the others so that we could estimate all chronic responses when not all of them were measured or reported. In addition, it is necessary to extrapolate from acute lethal effects (LC50s) to chronic responses (Figure 7-2). Therefore, regression models were developed that related each of the chronic responses to an acute response of the same species to the same chemical. With these models, all 5 toxic responses could be estimated from any chronic response or an acute LC50.

When multiple extrapolations are required, one must perform extrapolations in sequence. For example, if the available toxicity datum for a chemical is a fathead minnow LC50 and the assessment endpoint is rainbow trout production, one must extrapolate between minnows and trout using the Cypriniformes-Salmoniformes model and then use the estimated rainbow trout LC50 to estimate effects on the chronic responses. The only trick to this is to carry the uncertainties appropriately through the two models so that each response parameter is estimated as a distribution for use in the Monte Carlo analysis of the population model.

This approach is purely empirical. An alternative approach to applying toxic responses to population models is to model the toxic mechanism in the individual organism. One may assume, for example, that a narcotic chemical slows metabolism and feeding. Then the model can be calibrated for a particular narcotic chemical by adjusting the metabolism and feeding parameters so that the organism dies at the

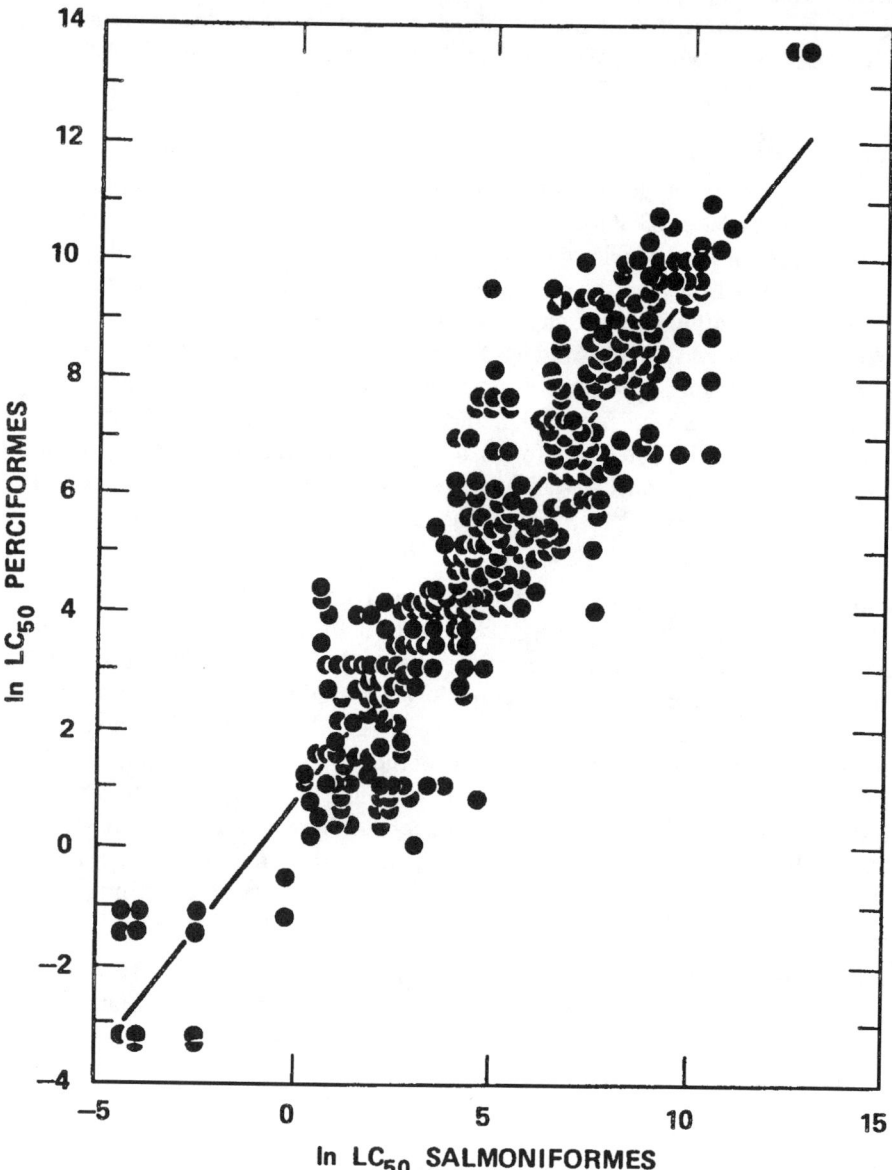

Figure 7-1 Example of a taxonomic regression. This model would be used if the test species was a member of the family *Salmoniformes* and the endpoint species was a member of the family *Perciformes*.

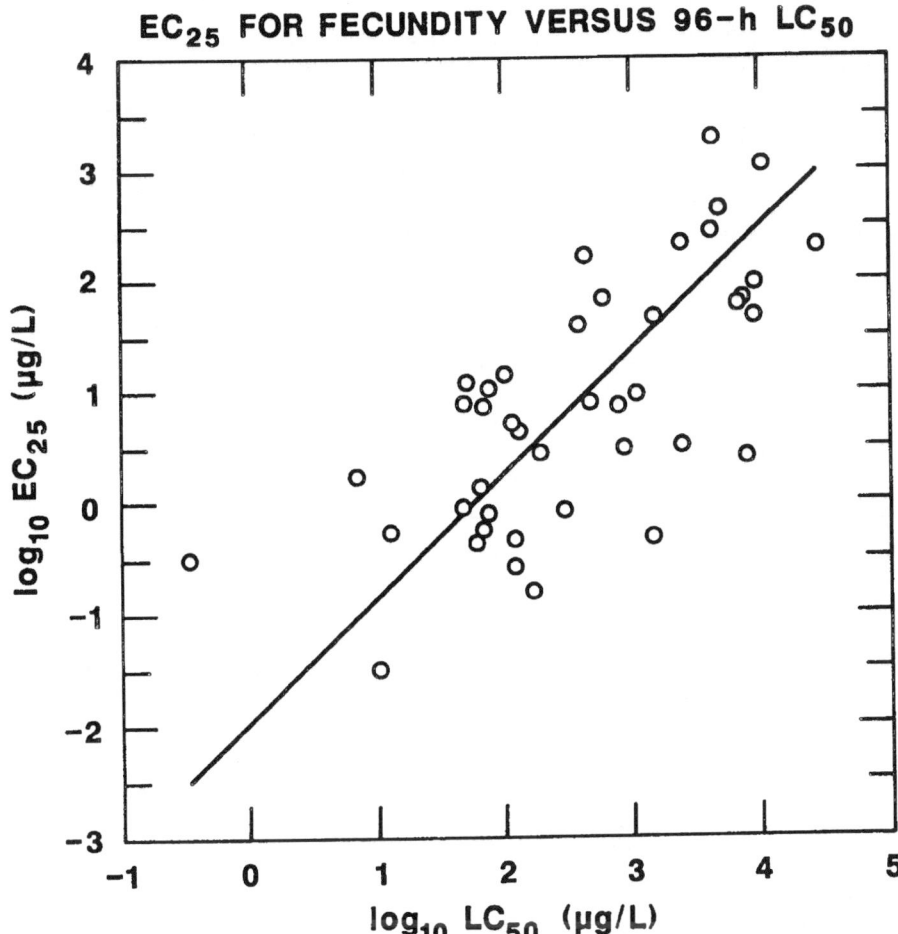

Figure 7-2 Regression model to estimate a chronic quartile effective concentration (EC25) for fecundity from an acute median lethal concentration (LC50)

lethal concentration. The approach may be elaborated to include seasonal, habitat, and inter-individual differences in susceptibility (e.g., fat reserves), as long as the mechanism is known. This mechanistic approach has been applied to *Daphnia* populations (Hallam and Clark 1983; Hallam et al. 1990). Mechanistic toxicology is certainly the state of the science in population modeling, but current knowledge of mechanisms of action is seldom sufficient, even for mammals.

Toxicity-test design places constraints on the ability to model effects on populations. Life-cycle tests provide the information needed to model the effects of chronic exposures. However, for vertebrates, those tests are expensive and lengthy. To reduce cost and time requirements, fisheries toxicologists have reduced the chronic tests from life-cycle tests, to early life-stage (eggs and larvae) tests, to 7-d larval tests. Although it has been claimed that these tests are as sensitive as life-cycle tests, analysis of published results indicates that, in practice, they are not (Suter et al. 1987; Suter 1990). In particular, fecundity is, on average, more sensitive to toxic effects than any of the early life-stage responses of fish. Therefore, it is not sufficient to include only early life-stage effects in population models. The body of life-cycle tests, most of which were conducted in the 1970s, must serve to estimate the sensitivity of adults when only early life-stage responses are available. Without those tests, modeling of truly chronic exposures would not be possible.

Modeling Problems and Solutions

Our strategy for developing population-level assessment methodologies involved combining a rigorous statistical analysis of toxicity-test data (discussed above) with relatively simple, but well-established, population-assessment techniques. Hence, we employed scientific concepts developed by theoretical population biologists nearly 100 years ago and empirically oriented modeling techniques developed by fisheries managers over the past 4 decades .

So-called "demographic" models that describe population properties such as abundance, age structure, and extinction probability in terms of the survival and reproduction of individual organisms were the obvious choices. These kinds of models are the quantitative foundation of both resource management and conservation biology (Caswell 1989; Barnthouse 1993). Integrating toxicology with population biology required the development of methods for linking laboratory-derived toxicity data to the mortality and reproduction parameters of typical population models. For fish, this task is conceptually straightforward. Results of acute and chronic lethality tests and reproduction tests can be translated directly into reductions in the survival and reproduction parameters of age-structured fish population models (Barnthouse et al. 1987; Barnthouse et al. 1988).

One complicating factor is that the fish species of greatest interest for risk assessment, (primarily species of recreational or commercial interest or endangered/threatened species), are usually not suitable for toxicity testing. Effects on species of interest usually must be estimated from data on standard test species such as fathead minnow, rainbow trout, or sheepshead minnow. The empirical regression models described by Suter et al. (1983) can be used to predict acute and chronic effects benchmarks for any species of interest from data derived from any of the common test systems. For purposes of population-level assessment, however, a full dose–response relationship rather than a simple LC50 is required. Barnthouse et al. (1987) demonstrated a method for extending the method of Suter et al. (1983) to obtain estimates of the parameters of the logistic concentration–response model. Uncertainties inherent in each step of the extrapolation are accumulated, and the result expresses an "expected" concentration-response relationship with a 90% confidence band.

We initially applied the above approach to a simple index of lifetime reproductive potential, which is defined as the expected number of eggs produced by a female fish over her reproductive lifetime. This index provides a direct means of comparing effects of chemicals to effects of other stresses, e.g., fishing. The reproductive-potential index is now widely used by fisheries managers to identify populations that are at risk of decline because of overfishing. The index does not, however, provide numerical estimates of population abundance. In a later paper (Barnthouse et al. 1990), we coupled our toxicity extrapolation equations to matrix projection models of 2 well-studied fish populations, the gulf menhaden (*Brevoortia patronus*) and the Chesapeake Bay striped bass (*Morone saxatilis*). These models had each been developed and tested using more than 20 years of field observations. We used them to investigate potential impacts of chemicals on the long-term abundance, extinction risk, and yield of populations exposed to varying degrees of environmental variability, harvest intensity, and chemical exposure.

Results

Our initial objective was to test the then-common assumption that "no-effects" concentrations that were derived from the application of conventional statistics to laboratory data were measures of actual biological effects thresholds. Given the relatively low statistical power of many test systems, it seemed possible that biologically significant effects on real populations might occur at exposure levels too low to produce statistically detectable effects in the laboratory. We translated individual-level data for a variety of chemicals into population-level effects using the reproductive-potential index (Barnthouse et al. 1987). We found that, even under the most favorable conditions (i.e., a full life-cycle test performed on the species of interest, with no inter-species or acute-chronic extrapolations), no-effect concentrations from the laboratory often corresponded to greater than 50% reductions in reproductive potential. Fishing a population at or above its level of maximum sustained yield would produce a similar reduction in reproductive potential. This result does not imply that real populations exposed to chemicals at or near a no-observed-effect concentration (NOEC) would necessarily suffer adverse effects because our analysis included only a few of the many factors that differ between laboratory test systems and real ecosystems. However, our analysis did demonstrate that 1) no-effect concentrations measured in the laboratory are in part a function of test design (e.g., the specific concentrations tested, the number of replicates, and the variability in responses among replicates), and, therefore, are not measures of actual biological effects thresholds and 2) potentially significant population-level effects can occur at or below no-effect concentrations. We subsequently evaluated the contributions of different components of a typical life-cycle test design to uncertainty concerning population-level effects (Barnthouse et al. 1988) and found that uncertainty inherent in the tests used to measure fecundity were by far the most important.

Analyses of the menhaden and striped bass models focused more on ecology than on toxicology (Barnthouse et al. 1990). These 2 populations differed dramatically in terms of life history and population status. The gulf menhaden has a maximum life span of only 5 years; the size of the gulf menhaden population has fluctuated within a relatively narrow range for several decades. Striped bass, in contrast, do not reach sexual maturity until at least 5 years of age and can live as long as 30 years. Annual variability in abundance is high, and at the time our work was performed the Chesapeake Bay population was at an extremely low level. Based on previously published results, we expected that

- simulated chemical exposure would be more likely to cause extinction of striped bass than of gulf menhaden,
- simulated yields of striped bass would be more sensitive to chemical-related effects than would yields of menhaden, and
- chemicals would have greater relative effects on heavily fished populations than on lightly fished populations.

We found that these predictions were qualitatively accurate but that the magnitudes of the differences were surprisingly small. Uncertainties related to differences between life-history types and differences in fishing intensity were far smaller than uncertainties related to the type of toxicity data used to estimate effects of chemicals on the modeled populations.

Recommendations

The following recommendations are derived from the experience described above. The first set of recommendations deals with improvements in toxicity-test design; the second set deals with steps needed to increase the understanding and use of population-level assessment tools in regulatory assessments.

Improving toxicity-test design

Toxic effects should be considered in terms of the life cycle of the endpoint species and not simply in terms of the duration of the test. If the test does not include the entire life cycle, extrapolation models must be developed to estimate the effects that would occur if the exposure was extended to the entire life cycle and the associated uncertainties.

The population models that we developed for fish were based on the assumption that exposures were continuous and constant. This assumption would be appropriate for some wildlife scenarios such as populations resident on an area with contaminated soils. However, for scenarios such as pesticide applications, it will be necessary to develop models that incorporate temporal dynamics.

The development and verification of extrapolation models require a large body of results from standard toxicity tests. The chronic toxicity tests conducted for avian and mammalian wildlife and for human health are poorly standardized. As a result, variance resulting from test protocols will tend to obscure differences that are due to taxonomy, size, life stage, etc. For this and other reasons, it is highly desirable that standard chronic tests be embraced by wildlife toxicologists.

The standard chronic test protocols should be designed to support risk assessment in general and population modeling in particular. They should include a sufficient range of doses to establish dose–response relationships. This need not drive up the cost of testing, since little replication is required for regression analysis. Numerical results should be presented for all replicates to allow estimation of variance and for multiple time intervals to allow estimation of temporal dynamics. As far as practicable, all life stages should be included.

Extrapolation models other than allometric scaling should be developed, compared, and validated. Clearly a deer is not a large rat, and a fetal deer is not a small adult deer. This point also has been made by Fairbrother and Kaputska (1996). For a recent review of alternative approaches to ecotoxicological extrapolation modeling see Suter (1998).

Overcoming impediments to the use of population modeling

It has been our experience that the modeling of contaminant effects on populations has been scientifically accepted but has not been incorporated into regulatory practice. We believe that there are 2 major problems, neither of which has to do with the models themselves.

The first problem is the training and experience of the risk assessors and risk managers. Most ecological risk assessors are toxicologists, chemists, or general field biologists. They are very familiar with the design and interpretation of test systems and field surveys but have little quantitative training beyond basic statistics. They are skeptical of types of analysis that they do not understand or that appear to go beyond a strictly limited interpretation of the test data. The problem is compounded when Monte Carlo analyses or other uncertainty analysis techniques are used. The problem carries over to risk managers who are unlikely to have sufficient understanding of the issues to demand or even understand a population analysis.

The second problem is the risk management implications of population-level assessment approaches. When fish, wildlife, and forest resources are managed, a certain level of harvest is accepted as long as the population maintains itself in the long run. The idea that effects of chemicals on populations can be assessed using the same methods that are used to manage fish, wildlife, and forest resources implies that there might be acceptable levels of population impairment, at least on a short-term basis. However, the acceptable level of effects is not the same. Conventionally, the effects on populations that are induced by harvesting would not be acceptable if they were induced by pollutants. Many environmental advocacy organizations and many environmental regulators categorically reject this notion and argue instead for the "precautionary principle" of minimizing risks from chemicals to the greatest possible extent. Hence, a standard has been applied to regulation of chemicals that is qualitatively different from those applied to regulation of natural resource harvesting and to other human activities (e.g., recreation, power plant operation, land use, water management, and transportation).

Without an awareness of the policy implications, the practice of ecotoxicology is pushing regulatory practice toward a precautionary approach. Increasingly, biochemical endpoints are used in place of the traditional endpoints of survival, growth, and reproduction. Enzyme induction and other biomarkers are useful for diagnosis of the causes of overt effects and for estimating levels of internal exposure. However, when biomarkers are used as endpoints for assessments and then as bases for regulatory or remedial actions, the level of concern is pushed down 1 level of biological organization. Increasingly, our eyes are not just on the sparrow, they are on the sparrow's mixed function oxidase levels. However, we have no means of relating the enzyme level to effects on valued properties of populations or ecosystems.

We have the following recommendations to move forward with population-based management of chemical contaminants:

- Population-level risk assessments must meet the highest possible standards of scientific credibility because of the skepticism of many risk assessors and managers. Both the population biology and the toxicology must be defensible to a degree not required of conventional risk assessments.
- We must seek high-visibility successes, e.g., regulatory problems for which a population-level assessment approach is actually implemented and adopted and that provides a clearly superior solution.
- We must spend time educating the current generation of assessors, not by writing papers and giving seminars, but by sitting down with them and helping them figure out how to apply population-level concepts to their day-to-day activities.
- University programs in ecology and environmental management should ensure that all of their graduates have a basic knowledge of mathematical and statistical modeling sufficient to write a basic chemical exposure model, fit a dose–response function, and parameterize a basic population model. Mathematics is the language of science, but many environmental science graduates are quantitatively illiterate.
- Regulatory agencies have not been clear about what components and properties of the environment they will protect. In the absence of guidance,

risk assessors need to be clear themselves about what they are estimating. It is not sufficient to say that the estimated dose exceeds the no-observed-adverse-effect level. What do they expect to happen at that dose? Attempts to answer that question will reveal the need to work out the implications of observed toxic responses, including population implications.

- Multi-level assessment is a good strategy in the absence of guidance concerning levels of organization to be protected. That is, for an individual assessment, present the estimated effects on all potentially relevant levels of organization. For example, if assessing risks from an organophosphate, one might estimate that w individuals will experience prescribed cholinesterase reductions, x individuals will be killed within a region, which includes $y\%$ of the regional population of the most affected species, and that would result in a $z\%$ reduction in the breeding population the following year. Such a presentation could be a better basis for decision-making and would serve to educate the decision-maker concerning the value of population assessments.

References

Baril A, Jobin B, Mineau P, Collins BT. 1994. A consideration of inter-species variability in the use of the median lethal dose (LD50) in avian risk assessment. Canadian Wildlife Service: Hull (PQ No. 216).

Barnthouse LW. 1993. Population-level effects. In: Suter GW, editor. Ecological risk assessment. Boca Raton FL: Lewis. p 247–274.

Barnthouse LW, Suter II GW, Rosen AE. 1988. Inferring population-level significance from individual-level effects: An extrapolation from fisheries science to ecotoxicology. In: Suter GW, Lewis M, editors. Aquatic toxicology and hazard assessment, 11th Symposium. Philadelphia PA: ASTM. p 289–300.

Barnthouse LW, Suter II GW, Rosen AE. 1990. Risks of toxic contaminants to exploited fish populations: Influence of life history, data uncertainty, and exploitation intensity. *Environ Toxicol Chem* 9:297–311.

Barnthouse LW, Suter II GW, Rosen AE, Beauchamp JJ. 1987. Estimating responses of fish populations to toxic contaminants. *Environ Toxicol Chem* 6:811–824.

Barton A, Berish C, Daniel B, Ells S, Marshall, T, Messer, J, Powell M, Rice M, Sergeant A, Serveiss, V, Sunzenauer I, Whitworth M. 1997. Priorities for ecological protection: An initial list and discussion document for EPA. Washington DC: U.S. Environmental Protection Agency. EPA/600/S-97/002.

Caswell H. 1989. Matrix population models. Sunderland MA: Sinauer Associates, Inc.

[USEPA] U.S. Environmental Protection Agency. 1998. Guidelines for ecological risk assessment: Risk assessment forum. Washington DC: U.S. Environmental Protection Agency. EPA/630/R-95/002F.

Fairbrother A, Kaputska LA. 1996. Toxicity extrapolation in terrestrial systems. Sacramento CA: California Environmental Protection Agency.

Hallam TG, Clark CE. 1983. Effects of toxicants on populations: A qualitative approach. 1: Equilibrium environmental exposure. *Ecolog Model* 18:291–304.

Hallam TG, Lassiter RR, Li J, McKinney W. 1990. Toxicant-induced mortality in models of *Daphnia* populations. *Environ Toxicol Chem* 9:597–621.

Mayer Jr FL, Ellersieck MR. 1986. Manual of acute toxicity: Interpretation and data base for 410 chemicals and 66 species of freshwater animals. Washington DC: U.S. Fish and Wildlife Service. Resource Pub. 160.

Suter II GW. 1990. Seven-day tests and chronic tests. *Environ Toxicol Chem* 9:1435–1436.

Suter II GW. 1998. Ecotoxicological effects extrapolation models. In: Newman MC, Strojan CL, editors. Risk assessment: Logic and measurement. Ann Arbor MI: Ann Arbor Press. p 167–185.

Suter II GW, Rosen AE, Linder E, Parkhurst DF. 1987. Endpoints for responses of fish to chronic toxic exposures. *Environ Toxicol Chem* 6:793–809.

Suter II GW, Vaughan DS, Gardner RH. 1983. Risk assessment by analysis of extrapolation error, a demonstration for effects of pollutants on fish. *Environ Toxicol Chem* 2:369–378.

Troyer ME, Brody MS. 1994. Managing ecological risks at EPA: Issues and recommendations for progress. Washington DC: U.S. Environmental Protection Agency. EPA/600/R–94/183.

CHAPTER 8

Group A Discussions of Endpoint Selection, Study Design, and Extrapolation

John B. French, Jr., Steven P. Bradbury, Hank Krueger, Elizabeth McGee, Bradley E. Sample

A mechanistic basis for understanding effects of contaminants is gained at lower levels of organization, whereas ecological relevance is gained at higher levels of biological organization. A major issue regarding these levels is the ability to define them scientifically, with definitions becoming increasingly difficult and less operational from population to community to ecosystem. Furthermore, environmental protection based on harm to populations has occurred only rarely in the U.S. and is almost never based on harm to communities or ecosystems, despite references to them in environmental legislation. That fact, and the current emphasis on land-management based on ecosystems, provided some of the motivation for this symposium.

Among environmental concerns, the protection of individual organisms is probably the most widely held value for society at large. Outside the scientific community, the welfare of higher levels of organization is understood and valued by progressively fewer persons as the level of organizational complexity increases. Consequently, scientific efforts to document the effects of contaminants on populations, communities, and ecosystems might not be effective in promoting environmental protection. However, given a positive correlation between degree of scientific understanding and degree of societal concern, it is possible that greater scientific understanding of these higher levels of organization will be translated into greater societal appreciation and valuation.

Session 1: Endpoint Selection and Study Design

Endpoint selection

A discussion of endpoint selection must be placed within the context of environmental management issues and associated legislative mandates. Different types of endpoints may be more or less appropriate depending on the management issue. Management issues can include

CHAPTER PREVIEW

Session 1: Endpoint selection and study design 189
Session 2: Extrapolation 194

- assessing, with known confidence, the nature and extent of the condition of wildlife populations and their associated communities and ecosystems;
- establishing relations between natural events or human activities (stressors) and the condition of wildlife populations and their associated communities and ecosystems, with a known level of confidence;
- diagnosing the causes of adverse effects in populations, communities, and ecosystems; and
- establishing quantitative stressor-response profiles for populations, communities, and ecosystems.

Communities and ecosystems

Although monitoring and assessing wildlife at the community and ecosystem levels of organization are important for evaluating the integrity and health of the terrestrial environment, we did not identify community or ecosystem endpoints that are indicators specific to the effects of contaminants. Part of our inability to identify endpoints was conceptual, as communities and ecosystems are very complicated, and we have not yet conceived of and measured all the properties that emerge at these levels; another part was operational, in that there are relatively few of these entities and the entities are large in scale. Hence, small sample sizes prevent reasonable confidence levels from being attached to conclusions; often, this lack of confidence is described as the unpredictability or stochasticity of communities or ecosystems. Measuring the effects of contaminants on communities or ecosystems would benefit from the development of such indicators, if they exist.

Populations

The following endpoints can be used to assess the condition of wildlife populations in monitoring and assessment programs:
- population trends,
- age structure,
- "harvestable numbers" (visible numbers),
- age-specific survival,
- age-specific fecundity,
- abundance,
- sex ratio, and
- prevalence and incidence of disease.

Where possible and appropriate, sampling for these endpoints must include a design that can identify metapopulation characteristics in order to facilitate interpretation of results at a range of spatial scales. The extent to which media contaminant concentrations or organismal contaminant exposure can be quantified and linked to population attributes determines the likelihood of formulating hypotheses of cause and effect. Inclusion of organismal or sub-organismal biomarkers of effects increases the means of diagnosing causes of population impairment.

Organisms

Evaluating the effects of contaminants on individual organisms will remain important. Through a combination of partial life-cycle, full life-cycle, or multi generational studies, age-specific mortality and fecundity can be used to establish dose–response relationships for organisms that can be used to qualitatively or quantitatively infer or predict population responses. Many of these endpoints are included in standard Federal testing requirements. Most techniques have been developed for mammals and birds, whereas techniques for amphibians and reptiles require further development and refinement. A few terrestrial invertebrate species have been used in toxicological studies, but more studies are needed to support more-integrated assessment of ecosystems. Making experimental studies on individuals more useful for interpretation of field studies, especially monitoring and survey studies, should be a goal for all investigators.

Study Design

In our discussion, 3 types of studies were identified: survey and monitoring, experimental, and modeling. These study types should be considered complementary to each other. They are best used in an integrated approach to monitor and assess condition, to diagnose causes of impairment, and to predict future status of populations, communities, and ecosystems. Concern was expressed about the inadequate number of field surveys and monitoring studies being performed; it is from incidents observed in the field that many contaminants problems have been identified. Indeed, in the history of contaminant-related issues in animal populations, many of the problems (e.g., dichlorodiphenyltrichloroethane [DDT], polychlorinated biphenyls [PCBs], lead shot, selenium) were first identified as field effects. Field monitoring is also one of the only ways to evaluate the accuracy of our risk assessments.

Survey and monitoring studies

Survey and monitoring studies are almost always observational field studies that are distinguished by the absence of investigator control over which subjects are exposed to contaminants and which are not. Other important aspects of survey and monitoring activities include
- Goals of the study, which can include one or more of the following: 1) to establish "wildlife condition" (i.e., the status and health of the population); 2) to establish the associations between condition and possible stressors; 3) to suggest cause and effect; 4) to evaluate impacts following contaminant releases or subsequent mitigation action; or, 5) to establish baseline conditions in areas of concern or interest.

- Endpoints, or what to measure in the study. Most surveys measure characteristics of individual organisms and, when aggregated, draw conclusions about populations or species. There is rarely an explicit definition of the population in question. A definition would be helpful and is crucial if true population characteristics are measured (see above) or metapopulation structure is evident. As previously noted, there is no consensus on the definition or use of emerging community or ecosystem properties that might be measured and compared in a contaminants survey. It is sometimes useful to compare traditional descriptors of communities or ecosystems (e.g., number of species, species importance, measures of diversity across communities with differing levels of contamination), but this approach suffers from the difficulty of finding replicates that are the "same" except for the presence of contaminants. Measures of contamination are also needed to interpret the community and ecosystem condition in the context of wildlife populations and their potential exposure. Examples include contaminant concentrations in soil, water, or air and the spatial and temporal pattern of this contamination. Regardless of the level of biological organization of effects, organism-level biomarkers of exposure and actual concentrations measured in organisms can provide useful information to evaluate the strength of the association of effects with contaminant exposure.
- Sampling design of the study. This should be determined by the goals of the study and should answer the question "compared to what?". Detection of trends across space and time and associations with stressors at state, regional, and national scales has typically required the use of sample surveys because the populations of interest are too large to count. For example, the National Agricultural Statistical Survey (Cotter and Nealon 1987), the Forest Inventory Assessment (Hazard and Law 1989), the National Wetlands Inventory, and the USEPA Environmental Monitoring and Assessment Program (Larsen et al. 1994) all employ probability-based sample surveys. The consistent use of a common design over time permits an unbiased estimate of the status, extent, change, and trends in indicators of ecological condition with known confidence. These designs can provide rigorous information on wildlife population attributes, habitats, and contaminant trends at large spatial and temporal scales to establish associations between stressors, land-use patterns, and ecological condition. Often there is a need to determine the effects of a local contamination event on a wildlife population or particular habitat, and the obvious design is to compare the affected area with a clean area. For comparison of populations, communities, or ecosystems, problems often occur when looking for sampling replicates within the contaminated area and trying to locate uncontaminated (reference) sites. Sometimes a study along a diminishing gradient of contamination can be useful. Also, multivariate analyses should be considered because they can provide descriptions of many attributes of the habitat at once.

Experimental studies

Experiments in which treatments are selectively applied on individual organisms have been and will remain a mainstay of wildlife toxicology. Experiments on populations of vertebrates and higher levels of organization are difficult but are important for measuring directly, rather than by extrapolation, the effect of contamination. Important aspects of experimental studies include

- Goals of the study, which can include one or more of the following: 1) to elucidate mechanisms of contaminant action, 2) to demonstrate cause-effect relations and establish stressor-response relations, 3) to confirm field observations, or 4) to provide the basis for extrapolation between species and across levels of biological organization.
- Endpoints of the study. Measurements include organismal and population characteristics. There is a great need to link exposure measured on individuals with endpoints of effect measured at higher organization levels.
- Experimental design of the study. Design of experiments in which the subjects are populations or communities is an area that could benefit from some imaginative thinking. Designs in use now include mesocosm or enclosure studies and field manipulations; the latter are often limited by (reasonable) prohibitions on releasing contaminants into the environment. Here, novel methods of dosing are needed to both reduce the amount of chemical used and improve measurement of exposure of individual animals. Other considerations include the ability to detect stressor-response profiles in the field. These studies should address relevancy to exposure conditions in the field and address habitat constraints that influence chemical fate, transport, bioavailability, and exposure to a contaminant.

Modeling studies

Overall, we believe that population models are under-utilized in assessing the effects of contaminants on terrestrial populations. Population models represent a quantitative and useful method for extrapolating from effects on individuals to effects on populations. However, before these models can become more widely accepted and applied, the validity and limitations of their assumptions must be established. In addition, validation of the predictions of ecological models is needed and generally consists of an objective assessment of how well the model output fits the data (Jorgensen 1986). Field validation can involve comparing predictions of population models (e.g., changes in population structure or age at first reproduction) with data from populations in a controlled setting (e.g., pesticide field studies), or by monitoring exposed field populations. Field validation is crucial to the development and application of scientifically credible population models.

Session 2: Extrapolation

Extrapolations of contaminant effects are possible within and between species and from individuals to populations, but extrapolations are more difficult from populations to communities and are extremely difficult from communities to ecosystems. The difficulty with higher-level extrapolations is the result of a traditional focus of toxicologists on individuals and the lack of well recognized descriptors of community and ecosystem function. In addition, as system complexity increases, uncertainty increases, and our ability to extrapolate contaminant effects on populations to higher levels of organization becomes more difficult.

It was proposed that we might be approaching the extrapolation problem from the wrong direction. Instead of asking how contaminants affect communities and ecosystems, a more fruitful question may be to ask how ecosystems or communities affect contaminants. This line of questioning emphasizes exposure more than effects, but they are very closely related. It also brings into the discussion the environmental fate of contaminants. When contaminants are released into the environment, they are altered through physical degradation and metabolism. Physical degradation involves abiotic factors of ecosystems such as pH, temperature, photolysis, and hydrolysis. Metabolism involves chemical alteration by organisms; microbial communities (decomposers) play a major role.

In product testing, soil metabolism and biodegradation studies are routinely required for registration. There also are many models available to evaluate the environmental fate of contaminants in ecosystems. Thus, while we cannot currently extrapolate effects from individuals and populations to ecosystems, a better understanding of the fate of contaminants in an ecosystem is critical to determining when and where effects are likely to occur. Once the spatial and temporal distribution of a contaminant is understood, we can better determine what types of habitats and populations may be at risk.

Extrapolation within individuals

The sensitivity of molecular and cellular measures makes them useful as screening tools for identifying individuals exposed to contaminants. Extrapolation within individuals is performed when information on molecular or cellular effects of contaminants is used to estimate the effects on organs or the whole animal. Methods for intra-individual extrapolation are generally observational, relying on knowledge of the mechanisms of toxic action of a chemical coupled with knowledge of animal physiology to explain effects on the whole animal. Using molecular or cellular effects data to draw conclusions concerning effects to whole animals should be performed with caution because effective repair mechanisms and redundancy within organ systems can preclude the expression of these effects; negation or reduction of effects are possible. While intra-individual extrapolation may be used to generate estimates of potential effects to individuals, the variability of expression of molecular and cellular effects often precludes reliable extrapolation to higher levels of organization.

Extrapolation within species

Toxic responses within a given species may vary according to sex, life stage, condition, or reproductive state of an individual; the prior exposure history among individuals; exposure route (dermal, oral, inhalation); and genetic variation within the population, with isolated subpopulations likely to have different sensitivities. There are no off-the-shelf methods for intraspecies extrapolation, which generally is performed empirically and is based on direct observations. Intraspecies variability is addressed by evaluating what is presumed to be the most sensitive subgroup within the species. This is determined to a large degree by the nature of the contaminant of concern. For example, if a chemical is known to cause female reproductive toxicity, data-collection efforts should be focused on females, particularly those of reproductive age. Similarly, selected subgroups of a population may be known to be sensitive. For example, it is generally accepted that younger life stages of most species are more sensitive to contaminant effects than are adults. This sensitivity has been attributed to the higher metabolic rates, and greater food and water consumption of younger life stages by focusing on the most appropriate portion of the receptor population (e.g., most sensitive, most exposed), the need for extrapolation is minimized.

Extrapolation between species

There is an obvious need to extrapolate from the few species tested in the laboratory to the large number of species in an ecosystem. For example, laboratory studies with mallards (*Anas platyrhynchos*) and bobwhite quail (*Colinus virginianus*) are typically used to evaluate effects on all birds in an ecosystem. The justification for such an extrapolation is that similar amounts of test substance per unit weight of animal should result in similar toxicological responses among species, assuming that the anatomy and physiology of birds are similar. Complicating factors are many, and include differences in routes and timing of exposure and in the actual dose that different species receive.

Interspecies extrapolation has been performed using multiple approaches, including allometric scaling, uncertainty factors, phylogeny, dietary habits, and physiologically based pharmacokinetic (PBPK) modeling. Allometric scaling is based on the observation that toxicity and other biological attributes tend to vary directly with body weight or a power of body weight. If a toxicity value for a test species and body weights for both the test species and the wild species of interest are known, a toxicity value for the wild species may be estimated. Uncertainty factors consist of values, generally ranging from 1 to 100, that are use to multiply results from a test species to generate an estimated toxicity value for a wild species. Because physiological responses are generally similar among phylogenetically related species, phylogeny has been employed for interspecies extrapolation. Similarly, interspecies extrapolation may be based on dietary habits. For both phylogenetic and food-habit–based extrapolation, 2 species that are similar taxonomically or in food habits are assumed to display comparable toxicity responses. Phylogeny and food habits are frequently used as a basis for selecting particular uncertainty factors. An approach that is currently in development is the use of PBPK models. In general, allometric models are used to extrapolate the various rate coefficients needed for the PBPK models with the resulting outputs being species-specific toxicity estimates.

A critical concern with most interspecies extrapolation methods is that they have not been validated. Validation is critical to quantifying the nature and magnitude of uncertainty in the estimates obtained from each extrapolation method. Multiple validation approaches may be used. For example, data from field studies may be compared to results obtained from laboratory tests. However, caution must be used with this approach because of the lack of control over confounding influences (e.g., weather, interspecies interactions, etc.) affecting field observations. Another alternative is to exclude data prior to development of a given extrapolation model and then apply the model to the excluded data. If the data are independent, this is a good approach. A variation on this approach is to use existing models to estimate potential toxicity for new chemicals. If the model is appropriate, then the estimated values should approximate actual values.

Extrapolation among the members of a major taxon such as birds is difficult, but extrapolating from laboratory data on mammals or birds to other major taxa, such as reptiles or amphibians, is suspect. There should be an explicit physiological basis for doing so. More laboratory and field-related contaminant information is needed to understand better the effects of contaminants on reptiles and amphibians. Perhaps current investigations into amphibian declines and frog deformities will improve the situation for amphibians.

Extrapolation from individuals to populations

A variety of population-analysis approaches have been applied to toxicological problems in the last few decades (see Barnthouse 1993 for a review). One of the earliest applications was the use of the intrinsic rate of population growth, r, to interpret aquatic life-cycle toxicity tests (Daniels and Allan 1981; Gentile et al. 1982). This approach has the advantage of combining information on survival and reproduction into a single index; hence, it is more representative of population status than any single endpoint. Estimates of r are obtained from life-table analysis of laboratory toxicity data and, as a result, they typically involve full, or partial, life-cycle exposures. Consequently, this method is of limited utility in evaluating population effects on vertebrates that have comparatively long lives.

An alternate, yet theoretically similar, approach uses matrix projection models to evaluate population effects (Caswell 1989). Like the life-table method, these demographic models integrate multiple endpoints, including survival, growth, and reproduction, to project changes in population parameters such as abundance, growth rate, or probability of extinction. Projection matrices can accommodate organisms classified by age, life stage, or size (Lefkovitch 1965; Leslie 1945). Toxicological effects on individuals, such as age-specific mortality and impaired reproduction, are simulated by altering the appropriate parameters in the population model. For example, Samuels and Ladino (1983) used a stage-based matrix model to evaluate potential effects of oil spills on seabird populations. Tipton et al. (1980) used a stage-based matrix model to quantify the relationship between pesticide exposure, cholinesterase inhibition, and population survival of quail in field pesticide tests. In addition, matrix models have recently been used as a tool for managing endangered species populations (Beissinger and Westphal 1998). The matrix modeling approach is more flexible than the life-table method, as it allows an unlimited number of exposure scenarios to be simulated, not just those evaluated in the laboratory. In addition, density dependence and stochastic variation can be introduced in to the model to provide ecological realism. The weakness of this approach is the large amount of demographic information needed to develop the models; hence, there have been few applications in wildlife toxicology (Slade 1993).

Single population models, like those described above, do not consider the effects of spatial factors, e.g., habitat heterogeneity and movement among subpopulations, on overall population viability. Metapopulation models were developed to address spatial variation. In these models, metapopulations are spatially distributed into habitat patches, and dispersing individuals serve as links among these patches. Such models are multi-habitat versions of single-population models in which demographic rates are patch (habitat)-specific, and dispersal rates among patches are determined by patch size and distance. Although intuitively appealing because of their explicit consideration of the heterogeneity of habitat and toxicant distribution, these models can be quite complex. Hence, their application to assessing toxicological effects on terrestrial vertebrates has mostly been theoretical (e.g., Maurer and Holt 1996).

Extrapolation between field and laboratory

In the area of product testing (e.g., agricultural chemicals and pharmaceuticals), an abundance of laboratory data, some field data (exposure related), and models exist to use for the evaluation of potential effects of a test substance when it is released into the environment. Product testing uses large, laboratory-generated databases and extrapolates to the field to determine product safety or unacceptable risk to animal populations. In product testing we start with a known test substance, conduct lab tests to understand potential effects, and then extrapolate what effects may occur in the field through the process of ecological risk assessment.

A different type of problem can also benefit from lab-to-field extrapolation, but here field effects may drive the process. For example, when dead or deformed animals are found or other incidents of morbidity are reported, we often need to deduce which contaminants caused the problem. Additional field work (e.g., collection of samples for analysis), as well as laboratory work, may be needed to identify the contaminants. Once identified, there may be an adequate amount of existing laboratory data to draw upon, or additional data may be needed from laboratory studies to better relate the effects to contaminants identified in the field.

Structure-activity relationships

Structure-activity relationships (SARs) are based on the principle that structurally similar chemicals should have similar biological activity. Structure-activity relationships relate a specifically defined toxicological activity of chemicals to their molecular structure and physico chemical properties. Once such a relationship is established, it can be used to predict the activity of untested chemicals. Within the last decade, SAR applications in the field of pharmacology and mammalian toxicology, as well as ecotoxicology, have steadily increased (Bradbury 1995; Bradbury et al. 1998).

Structure-activity relationships are useful when ecological risk assessments of chemicals are confronted with a lack of toxicity data for the chemical of concern, a typical situation. In aquatic toxicology, quantitative structure-activity relationships are available for predicting the acute and, in some instances, sub-chronic toxicity of chemicals when little or no empirical data are available. In addition, "analog-selection" techniques use data on structurally similar chemicals to estimate risk levels for compounds on which no data are available (Bradbury 1995).

Because of the toxicokinetic complexity associated with oral exposure of xenobiotics, the development of SAR techniques to predict all *in vivo* endpoints for wildlife is unlikely. However, SARs are being developed that predict the potential for compounds to interact with specific molecular receptors, and these models could be used in "screening" assessments for risk to wildlife. For example, SARs can predict the potential for compounds to bind to specific hormone receptors, when rigorous *in vitro* toxicological knowledge bases have been established, and thus predict the endocrine activity of the contaminant (Bradbury et al. 1998). Because vertebrate hormones are evolutionarily conserved, it is reasonable to assume that SARs based on hormonal binding or other endpoints could be used to assess hazards to a wide variety of vertebrates. Challenges to improve predictive models for "chemical" extrapolations in wildlife will center on the uncertainties about mechanisms of chemical toxicity and xenobiotic metabolism.

Multiple stressors

Typically in human-health and ecological risk assessments, mixtures are evaluated assuming additivity of constituents that share a mode of action with no account taken of chemicals with different modes of action. For example, many polyhalogenated aromatic hydrocarbons have been shown to produce toxic effects that are similar to those associated with exposure to 2,3,7,8-tetrachlorodibenzo-p-dioxin (TCDD). Specific polychlorinated dioxins (PCDDs), furans (PCDFs) and biphenyls (PCBs) are believed to exert their toxic effects as a result of binding to the aryl hydrocarbon receptor. Based on their mechanistic similarity to TCDD, toxic equivalency factors have been calculated that express the toxicity of individual PCB, PCDF, and PCDD congeners relative to the toxicity of TCDD, thus allowing estimation of their combined effect in mixtures (Van den Berg et al. 1998).

The further development of credible models for the action of chemical mixtures requires improved understanding of toxic mechanisms to determine when additivity assumptions are reasonable or when synergism or antagonism can be expected. Also, the ability to identify and quantify causes of wildlife toxicity in samples with unknown composition is critically needed in order to reduce uncertainties in assessing chemical mixtures. This reduction might be accomplished through the use of toxicity identification and evaluation techniques, most likely based on in vitro assays.

Summary

Our discussion highlighted the disparity between the perceived need for ecologically relevant information at the population, community, and ecosystem levels and the small number of tools (operational concepts and datasets) currently available to produce that information. This lack of tools is partially operational ("the science is not there yet") and partially historical (the study of individual organisms has been primary in wildlife toxicology). Also, it is unclear whether harm from pollution at higher levels of organization will be a more or less sensitive indicator than organismal harm, or whether the changes associated with pollution in populations, communities, and ecosystems can be distinguished from changes due to other influences. Promising areas of research include

- developing practicable indicators of function or health of populations, communities, and ecosystems;
- evaluating the susceptibility of population parameters to changes caused by exposure to contaminants;
- identifying species for testing that are low in terrestrial food webs but have great trophic and ecological importance, especially invertebrate species;
- compiling meta-analyses of spatially or temporally diverse datasets to identify broad associations between stressors, land-use patterns, and ecological conditions;
- developing and validating models that are useful for interspecies extrapolations, especially PBTK models;
- applying current types of population models (simple aggregated models or stage- or age-based models) to extant datasets to determine usefulness of the

models for evaluating population consequences of exposure to contaminants; and
- developing SARs for multiple modes of action of contaminants to use as a tool for identifying chemicals with a high potential for producing biotic harm.

References

Barnthouse LW. 1993. Population-level effects. In: Suter GW, editor. Ecological risk assessment. Chelsea MI: Lewis. p 247–274.

Beissinger SR, Westphal MI. 1998. On the use of demographic models of population viability in endangered species management. *J Wildl Manage* 62:821–841.

Bradbury SP. 1995. Quantitative structure activity relationships and ecological risk assessment: An overview of predictive aquatic toxicology research. *Toxicol Let* 79:229–237.

Bradbury SP, Mekenyan OG, Ankley GT. 1998. The role of ligand flexibility in predicting biological activity: Structure-activity relationships for aryl hydrocarbon, estrogen and androgen receptor binding affinity. *Environ Toxicol Chem* 17:15–25.

Caswell H. 1989. Matrix population models. Sunderland MA: Sinauer. 328 p.

Cotter J, Nealon J. 1987. Area frame design for agricultural surveys. Area frame section. Washington DC: Research and Applications Division, National Agricultural Statistical Service, U.S. Department of Agriculture.

Daniels RE, Allan JD. 1981. Life table evaluation of chronic exposure to a pesticide. *Can J Fish Aquat Sci* 38:485–494.

Gentile JH, Gentile SM, Hairston NG, Sullivan BK. 1982. The use of life-tables for evaluating the chronic toxicity of pollutants to *Mysidopsis bahia*. *Hydrobiol* 93:179–187.

Hazard JW, Law BE. 1989. Forest survey methods used in the USDA Forest Service. EPA/600/3–89/065. Corvallis OR: USEPA, Office of Research and Development.

Jorgensen SE. 1986. Fundamentals of ecological modeling. New York NY: Elsevier.

Larsen DP, Thorton KW, Urquhart NS, Paulsen SG. 1994. The role of sampling surveys for monitoring the condition of the nation's lakes. *Environ Monitor Assess* 32:101–134.

Lefkovitch LP. 1965. The study of population growth in organisms grouped by stages. *Biometrics* 21:1–18.

Leslie PH. 1945. On the use of matrices in certain population mathematics. *Biometrika* 33:183–212.

Maurer BA, Holt RD. 1996. Effects of chronic pesticide stress on wildlife populations in complex landscapes: Processes at multiple scales. *Environ Toxicol Chem* 15:420–426.

Samuels WB, Ladino A. 1983. Calculation of seabird population recovery from potential oil spills in the mid-Atlantic region of the U.S. *Ecol Model* 21:63–84.

Slade NA. 1993. Models of structured populations: Age and mass transition matrices. In: Kendall RJ, Lacher TE, editors. Wildlife toxicology and population modeling. Boca Raton FL: CRC Press. p 189–200.

Tipton AR, Kendall RJ, Coyle JF, Scanlon PF. 1980. A model of the impact of methyl parathion spraying on a quail population. *Bull Environ Contam Toxicol* 25:586–593.

Van den Berg M, Birnbaum LS, Bosveld ATC, Brunstrom B, Cook P, Feeley M, Gisey JP, Hanberg A, Hasegawa R, Kennedy SW, Kubiak T, Larsen JC, van Leeuwen FXR, Liem AK, Nolt C, Peterson RE, Poellinger L, Safe S, Shrenk D, Tillitt D, Tysklind M, Younes M, Waern F, Zacharewski T. 1998. Toxic equivalency factors (TEFs) for PCBs, PCDDs, PCDFs for humans and wildlife. *Environ Health Perspect* 106:775–792.

Group A Leaders
John B. French, Jr., Steven P. Bradbury, Hank Krueger, Bradley E. Sample

Group Members
Stanley H. Anderson, Russell J. Hall, Gary H. Heinz, Almira Hoogesteyn, Kelly Lippenholz, James MacMahon, John McCarty, Elizabeth McGee, Barnett Rattner, Anne L. Secord, John R. Skalski, Mark Woythal

CHAPTER 9
Group B Discussions of Endpoint Selection, Study Design, and Extrapolation

James Chapman, Steve Sheffield, Rick P. Brown, Glenn W. Suter II

Session 1: Endpoint Selection and Study Design

This group discussed the strengths and limitations of population-level field studies and shared opportunities for improving experimental designs. A discussion of population modeling and establishing the spatial extent of study populations followed. An evaluation of currently used endpoints and statistical procedures concluded our exchange.

Can we design better and more efficient experiments and field surveys?

The principles of study design were initially reviewed (Figure 9-1), and it was established that it is essential to invest significant effort in the up-front development of the appropriate design by establishing the study goals. This must include addressing data-analysis issues, defining what types of statistical inferences need to be made, and identifying decision criteria clearly. Based on these considerations, the appropriate response variables to be assessed can be identified and a determination can be made regarding the geographical boundaries that define the experimental unit (i.e., site-specific, local, or regional scale).

A major concern in the conduct of field studies is low resolution that is due to high natural variability, short study duration, and geographical scale and boundary issues. For studies of pollution or accident events, there also is limited opportunity for replication, inadequate understanding of reference conditions, and limited baseline data. The resolution problem is a key issue in evaluating the meaning and reliability of "no-adverse-effect" findings when demonstration of population- or higher-level effects is the standard for determining the significance of an environmental impact.

CHAPTER PREVIEW

Session 1: Endpoint selection and study design 203
Session 2: Extrapolation 210

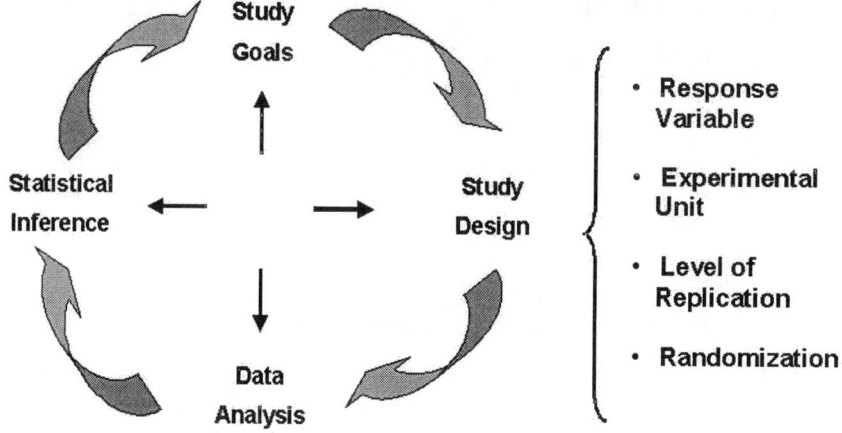

Figure 9-1 Principles of study design.

In general terms, 2 study designs can be considered: observational and manipulative. Observational studies are opportunistic and can be used to develop correlative relationships. In these studies the researchers need to be cognizant of the potential effect of confounding factors that influence data interpretation. Manipulative studies are premeditated investigations that offer greater control over the influence of confounders. This type of study often can be designed and used for the purpose of establishing cause-effect relationships.

Manipulative study designs can be further divided into direct and indirect assessments. In a direct assessment, a specific measure of response, as may be obtained from a population survey, can be examined. Whereas potentially confounding factors can be identified and their influence limited to some extent, these factors still can play a significant role in changes in the response measure and subsequent data interpretation. Recognizing these limitations, and understanding their influence, is key to being able to conduct a population-level field study successfully and to provide useful information for risk management decisions.

An indirect assessment often uses individual-level responses to assess a potential effect on a population. This is a model-based assessment that is burdened by the default assumptions inherent in the selected model. In this type of experimental design, a careful analysis and incorporation of epidemiological tools can be applied to evaluate effects at the population, community, and ecosystem levels.

The low resolution of population-level field studies that results from natural variability can be effectively addressed by increasing study duration or number of observations; however, biologically appropriate temporal scales might conflict with regulatory time frames and funding limitations. Population modeling is not temporally constrained, but model validation is. Also, input data for population modeling are available for relatively few species, a situation that is unlikely to change appreciably because research on natural history is poorly funded. Increasingly, this forces the researcher to depend upon the reasonableness and accuracy of default assumptions in the model.

The investigation of higher-level effects (i.e., at the population, community, and ecosystem level) is valuable because it provides strong evidence for taking action when adverse impacts are demonstrated. Studies of higher-level effects should be integrated with exposure assessments and evaluations of contaminant fate and transport. The role and utility of investigations of higher-level effects are related to the geographic scale of the stress under consideration. Studies of higher-level effects are often more problematic for assessments of contaminant effects in spatially restricted areas, such as hazardous waste sites or localized spills, than for products or contaminants that are widely distributed, such as pesticides, air pollutants, and regionally or globally transported chemicals. In other words, investigation of higher-level effects works best as an integral component of a suite of tools in a weight-of-evidence approach to risk assessment.

Discussion of the optimization of experimental designs centered on the role of hypothesis testing in risk assessment and differing concepts of the goals of risk assessment studies. One view emphasized the importance of hypothesis testing in retrospective risk assessment, particularly for demonstrating differences between affected and reference locations. A primary need is an appropriate experimental design for assessment of the impact of accidental spills or releases. This type of site-specific assessment often comprises most of the work performed in the field, yet it is not a true experiment. Often, spatial and temporal baseline information is lacking and there is no replication or randomization. In this regard, it is best characterized as an experience, not an experiment. The analyses conducted are based on individual-level responses, such as death and reproductive impairment, that occur during or after the event. A long-term commitment of resources is needed to complete the assessment successfully. The primary problem is that power calculations usually show that very large effects are needed to attain statistical significance, i.e., the usual designs have low power to discern differences between affected and reference locations.

An alternate view is that risk assessment is not an experimental science and, therefore, hypothesis testing is not an appropriate approach. Estimates of the nature and magnitude of effects are needed for risk assessments, which are best addressed by multiple lines of evidence. This means that no single line of evidence need prove whether significant adverse effects exist. Risk assessors are more interested in exposure–response relationships, so replication is not as important here as it is as for hypothesis testing. Hypothesis testing is inappropriate for risk assessment purposes because it has little to do with biological relevance, and hypothesis testing does not reveal the shape of the dose–response curve. The role of statistics is mainly for evaluating the precision of estimates (i.e., confidence limits).

Modeling population effects

Different views were expressed regarding the magnitude of problems associated with population modeling in ecological risk assessments. In one view, population modeling is not a problem for risk assessment purposes because population response curves are most often quite steep. Usually there is either no discernable adverse effect on an ecological receptor (an identified species or population) or the receptor is heading for extirpation. It is unlikely that a study will occur while the population of the receptor is undergoing rapid change. Because the risk assessment is most effective in examination of large effects or no effects, intermediate effects become trivialized.

An alternate view was that a risk assessment is superfluous if a population is already at the bottom of a population cycle. The objective is to detect the potential for adverse effects before or just at the declining portion of the curve, not after a population has crashed. Furthermore, risk assessment implies that some level of population effect is acceptable and it therefore runs counter to the concept of the precautionary principle.

Overall, however, the group agreed that the best approach may be to incorporate parameters from individual-level studies into population models to project higher-level effects. However, few species are sufficiently characterized for effective population modeling. While broad life-history categories (e.g., r versus K) are available for most species, there is a general lack of information for the data requirements of population models, and there is no single, measurable parameter that is highly predictive of population-level effects. Unfortunately, more detailed information is not forthcoming for most species because natural history is now a low academic and funding priority.

The utility of population modeling was questioned based on the argument that models have not been successful for ensuring sustainable fish harvests. The appropriateness of the fisheries example was in turn questioned because these models project the next year's effects, which are highly stochastic. Ecological risk assessments usually project long-term effects, which are less stochastic than estimates of annual change. It was also pointed out that the failure to sustain fish harvests resulted as much, or more, from a failing of political will as from the inadequacy of the population models. It does serve as a sobering reminder that even sustained modeling efforts do not necessarily result in successful management policies.

Modeling can be successful if toxicological information is correct, but such information is often not sufficiently characterized or is misinterpreted. Because the model output is only as good as the toxicological input, we must consider what input is most useful (i.e., reproduction, lethality, gross toxicological effects, etc.). The exposure side of the equation often is poorly understood because of the difficulties in measuring it in the field and relating it back to the toxicological dose. In addition, the time course (kinetics) of toxicological effects usually is not known. Another complication is that the observed manifestation of a toxic effect can change depending on nutritional status and the history of the receptors. For example, migrating birds may be more susceptible because of loss of fat reserves.

A complicating factor that often is not included in population modeling is biotic interaction, which can produce counterintuitive results in the field. For example, ultraviolet (UV) exposure has been shown to reduce algal productivity in laboratory experiments, but increased UV exposure in a field mesocosm study resulted in increased algal production. The explanation is that algae-consuming chironomid larvae are more susceptible to UV than are their algal prey (Bothwell et al. 1994). Several opinions were expressed in the group regarding the significance of biotic interactions, from concluding that whole-community studies should be required since single populations cannot be reliably abstracted, to having low concern because the major interactions are understood and incorporating them into models is straightforward. While the interactions may be known, toxicological information is usually available only for a limited number of potential receptor species, thus limiting the ability to predict outcomes of biotic interactions in modifying population responses to contaminant stress.

Group members agreed that the discussions on population modeling demonstrated the importance of determining the scale and level of effect considered significant before performing an ecological risk assessment. Also in evidence were the differences in concepts and methods between research biologists and risk assessors.

Defining population boundaries

The most important decision in a population-level assessment is the delineation of the population boundary because this strongly influences the resolution of the assessment and interpretation of the results. For example, one of the early field studies of 1,1,1-trichloro-2,2-bis(p-chlorophenyl)ethane (DDT) showed 70% mortality of resident American robins (*Turdus migratorius*) in towns with regular spraying for Dutch elm disease beetles vector control (Wurster et al. 1965). Whether this is a population-level effect depends on the geographical scale considered, i.e., the population boundary. If the boundary corresponds to a town, an adverse effect on population size was demonstrated between sprayed and reference towns throughout the nesting season. If the boundary corresponds to a region (state or larger), there probably was no demonstrable population-level effect because robin populations generally increased in North America during the 20th Century as a result of favorable habitat changes. Therefore, the boundary (scale) decision can bias the likelihood of detecting population-level effects in response to spatially limited contaminant exposure.

Group members agreed that delineation of population boundaries depends on the study objectives and that, in most cases, this means that the population will be operationally defined. For example, the appropriate population boundary for Superfund sites is the site boundary because the objective is to determine the likelihood of local impacts. Impact to the North American population is not a relevant question in this context. However, it was also noted that an operational approach could result in absurdly small populations. The appropriate limit to "small" was not discussed.

The population-boundary question is a less-pressing issue for assessments of widely distributed contaminants or products intended for widespread use. In these situations there is greater overlap between the exposed (operationally defined) population and a more conventional definition of population (on a genetic or ecological basis).

Although adverse population-level impacts provide powerful justification for taking regulatory action, that action should not be justified solely on the basis of higher-level effects. One reason is that a requirement to demonstrate higher-level effects may be out of step with public opinion. The public often is more concerned with mortality and morbidity than with population size, e.g., the expenditure of resources to save beached whales or rehabilitate wildlife in an oil spill. In terms of hazardous substances, the public often focuses more on toxicity than on higher-level responses.

A major concern for all was the level of effort required to characterize higher-level effects adequately in field studies. To borrow an aquatic example, the level of effort for reliably detecting at least a 2-fold change in abundance of freshwater lake fish populations was estimated to be 14 years of monitoring, distributed evenly before and after the impact under investigation (Lester et. al. 1996). Other measures, such as instantaneous growth rate, may be more appropriate, but in any case, the level of effort to discern effects from the noise of natural variability may exceed regulatory time frames or available funding.

Although population-level effects are of particular concern, it is not necessary to demonstrate effects above the organism level in order to take regulatory or remedial action. In the diazinon case, the courts ruled that use of a pesticide could be restricted because it caused mass mortality of individual birds, even though there were no demonstrated effects on populations (Troyer and Brody 1994). Effects on ecosystems are also of concern, but they have been hard to interpret. The USEPA's Office of Pesticide Programs dropped its practice of requiring aquatic mesocosm tests largely because it had difficulty determining what constituted an unreasonable adverse effect on an aquatic ecosystem.

Endpoints

Because highly toxic substances are no longer commonly developed, the current focus of industry for assessing the toxicology of new products is on sublethal effects. A major contemporary issue is that of evaluation and interpretation of genetic effects. Most genotoxic studies have been on aquatic organisms, but terrestrial research activity is likely to increase in the future. There was a consensus that genotypic effects are measurable but that their application in risk assessment is uncertain. For example, microsatellite deoxyribonucleic acid (DNA) mutations can be shown to increase with contamination, but the effects of these mutations are unknown. Scientists at Oak Ridge National Laboratory have studied the effects of radiation on chironomids in an affected creek and river. A portion of the exposed larvae exhibited gill abnormalities and chromosomal aberrations (Nelson and Blaylock 1963), but, despite these effects, larval abundance was not depressed (Laws 1981). Gradual increases in exposure to radiation may increase fecundity or survival compared with populations exposed to background radiation (Luckey 1991).

An issue related to genetics is the interpretation of adaptation to contaminants. Interpretation of the significance of adaptation depends on one's viewpoint. Genetic adaptation is evidence of stress on individuals, without which natural selection will not occur. Many cases of adaptation to stress have been shown to involve costs that, in the absence of the stressor, are disadvantageous to the adapted population relative to unadapted populations. However, an alternate view is that adaptation is evidence of reduced overall impact to the exposed population, therefore reducing the justification for regulatory action.

Statistical considerations

Differences of opinion concerning the role of statistics in risk assessment, and therefore the optimal design of studies, are related to the main objectives of the studies. If the primary objective is to demonstrate whether effects differ between exposed and reference sites, emphasis is placed on the factors influencing the power of the design. If the primary objective is to evaluate dose–response relationships, emphasis is placed on the precision of the curve estimate. The number of dose levels is minimized in the former approach in order to maximize replication, while replication is minimized in the latter to maximize the number of dose–response points for curve-fitting.

The difference between statistical significance and biological relevance is important. It is necessary to have an objective method for concluding that differences exist, but the biological interpretation of the results must be done in the context of the situation that was measured. The distinction between individuals and populations in epidemiological studies is often unclear because individuals are sampled in both situations.

Session 2: Extrapolation

Extrapolation is the use of experimental or observational evidence of contaminant effects on organisms at 1 level of organization to estimate effects on other organisms either at the same level of organization or at higher levels of organization.

In the field of wildlife toxicology, extrapolation is employed mainly as a tool in ecological risk assessment. There are many deficiencies in the toxicological data for most animal taxa and there is much uncertainty associated with extrapolation in wildlife toxicological studies. As a result, interspecies and dose–response uncertainty factors must be used to estimate responses in the species of concern under varying exposure regimes. Current techniques to quantify these extrapolation uncertainty factors typically are based on empirical assessments of toxic effects and allometric relationships. Under complex exposure and effects scenarios and for animals of similar mass, the predictive power of these approaches is minimal. There is a great need to develop methods and models (e.g., physiologically based pharmacokinetic [PBPK] models) for inclusion into a biologically based approach to reduce these uncertainties.

Discussion

Our approach for examining extrapolation was the systematic examination of each possible extrapolation between ecological levels, including

Suborganism⇒ Organism⇒Population⇒Community or Ecosystem⇒Landscape, Region, or World
⇓ ⇓ ⇓
Organism   Population   Community or Ecosystem

Suborganism ⇒ Organism

Because suborganismal effects are seldom considered important in themselves, extrapolation between these 2 levels has been an important practice. However, most reports of suborganismal effects still are not explicitly extrapolated to higher levels. Associations between suborganismal and organismal responses in the laboratory have been used to estimate toxic effects on organisms in the field or to determine the cause of observed organismal effects. Examples include cholinesterase (ChE) inhibition that results from anti-ChE insecticides as an indicator of death and aminolevulinic acid dehydratase levels (ALAD) as predictors of lead toxicosis.

Reliable extrapolation depends on an understanding of the mechanistic connections between the suborganismal and organismal responses. For example, tumors, lesions, and deformities usually are not specific enough to be used for diagnosis of causation without supporting information (e.g., contaminant data or interpretive experimental studies). There is little information about the occurrence and causes of these gross pathologies in terrestrial vertebrates (with the possible exception of selenium). They are, however, important because they are unacceptable to the public. Standard methods should be developed for quantifying and reporting the normal and abnormal occurrence of gross pathologies such as tumors, lesions, and deformities.

It is possible to establish suites of suborganismal effects (instead of 1 or a few) that could be used for efficient extrapolation to the organismal level for specific classes of contaminants. When extrapolating between suborganismal and organismal effects, we need a solid diagnosis of cause and effect. The suites of suborganismal effects should be related to the health and function of organisms (e.g., body weights, incidence of disease, behavior, reproductive condition).

Suborganism⇒Population

Extrapolating of effects from the suborganism to the population is the level of extrapolation that is being sought by the USEPA. In the 1998 USEPA National Center for Environmental Research and Quality Assurance request for proposals for endocrine disruption, research was requested to estimate effects of contaminants on endocrine function that affect species at the population level. If the mechanisms by which physiological effects influence population processes can be established, we could skip the organismal level and extrapolate directly to the population level. An example of this is the examination of ChE activity in avian species in the eastern Canada forest-spray studies. There were indications that the percent of ChE inhibition in sampled individuals was related to the percent of the population showing altered behavior, but the range of inhibition in which altered behavior was detected was narrow (40% to 60%) and overlapped the lethal threshold of 50% inhibition (Busby et al. 1990; Grue et al. 1991). Such studies could lead to an estimate of population-level effects. Similarly, the degree of eggshell thinning can be used to estimate the degree of reproductive failure and subsequent population decline (Peakall 1993; Blus 1996).

Organism ⇒ Organism

Standard toxicity tests measure effects such as mortality, growth, and fecundity, on individuals. These effects are commonly extrapolated from the test species to species of interest. For terrestrial wildlife, this extrapolation is most commonly made by allometric scaling or by assuming that effects, bounded by some uncertainty, are the same.

Organism ⇒ Population

Extrapolation from the organism to the population can use demographic, individual-based, or metapopulation models. Although population-modeling techniques are well-validated (i.e., the limits of their predictive power are understood), the use of toxicity-test data with population models to predict effects of field exposures on populations has not been validated. Of special interest is the potential for differences in effects between "source" and "sink" components of an affected population.

Population models are designed for ecologically defined populations. They do not work well with populations that are artificially defined by local site boundaries because the demographic responses of small, artificially defined populations are dominated by exchange with the surrounding population rather than by internal responses.

Population ⇒ Population

In contaminants investigations, effects on local populations of a species are often extrapolated to a larger population. An example of this type of extrapolation is field studies of pesticide effects, in which effects observed in a few fields are extrapolated to regional populations. Christmas Bird Counts, Breeding Bird Surveys, and Migratory Bird Counts are examples of established, local population counts that are the basis for extrapolation to overall population estimates for avian species.

Different species can experience the same type of effect from environmental contaminants, which provides us with the opportunity to extrapolate between populations of those species. Thus, a well-established relation between contaminant exposure and effects on populations of 1 species can be used to predict the responses of populations of similar species. Examples of this include population declines of raptor species (e.g., falcons, hawks, eagles) exposed to DDE (Blus 1996; Peakall 1993) and reproductive failure in avian species resulting from selenium toxicity (Ohlendorf and Hothem 1995; Ohlendorf 1989). Obviously, extrapolation among species is most believable when the mechanisms of exposure and toxicity causing the observed effects are understood.

However, many questions remain regarding extrapolation between populations of different species. How closely related must terrestrial vertebrates be in order for one to extrapolate between populations reliably? Do we extrapolate between species in the same genus, family, or even order? Should extrapolation mostly be based on phylogenetic or non-phylogenetic (similarity of function) relationships? There are many combinations of species, such as mink (*Mustela vison*) and river otter (*Lutra canadensis*), that can be considered for population-to-population extrapolation because they have similar habitats and food habits and are taxonomically close.

Organism ⇒ Community or Ecosystem

Extrapolation from organismal to community or ecosystem effects can be performed using ecosystem models and the results of conventional organism-level toxicity tests. However, in these models the relationship between test parameters and model parameters is generally not as direct as that for population models. Species-sensitivity distributions have been used to estimate the proportion of aquatic species that will be affected at a particular exposure level. However, this approach is not practical for terrestrial vertebrates because 1) few species have been extensively tested with environmental contaminants and 2) they tend to be concentrated in a few taxa and therefore are not representative of natural communities.

Effects on organisms of individual species may be used to estimate effects on ecosystems if the role of the organism in the ecosystem is understood. Certain species, such as "keystone" species, or species that have important ecosystem function (i.e., affect many other species in the ecosystem), are critical species to consider for such extrapolations. Those species have the highest probability of having an ecosystem-level effect following contaminant exposure. Examples of such species are prairie dogs (*Cynomys* sp.), elephants, and zebra mussels (*Dreissena polymorpha*).

Population ⇒ Community or Ecosystem

In theory, population effects data could be used in community or ecosystem models, but such data are rare and therefore are unlikely to be available for a significant fraction of the species in an ecosystem model. Extrapolation at this level can be accomplished empirically using experimental approaches, such as mesocosms, where multiple levels of ecological organization can be analyzed simultaneously. Examples of extrapolation at this level are earthworm studies looking at nutrient dynamics (Makeschin 1997) and a study looking at insects and grey partridge (*Perdix perdix*) in the UK (Potts 1986).

Community or Ecosystem ⇒ Community or Ecosystem

There are no models in current use for extrapolating effects on wildlife among communities, ecosystems, or geographic regions in part because wildlife endpoints are usually defined at the population level rather than at other biological or spatial levels. In addition, the complexity of interactions at higher ecological levels or large geographic areas makes extrapolation difficult. One might use expert judgment to suggest the equivalence of effects at these different levels, but that approach is always questionable because the methods and assumptions are not explicit. Alternatively, a mechanistic or quasi-mechanistic approach, supported by analogy to similar situations, might be used to estimate community- or ecosystem-mediated effects on populations or feeding guilds. For example, reductions in abundance of prey species might be expected to reduce predator populations as a result of trophic requirements. The applicability of that mechanism to a particular situation might be supported by analogy to documented events at other locations. Examples of this include the reduction in otters and other piscivores in the Clark Fork River following metal-induced reductions in fish abundance (see Chapter 1) and pesticide-induced reductions in insect populations and plant cover that affected granivorous birds (see Chapter 5).

However, such extrapolations must be made with care. Many wildlife species have greater dietary flexibility than otters or are limited by factors other than trophic resources. In addition, because reported instances of effects on wildlife mediated by higher-level interactions are based on observations rather than on controlled and replicated experiments, the presumed mechanism might not be analogous. For example, the low abundance of Clark Fork otters might be due to direct toxic effects on the otters rather than reduced prey abundance; therefore, an extrapolation to other cases of reduced prey abundance would be erroneous.

Population ⇒ Landscape, Region, or World
Landscape ⇒ Region or World

It was believed that extrapolation from these levels of biological organization to these large and complex geographic areas was not feasible for terrestrial vertebrates.

References

Blus LJ. 1996. DDT, DDD, and DDE in birds. In: Beyer WN, Heinz GH, Redmon-Norwood AW, editors. Environmental contaminants in wildlife: Interpreting tissue concentrations. Boca Raton FL: Lewis. p 49–71.

Bothwell M, Sherbot D, Pollock C. 1994. Ecosystem response to solar ultraviolet-B radiation: Influence of trophic-level interactions. *Science* 265:97-100.

Busby DG., White LM, Pearce PA. 1990. Effects of aerial spraying of fenitrothion on breeding White-throated Sparrows. *J Appl Ecol* 27:743–755.

Grue CE, Hart ADM, Mineau P. 1991. Biological consequences of depressed brain cholinesterase activity in wildlife. In: Mineau P, editor. Cholinesterase-inhibiting insecticides. Their impact on wildlife and the environment. Amsterdam, The Netherlands: Elsevier. p 151–209.

Laws E. 1981. Aquatic pollution. New York: Wiley. 482 p.

Lester N, Dunlop W, Willox C. 1996. Detecting changes in the nearshore fish community. *Can J Fish Aquat Sci* 53(Suppl. 1):391-402.

Luckey TD. 1991. Radiation hormesis. Boca Raton FL: CRC Press. 306 p.

Makeschin F. 1997. Earthworms (*Lumbricidae*: *Oligochaeta*): Important promoters of soil development and soil fertility. In: Benckiser G, editor. Fauna in soil ecosystems: Recycling processes, nutrient fluxes, and agricultural production. New York NY: Marcel Dekker. p 173–223.

Nelson D, Blaylock B. 1963. The preliminary investigation of salivary gland chromosomes of *Chironomus tetans*[sic] Fabr. from the Clinch River. In: Schultz V, Klement Jr. W, editors. Radioecology. New York NY: Reinhold Publ. and Washington DC: American Institute of Biological Sciences. p 367–372.

Ohlendorf HM. 1989. Bioaccumulation and effects of selenium in wildlife. In: Selenium in agriculture and the environment. SSSA Special Publ. No. 23. Madison WI: American Society of Agronomy and Soil Science Society of America. p 133–177.

Ohlendorf HM, Hothem RL. 1995. Agricultural drainwater effects on wildlife in Central California. In: Hoffman DJ, Rattner BA, Burton, Jr. GA, Cairns, Jr. J, editors. Handbook of ecotoxicology. Boca Raton FL: Lewis. p 577–595.

Peakall D. 1993. DDE-induced eggshell thinning: An environmental detective story. *Environ Rev* 1:13–20.

Potts GR. 1986. The partridge: Pesticides, predation, and conservation. London: Collins.

Troyer ME, Brody MS. 1994. Managing ecological risks at EPA: Issues and recommendations for progress. EPA/600/R-94/183. Washington DC: U.S. Environmental Protection Agency.

Wurster D, Wurster C Jr, Strickland N. 1965. Bird mortality following DDT spray for Dutch elm disease. *Ecology* 46:488–499.

Group B Leaders
Rick Brown, James Chapman, Steve Sheffield, Glenn W. Suter II

Group Members:
Peter H. Albers, Tim Bartish, Glen A. Fox, Melanie S. Hawkins, Gerry M. Henningsen, David J. Hoffman, Kathleen Jennings, Michael J. Mac, John T. Paul, Jr., Susan Roddy, Joseph P. Skorupa, Mark L. Watson

CHAPTER 10

Group C Discussions of Endpoint Selection, Study Design, and Extrapolation

Michalann Harthill, Donald W. Sparling, Joseph P. Sullivan, Harry M. Ohlendorf

Session 1: Endpoint Selection and Study Design

Discussions addressed 3 primary questions: 1) What are the positive or negative attributes of endpoints currently in use? 2) Should new endpoints be identified and measured? 3) Can better and more efficient experiments and field studies be designed?

Study populations

Prior to addressing the primary questions of this session, group members dealt with the thorny issue of developing a suitable definition of "population" for purposes of a contaminants investigation. Most often the population studied is not a true biological population, but a portion of the biological population; the term suggested for this portion was "study population." The study population, or "population of concern," will be defined by some combination of 1) a predetermined area of contamination, 2) stakeholders, 3) fiscal constraints, and 4) the investigating biologists. The term "study population" was suggested because it captures the notion that the specifics of each investigation determine the assemblage of animals to be studied.

Because contaminants investigations often are structured by circumstances beyond the control of investigators, it is important to be attentive to matters of randomness in sample selection, appropriateness of habitat within the study area, and representativeness of individuals selected for evaluation. For example, mobility and health of affected individuals, heterogeneity of habitat, and contaminant distribution can profoundly influence the outcome of a study. Members discussed the need for preliminary surveys and assessments of nutrient and contaminant movement within the affected community and between trophic levels before study initiation. Such matters are important to subsequent efforts to translate the results into estimated effects on the true biological population and proposals for population

CHAPTER PREVIEW

Session 1: Endpoint selection and study design   217
Session 2: Extrapolation   221

restoration. Factors that would determine the size and extent of the study population include
- scope of the assessment (e.g., a particular site or a general threat to many locations);
- spatial distribution of the contamination, physical characteristics of the contaminated site, types and mixtures of contaminants and their respective environmental fates, and observed or expected responses to exposure;
- characteristics of the species of concern and trophic level of concern;
- amount of movement of the species into and out of the contaminated area;
- statistical considerations for selection of habitat and organisms to be studied; and
- concerns of stakeholders.

Selection of species for study is influenced by several of the previously mentioned factors, i.e., scope of the assessment, characteristics of the contaminants, and affected habitats and species. Two examples of good study species were discussed. The black-crowned night heron (*Nycticorax nycticorax*) was suggested as a species for which considerable information has been collected and would be appropriate for use in monitoring. Another species proposed for monitoring was the mink (*Mustela vison*), although information on the population dynamics of free-living mink is limited.

Endpoints

The presentation in Chapter 3 on current endpoints for contaminants studies provided a base for the discussion. At and below the level of the individual organism, biomarker and bioassay response patterns have been associated with contaminant exposure (Melancon 1994). Although such molecular, cellular, organ, and organismal measures may constitute excellent warning signs of contaminant exposure, rarely are such responses causally linked to effects of contaminants at the population level (NRC 1987; Heinz 1996). A further dilemma is posed by the fact that few contaminant concentration thresholds, guidelines, or standards in multiple matrices, e.g., water, soil, sediment, and foods, have been developed for the protection of wild terrestrial vertebrates.

There was no agreement over whether there were adequate existing toxicological endpoints to evaluate contaminant impacts on populations. It was recognized that new endpoints are likely to be needed as new classes of chemicals appear in the environment. Also, many of the behaviors and interactions among conspecifics (e.g., mate selection), competitors, or other aspects of the environment have not been adequately characterized. Overall, the relations between biomarker responses and environmental stressors and between systemic responses and population responses need to be better understood so that we can maximize our use of this information.

Populations, communities, and ecosystems can be evaluated with endpoints that measure the vertebrates directly (e.g., survival, reproduction, movement, overall population change, species composition, species diversity) or that measure habitat (e.g., vegetation composition, nutrient dynamics, water and soil chemistry). Species presence or absence (population extinction) should be viewed as a measure of population resiliency resulting from a quantifiable type and mode of contamination. Determination of the mechanism responsible for measured changes at these levels of organization is difficult. Population measures have intuitive appeal to stakeholders and the general public, but changes in community and ecosystem characteristics may have unclear, hence unconvincing, associations with the contaminants of concern.

The only biomarker endpoint that was known to have been used in court decisions was acetylcholinesterase, which indicates exposure to and detrimental impacts from carbamate or organophosphate insecticides. Aminolevulinic acid dehydratase (ALAD) also is a reliable biomarker for lead exposure and can be used to diagnose degree of toxic response. Group members felt that other toxicological endpoints may be as reliable and have the potential for use in courts of law.

It is unlikely that any single measurement will be sufficient to evaluate the effects of a contaminant on populations, communities, or ecosystems. Measures of a number of responses spanning several levels of biological organizational will be necessary to discern effects and determine a cause-effect relationship. In essence, group members believe that a weight-of-evidence approach will be needed in most situations.

In every investigation, the selection of endpoints depends on the goals and special considerations of the study. Examples of considerations are

- scale of the study,
- contaminant or contaminant class,
- complexity of the contaminant stress, e.g., multiple contaminants, special circumstances,
- presence of non-contaminant stressors,
- whether effects are direct or indirect, and
- species and trophic dynamics.

Study design

A combination of studies, including both field and laboratory studies, will probably be required to evaluate the effects of contaminants on vertebrate populations, their communities, or the ecosystems within which they reside. The most desirable assessment of the effects of contaminants at these organizational levels would provide some measure of the likelihood of a detrimental impact. Many factors will contribute to the design of such studies, including

- scope of the study: site-based (e.g., Superfund Site) or threat-based (e.g., regulatory issue);

- goals of the study: research-oriented or management-oriented;
- the extent of potentially confounding circumstances, e.g., fate and bioavailability of contaminants and biological characteristics of exposed animals;
- temporal, spatial, and biological (community, trophic level) scales of the issue;
- available baseline data;
- characteristics of anticipated vertebrate responses to the contaminants;
- availability of reference populations, communities, or ecosystems;
- available resources to conduct the investigation; and
- species of concern.

The species of concern or the goals of the study might be influenced by political or societal concerns. Therefore, the study design may have to use high-profile species or a less direct line of inquiry to address the underlying biological issues. Whenever possible and appropriate, the investigator should work with species that have existing baseline information adequate for population modeling.

A shortcoming of many contaminant studies is their duration. Studies of short duration fail to capture the range of potential interactions with changing weather and water chemistry, contaminant combinations, natural fluctuations in plant and animal abundance, slowly developing physiological responses (e.g., immune system suppression), and transgenerational effects. Failure to address such interactions can lead to erroneous conclusions about the effects of the contaminant.

Appropriate reference sites or populations are needed for evaluation of presumed effects of contaminant exposure. Particular care should be taken to identify differences between contaminated sites and reference sites that preceded contaminant exposure or that are clearly unrelated to the potential effects of the contaminantion. Adequate reference sites are critically important when the study is not initiated until after the contaminant is present. Reference sites only approximate the original or current state of the affected site; biotic and abiotic dynamics are never exactly alike at different locations.

Research performed on populations or communities usually experiences the statistical consequences of small sample size. Replicates at these levels of organization are difficult to locate, and resources are often not available to work on more than a small number. Cooperation among research groups or government agencies to pool resources or research results would greatly increase the potential scope of such research. Also, results from published accounts of similar work or relevant information from long-term monitoring efforts could be used to supplement specific research at limited locations.

Session 2: Extrapolation

Discussions addressed 4 primary questions:
- How do we use information on contaminant effects to infer effects at higher levels of organization?
- How successful have we been at identifying effects at higher levels of organization?
- Are there better ways to assess "higher-level" effects?
- What can we learn from aquatics studies?

Initial discussion dealt with definitions of population, community, and ecosystem. The following working definitions were generally acceptable:
- Population: a group of organisms of the same species in the same area interacting and interbreeding.
- Community: a group of organisms of different species interacting in an area.
- Ecosystem: 1 or more communities plus their interactions with the abiotic components of the environment, typically involving a substantially larger physical area than do communities or populations.

Discussion

Group members agreed that cause-and-effect relations identified at the molecular or organism level can be useful in predicting higher-level effects. However, communities and ecosystems are so complex that it is usually impossible to extrapolate from populations to either of those levels, except when a keystone species is available, e.g., krill in the Arctic Ocean or wolves at Isle Royale. Such keystone species are not common and usually are restricted to simplified systems. However, community or ecosystem effects may result in indirect effects on important populations. For example, acidification may have a negative effect on loon populations because fish and invertebrate variety and abundance decline in acidified waters (community effect). Another example would be the effect of an herbicide on vegetative composition and structure, which could reduce food abundance or nesting habitat for birds.

Consensus was reached on the notion that, in general, it is better to measure community effects directly than to extrapolate population effects to the community. Once community effects have been measured, an evaluation of potential causative factors may be useful in understanding those effects. For example, evaluation of community metrics along with bioassays at the appropriate level (e.g., plant bioassays using soils from a site) can yield supportive information for weight–of–evidence arguments.

Potential endpoints for community effects include
- abundance estimates,
- number of links among populations,

- species richness,
- guild complexity,
- species or genetic diversity,
- presence or absence of sensitive species, and
- development of predictive ratios of taxa similar to the Ephemeroptera, Plecoptera, Trichoptera ratio used in aquatic ecology.

Group members thought it might be possible to develop procedures for assessment of disturbances in terrestrial communities caused by environmental contaminants. Such procedures might be similar to the Habitat Evaluation Procedures (USFWS 1980) and Habitat Suitability Index (USFWS 1981) models that are currently available or may include a modification of the Index of Biotic Integrity (USEPA 1995) that is currently used in streams and is being developed for wetlands. Specific endpoints were not identified for ecosystems but could include aspects of energy flow and nutrient cycling.

The environmental perspective of researchers is often different than that of resource managers and regulatory agencies. Whereas researchers frequently want to develop an understanding of how systems function, resource managers and regulatory agencies most often want answers to solve specific problems or sufficient information to allow for a decision. There are considerable differences between these viewpoints. At present, regulatory agencies that are involved in registration of new chemicals, such as the USEPA's Office of Pesticide Programs, have well defined protocols involving effects at the organismal level for registering chemicals. Such agencies may not have a current need for understanding higher-level effects. However, ignorance of higher-level effects does not nullify them, and it may be that regulatory agencies are taking too narrow an approach. States, tribes, and federal regulatory agencies that need to assess damage caused by contamination of environments should consider assessment of community and ecosystem effects. Considerable research would have to be conducted to develop tools or bioindicators for evaluating the effects of contaminants at these levels of organization.

With regard to aquatic studies, it was agreed that the field of aquatic toxicology, especially as it involves ponds and small lakes, is further along than the field of terrestrial toxicology in understanding how contaminants affect populations. An important difference between the 2 disciplines is that pond vertebrates are often confined to the system and have minimal opportunity, compared to terrestrial vertebrates, for emigration or immigration during the period of study. Complete restriction of terrestrial species is usually not feasible because of their high degree of vagility, i.e., flight and large home ranges. Thus, it is considerably more difficult to study population and community effects in birds and large mammals than in species that inhabit ponds. Another important difference between aquatic and terrestrial studies is that animals in aquatic environments are typically immersed in their contaminated environment, which allows for an accurate characterization of exposure, whereas many terrestrial animals can find refugia or move to uncontaminated locations. However, stream inhabitants can also move away from point-source contamination and can benefit from increases in stream flow. Other concepts that could be borrowed from aquatic toxicology include the indices mentioned above and comparisons among taxa with regard to sensitivity to specific toxicants.

References

Heinz GH. 1996. Selenium in birds. In: Beyer WN, Heinz GH, Redmon-Norwood AW, editors. Environmental contaminants in wildlife: Interpreting tissue concentrations. Boca Raton FL: Lewis. p 447–458.

Melancon, MJ. 1994. Bioindicators used in aquatic and terrestrial monitoring. In: Hoffman, DJ, Rattner, BA, Burton, GA, Jr, Cairns, J., Jr., editors. Handbook of ecotoxicology. Boca Raton FL: Lewis. p 220–240.

[NRC] National Research Council Committee on Biological Markers. 1987. Biological markers in environmental health research. *Environ Health Perspect* 74:3–9.

[USEPA] U.S. Environmental Protection Agency Index of Biotic Integrity can be searched at <http://www.epa.gov/OWOW/wetlands/wqual/genguide.html>.

[USFWS] U.S. Fish and Wildlife Service Habitat Evaluation Procedures, ESM 102. Division of Ecological Services, Washington DC. 1980.

[USFWS] U.S. Fish and Wildlife Service Standards for Development of Habitat Suitability Index Models, ESM 103. Division of Ecological Services, Washington DC. 1981.

Group C Leaders

Michalann Harthill, Joseph P. Sullivan, Harry M. Ohlendorf, Donald W. Sparling

Group Members

W. Nelson Beyer, Gary D. Brewer, Daniel D. Day, Hector Galbraith, Sarah Gerould, Mark S. Johnson, Peter M. Kareiva, Robert L. Lochmiller, David T. Mayack, John F. McLaughlin, Katharine C. Parsons, Janet Whaley

CHAPTER 11

Estimation of Population-Level Effects on Wildlife Based on Individual-Level Exposures: Influence of Life-History Strategies

Bradley E. Sample, Kenneth A. Rose, Glenn W. Suter II

With the exception of threatened or endangered species, or migratory birds, current policy at Superfund sites is that risk management decisions should be based on protecting populations (USEPA 1997). Although risk managers are concerned with populations, risk assessments for wildlife are typically based on effects on individuals. Toxicity data available for wildlife are, like those for humans, expressed as measures of effects on individuals. That is, toxicity tests determine the responses of individual organisms, e.g., mortality and fecundity, rather than responses of populations, e.g., age structure or population growth rate. Individual-level responses can be used to estimate population-level responses, but logistical requirements preclude population-level toxicity tests for wildlife. The relationship between toxic effects on individuals and their subsequent consequences on populations is complex. Wildlife population abundances fluctuate from year to year as a result of variation in a multitude of abiotic and biotic factors. How contaminants, which typically affect only certain biological processes and individuals during certain life stages, interact with these varying factors to affect long-term population dynamics is not intuitively obvious.

Existing risk assessment methods for relating toxic effects on individuals to population responses are based on overly simplistic assumptions about wildlife population dynamics. In many wildlife risk assessments, estimation of population-level effects are performed in an ad hoc and indirect manner by equating population exposure to contaminant levels averaged over large spatial areas (Sample, Hinzman et al. 1996) and by using toxicity endpoints (e.g., reproduction, survival) that are logically associated with population effects (Sample, Opresko et al. 1996). The assumption is that if the average exposure of individuals within an area exceeds threshold levels associated with impaired reproduction or reduced survival, then population-level effects are likely. While this assumption may be qualitatively true, the associated

CHAPTER PREVIEW
Methods 227
Results 235
Discussion 238

uncertainty is high and the magnitude of the population response is not quantified. Depending on the life-history strategy of a given species, comparable levels of reproductive impairment or reduced survival may result in dramatically different population-level responses. For example, species with high reproductive output (e.g., r-selected species such as voles) are generally short-lived and experience high mortality as a result of natural causes. These species tend to be resistant to episodic or localized contaminant exposures because of their high fecundity and rapid population turnover. In contrast, species with low reproductive output (e.g., K-selected species such as bats and most carnivores), are typically long-lived and experience low natural mortality rates. Populations of these species may be highly susceptible to relatively small decrements in their reproductive success because of their low fecundity and delayed maturation that is associated with their long lifespan. In the current approach to wildlife risk assessment, the significance of reproductive or survival effects experienced by individuals of different species is considered comparable. Differences in how abiotic and biotic factors affect population dynamics among species and among populations of the same species, and differences in life-history strategies among species, are largely ignored.

The disjunction between policy (protect populations) and practice (individual-level toxicity) in wildlife risk assessment may result in inappropriate remedial decisions. In the absence of appropriate population-level risk assessment methods, actions may be taken to minimize risks to individuals. While this level of protection is appropriate for humans, it would be considered excessively protective of wildlife by most resource managers and stakeholders. Attempts to estimate risks to populations based on simplistic or inappropriate assumptions and methods also can result in unnecessary remediation, thereby increasing remedial costs, or inadequate remediation and resulting in possible long-term detrimental ecological effects. The risks to wildlife that are due to toxic effects must be balanced against the risks of habitat loss resulting from remediation actions. In order to properly protect populations, and balance the benefits and costs of remediation, effects should be expressed in common units such as population abundance or production.

In this chapter, we use age-structured matrix projection models of hypothetical bird populations to compare how life-history strategies influence population responses to contaminant exposure. Our analyses illustrate how currently available modeling methods can be used to realistically express individual-level contaminant effects as population-level responses. The population model is configured for a short-lived, high-fecundity species whose annual abundance is highly variable and strongly coupled with environmental fluctuations (r-strategist) and for a long-lived, low-fecundity species whose annual abundance is more constant (K-strategist). Survival and fecundity are specified as density-dependent and vary randomly from year to year as a representation of environmental stochasticity. One-hundred-year simulations were performed, with various combinations of fecundity and mortality incrementally reduced to mimic the general effects of contaminant exposure. Long-term mean population abundances were compared between baseline (no contaminant) and reduced survival and fecundity simulations for the r- and K-strategy species. We envision this is as the first step. Future work will refine the models to reflect specific species and incorporate the toxicological effects of particular contaminants.

This research is part of a larger project to provide the U.S. Department of Energy (USDOE) with the means to appropriately assess risks to wildlife populations. Components of the project include development of improved approaches for interspecies extrapolation of toxicity data, development of dose-response models, and development of matrix-based population models for wildlife species. The goal is to integrate contaminant exposure–response data with species-specific population models to estimate the magnitude of population effects for a given level of exposure to a particular contaminant. This project builds on prior work at the Oak Ridge National Laboratory (ORNL) that has developed ecotoxicological screening benchmarks for wildlife (Sample, Opresko et al. 1996), wildlife exposure models (Sample et al. 1997), effects-based chronic toxicity test endpoints for fish (Suter et al. 1987), and integration of toxic effects with population models of fish (Barnthouse et al. 1987; Rose et al. 1993).

Methods

Life-history strategies

Life-history theory deals with the constraints among demographic variables and the manner in which these constraints shape strategies for dealing with different environments. The life cycle can be viewed as a progression of basic life stages (egg, nestling, juvenile, immature adult, sexually mature adult, senescent adult) linked by major developmental events. Life-history strategies are generalizations of the different combinations of traits and vital rates that make up the diversity found in life cycles. Life-history characteristics are important determinants of population-level changes in response to exogenous stressors, including contaminant exposure (Emlen and Pikitch 1989; Barnthouse et al. 1990).

Life-history patterns for vertebrates can be defined in terms of linear or multidimensional continua. The most commonly recognized pattern is the r–K continuum (Pianka 1970). The r-strategy refers to species characterized by individuals with short lives, early sexual maturity, large broods, and high annual reproductive effort. The K-strategy refers to species characterized by individuals with long lives, delayed sexual maturity, small broods, and low annual reproductive effort.

The relevance of the r–K dichotomy to real species has been questioned (e.g., Stearns 1977), and other continua have been proposed. For example, Winemiller and Rose (1992) proposed a triangular continuum for North American fishes. The 3 extreme strategies were denoted opportunistic, equilibrium, and periodic. These strategies were defined by differences in the values of juvenile survivorship, age of maturity, and fecundity. The equilibrium strategy is similar to the K-strategy. The opportunistic strategy is similar to r-strategy, except in having the smallest rather than the largest clutches. A periodic strategist exhibits delayed maturation, large clutches consisting of small eggs, low juvenile survivorship, and high recruitment every few years that sustains the population. We use the traditional r–K continuum as the basis for our analyses because of its simplicity and its wide useage in ecology. The exact continuum used is not important. What is critical is that a theoretically sound life-history framework is employed for analyzing population responses to contaminant exposures.

Population modeling

We configured an age-structured matrix model for representative hypothetical r-strategy and K-strategy bird species. By focusing on a particular taxon we ensure realism in our specification of life history characteristics and vital rates.

The age-based population can be compactly expressed in matrix notation (Caswell 1989):

$$n(t+1) = A \cdot n(t) \qquad \text{(Equation 11-1)},$$

where n(t) is a k × 1 vector, with each element being the number of individuals in each of the k age-classes at year t, and A is a k × k matrix:

$$A = \begin{bmatrix} s_0 f_0 & s_1 f_1 & s_2 f_2 & \cdots & s_{k\&1} f_{k\&1} & 0 \\ s_0 & 0 & 0 & \cdots & 0 & 0 \\ 0 & s_1 & 0 & \cdots & 0 & 0 \\ 0 & 0 & s_2 & \cdots & 0 & 0 \\ 0 & 0 & 0 & \cdots & s_{k\&1} & 0 \end{bmatrix} \qquad \text{(Equation 11-2)}.$$

The matrix A consists of a top row of annual fertilities and an off-diagonal of annual survival rates from the jth age class to the j + 1th age class. Thus, age 1 survival (s_1) is from age 1 to 2, and age 1 refers to the 365-d period coinciding with the individual's second year of life. We follow females only, assuming a fixed sex ratio in the population. Fecundity is therefore expressed as female eggs per female. This version of the A matrix assumes that individuals are counted just after breeding; pre-breeding census versions of the matrix would have slightly different meanings to the age-zero rates (Akcakaya et al. 1997). By repeated multiplication of the n(t) vector by the A matrix, annual abundances in each age class can be computed through time. Total population abundance (number of females) is the sum over age classes in each year. To better represent dynamics of field populations, we included 2 modifications to the A matrix incorporating stochasticity and density dependence.

The A matrix was specified for the r-strategy and K-strategy species. The r- and K-strategy species differ in their maximum age, fecundity, first-year survival, adult survival rates, and age at maturation (Table 11-1). Maximum ages of birds typically range from 2 to 3 years (y) to 15 to 25 y, with several extremely long-lived species exceeding 30 y (Table 1 in Botkin and Miller 1974; Table 26.4 in Newton 1989). Because avian vital rates have been reported as functions of body weight, we assigned an adult body weight of 10 g for our r-strategist and 500 g for our K-strategist (Figure 1 in Dobson and Hudson 1994). Annual adult survival rates corresponding to 10 g and 500 g were approximately 30% to 50% and 60% to 80%, respectively (Figure 1 in Dobson and Hudson 1994). Age at maturation was near 1 y for species with adult annual survival rates less than 60%, and between 2 and 4 y for species with adult survival rates between 60% and 80% (Figure 2 in Ricklefs 1973). We assumed that 100% of age 1 and older birds were mature for the r-strategist and that 50% of age 2 and 100% of age 3 and older were mature for the K-strategist. First-year survival, which we calculate as described below, ranged from 10% to 35% for an adult annual survival rate of 30% to 50% and from 15% to 60% for an adult annual survival rate of 60% to 80% (Figure 3 in Ricklefs 1973). Coefficients of variation (CV; 100 * [standard deviation/mean]) of annual population abundances ranged from 10% to 20% at the low end and from 60% to 70% at the high end (Table 6 in Ricklefs 1973). Annual fecundity in the A matrix is the product of the number of female eggs/female in each clutch times the number of clutches in a breeding season. Because estimating the number of clutches is problematic, we based our fecundity on lifetime reproduction, which has been reported for a variety of species (Newton 1989). Expected lifetime egg production is 8.75 female eggs/female (17.5 eggs/

Table 11-1 Life-history parameters for hypothetical r- and K-strategy avian species

Characteristic	r-strategist	K-strategist
Maximum age (years)	3	20
Fecundity (female eggs/female)	10 at age 1	2 at age 2
	15 at age 2	3 at age 3 & older
Adult weight (g)	10	500
Age-zero survival	0.11[a]	0.36[a]
Adult survival	0.5	0.7
Maturity	100% at age 1	50% at age 2
		100% at age 3 & older
CV of annual abundance (%)	60–70	10–20
Lifetime egg production (total eggs/female)	17.5	11.2

[a] Estimated by model

female) for our *r*-species and 5.6 female eggs/female (11.2 eggs/female) for our *K*-species. Species characterized as short-lived (5 y) in Newton (1989) had lifetime production values of 10 to 25 fledglings/female. For example, the house martin (*Delichon urbica*) weighs ~20 g and produces 10 to 20 fledglings over a 3-y lifespan (Bryant 1989). Species characterized as long-lived ($\geq$15 y) by Newton (1989) had maximum lifetime production values of ~10 fledglings/female. The short-tailed shearwater (*Puffinus tenuirostris*) weighs ~500 g and produces ~12 fledglings/female over a 20-y lifespan (Wooller et al. 1989). The lifetime production values we assumed for our *r*- and *K*-species should be higher than reported values because reported values are based on fledglings, which already account for some mortality of eggs.

Stochasticity was incorporated into the population model by randomly varying the values of age-specific fecundity and mortality rates in the A matrix. For each year of the simulation, multipliers of age-zero survival, adult-survival rates, and fecundity rates were independently generated from triangular probability distributions (defined by minimum, mode, and maximum values; Table 11-2). We used a triangular distribution because it is flexible, has easily understood parameters, and has finite tails (negative values and extreme positive values can be unrealistic). The mode for stochasticity multipliers was 1 for all distributions because generated random deviates were used to multiply the actual assigned fecundity and survival rates. Minimum and maximum values for the triangular distributions were determined so that predicted CVs of abundance were about 60% to 70% for the *r*-strategist and 10% to 20% for the *K*-strategist. Greater variation was imposed on survival and fecundity rates for the *r*-strategist to reflect the greater inter-annual variation and influence of density-independent (environmental) factors typical of *r*-type species.

Density dependence was incorporated by making age-zero survival and fecundity functions of adult abundance. In a synthesis of several literature reviews, O'Connor (1994) concluded that both theoretical and empirical studies indicate that density dependence is widespread, though not universal, within bird populations. Empirical evidence suggests that the first year of life is most likely influenced by population abundance (Ricklefs 1973); adult survival in most birds appears variable but in

Table 11-2 Distributions for inter-annual stochasticity multipliers for hypothetical avian *r*- and *K*-strategist species

Parameter	*r*-strategist			*K*-strategist		
	Minimum	Mode	Maximum	Minimum	Mode	Maximum
Fecundity	0.5	1.0	1.5	0.8	1.0	1.2
Age-zero survival	0.2	1.0	4.0	0.8	1.0	1.2
Adult survival	0.5	1.0	1.5	0.75	1.0	1.25

response to environmental factors not related to density (Dobson and Hudson 1994). Density dependence is more likely to affect the fecundity of adults where lowered energy intake resulting from competition for food or utilization of inferior habitats could lead to fewer viable young (Ricklefs 1973; Dobson and Hudson 1994). In some taxa where maturity is size-governed, age of maturity could be density-dependent via density-related growth rates. Because this dependence is not prevalent in birds, we treated age at maturity as fixed. Density-dependence multipliers of age-zero survival and fecundity were defined as functions of normalized adult (mature) abundance (Figure 11-1). Adult abundance was computed as numbers in each age class times the fraction mature by age. Adult abundance was normalized by dividing simulated values each year by the defined average abundance (1000 adults for the K-strategist and 20,000 adults for the r-strategist). Density dependence was made stronger in the K-strategy species by defining steeper slopes and more extreme maximum and minimum multiplier values.

Contaminant effects

Contaminant effects were imposed for both the r-strategy and K-strategy species by reducing the values of fecundity and juvenile and adult survival rates in the A

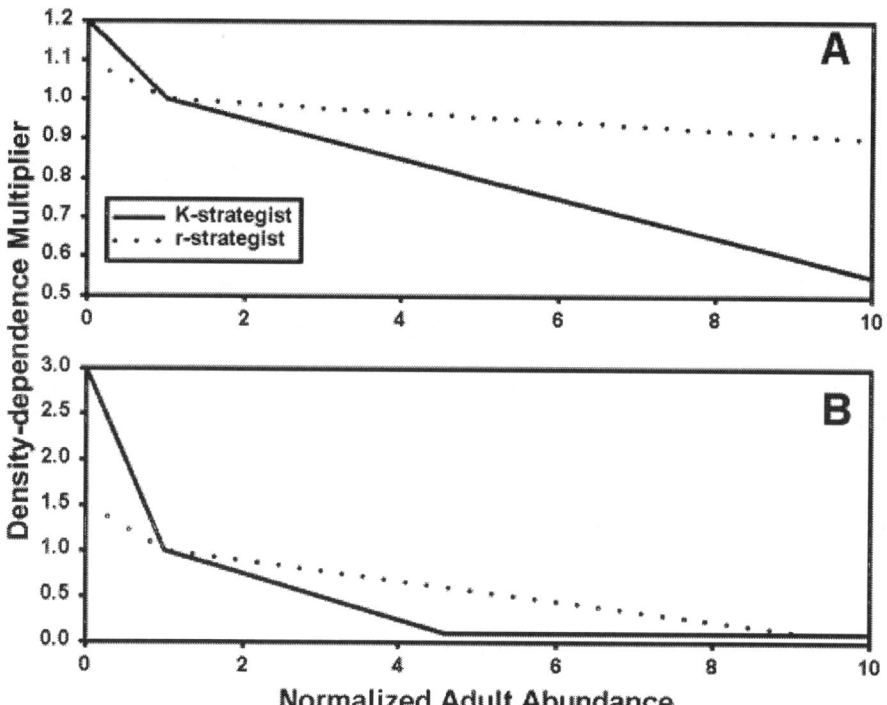

Figure 11-1 (a) Density-dependence multipliers for fecundity and (b) age-zero survival for hypothetical avian r- and K- strategist species

matrix. Fecundity and survival rates were reduced by specified percentages (e.g., 15% reduction was achieved by multiplying by 0.85). Unless otherwise stated, reductions were imposed for every year of the simulation. Age-zero survival was reduced because the juveniles have been shown to be more sensitive than adults to some contaminants (e.g., Grue and Hunter 1984; Grue and Shipley 1984). Fecundity was reduced to mimic fewer viable eggs as a result of contaminant exposure or less energy allocated by adults to reproduction. We also included reduced adult survival to contrast how life-history strategy influences population responses. Estimates of reduction in survivorship and fecundity that are appropriate to specific chemicals will be developed and applied in later stages of this research program.

Design of simulations

General

All simulations were for 100 y. Initial numbers in each age class were 15,000 for the r-strategist and 50 for the K-strategist. Total abundances were predicted for each year of the simulation.

The first 20 y of simulation results were discarded to minimize any effects of the arbitrary initial conditions. Discarding the first 20 y of each simulation was adequate to minimize effects of initial conditions. Predicted mean population abundances under baseline conditions (e.g., no contaminants) based on the last 80 y were within 5% of the mean abundances based on the last 50 y. Ten replicate simulations were performed for baseline conditions and each contaminant-effect condition. Replicate simulations differed in the random number sequence used to determine random deviates from the triangular distributions affecting survival and fecundity.

Baseline

The only free (unspecified) parameter value in the baseline models was age-zero survival. Age-zero survival was computed for the r-strategist and K-strategist so that, without stochasticity and density dependence, stable, long-term populations would be predicted. To determine initial values, stochasticity and density dependence were eliminated by setting their multipliers to 1 for all years. The resulting age-zero survival was determined to be 0.11 for the r-strategist and 0.36 for K-strategist. The lower age-zero survival needed for the r-strategist for stability is more the result of the higher fecundity and younger age of maturation than of offsetting the shorter lifespan (and fewer ages of reproduction).

Contaminant effects

A total of 14 contaminant-effects scenarios were simulated (Table 11-3). The first 5 scenarios involved reducing both age-zero survival and fecundity by increasing percentages (5%, 10%, 15%, 20%, and 25%). The sixth and seventh scenarios used 50% reductions in fecundity and age-zero survival, but they were imposed on an average of every 5 or 10 y. Years of contaminant effects were selected if a random number from a uniform distribution between 0 and 1 was less than 0.2 for the every 5 y and less than 0.1 for the every 10 y condition. The 8 scenario was to eliminate the egg production of the first mature age of breeding. For the r-strategist, who was fully mature at age 1, fecundity of age 1 females was assumed to be 0. For the K-strategist, whose first age of maturity (age 2) was only 50% mature, we set the fecundity of age 2 to 0 and reduced by one-half the fecundity of age 3 to remove the equivalent of a full year of breeding. Individuals surviving to age 2 in the r-strategist and age 4 in the K-strategist resume their usual fecundity. The remaining 5 scenarios were reductions in adult survival by 5%, 10%, 15%, 20%, and 25%.

For the 10% and 25% reductions in age-zero survival and fecundity scenarios, we also determined the fraction of the r-strategist population that would need to be affected to yield the same percent reductions in total abundance that were predicted for the K-strategist. The contaminant-effects simulations assume that all individuals are affected. If only a small portion of the r-strategist population needed to be affected in order to yield the same percentage reductions that the K-strategist population yielded, this would imply that the r-strategist was very sensitive to the contaminant-effects conditions.

In all contaminant effects scenarios outlined above, stochasticity was held constant (i.e., independent of contaminant exposure). To investigate population responses to varying stochasticity, 4 additional simulations were performed for the r-strategist species: baseline and 10% reduction in fecundity and age-zero survival (using stochasticity multipliers from Table 11-2), and 10% reduction in fecundity and age-zero survival with either reduced or increased stochasticity. Stochasticity was varied by specifying new values for the minimum and maximum values of the triangular distributions that multiplied fecundity and age-zero survival (Table 11-4). Contaminants can affect the variation in fecundity and survival rates, as well as their mean values. In our simulations, we allowed for both increased and decreased stochasticity in fecundity and survival rates for the sake of generality and because situations can be envisioned in which contaminants could act to increase or decrease variability.

Prediction variables

For the baseline simulations, we show annual total population abundance over time and report means and CVs of abundance computed for each of the 10 replicate simulations. The minimum, average, and maximum of the 10 means and the 10 CVs are reported. For each of the contaminant-effects scenarios, we computed the percent reduction in average total abundance from the mean baseline value. We first computed the mean abundance (over years 21 to 100) for each of the 10 replicate

Table 11-3 Contaminant exposure and effects scenarios and simulation results

Contaminant effect on	Magnitude of effect (% reduction)	Frequency of effect	Mean % total population reduction versus baseline		Mean CV (%)	
			r-strategist	K-strategist	r-strategist	K-strategist
Age-zero survival and fecundity	5	Annual	27	13	54	14
	10	Annual	50	22	55	13
	15	Annual	46	30	55	13
	20	Annual	80	39	56	13
	25	Annual	90	47	60	13
	50	Every 5 years	52	20	74	22
	50	Every 10 years	27	10	61	18
Fecundity of first age of breeding	r: 100% age 1 K: 100% age 2; 50% age 3	Annual	99	33	198	14
Adult survival	5	Annual	21	19	53	12
	10	Annual	40	31	54	12
	15	Annual	55	46	55	13
	20	Annual	68	65	54	18
	25	Annual	77	88	54	38

Table 11-4 Distributions for reduced and increased inter-annual stochasticity multipliers for hypothetical avian *r*-strategist species

Parameter	Reduced			Increased		
	Minimum	Mode	Maximum	Minimum	Mode	Maximum
Fecundity	0.75	1.0	1.25	0.25	1.0	1.75
Age-zero survival	0.6	1.0	2.5	0.0	1.0	6.0

simulations. We next computed the average of the 10 means. This was repeated for the 10 baseline simulations. Percent change was then computed:

$$100 * [(Y_c - Y_b) / Y_b] \quad \text{(Equation 11-3)},$$

where Y_c is the average for the contaminant effects scenario and Y_b is the average for the baseline condition. All of the contaminant effects scenarios were compared to the same average baseline value.

Results

Annual total abundances under baseline conditions exhibited dynamics that were expected for *r*-strategist and *K*-strategist bird species. Annual abundances were higher and more variable from year to year for the *r*-strategist compared to the *K*-strategist (Figure 11-2). Average abundance and average CV were 390,000 and 52.6% for the *r*-strategist compared to 4300 and 12.3% for the *K*-strategist. These results are expected because we specified greater stochasticity and weaker density dependence for the *r*-strategist than for the *K*-strategist.

Percent reductions in abundance for simulations involving any combinations of reduced fecundity and reduced age-zero survival resulting from contaminant exposure were larger for the *r*-strategist than for the *K*-strategist (Table 11-3, Figure 11-3). Reducing age-zero survival and fecundity together by fixed percents or by 50% every 5 or 10 y, or by eliminating egg production of the first breeding year, resulted in percent reductions for the *r*-strategist that were about twice those determined for the *K*-strategist. Even relatively small percent reductions in age-zero survival and fecundity (5% and 10%) led to significant reductions in long-term average abundances (> 20%) for the *r*-strategist. None of the contaminant-effects conditions caused > 50% reductions in the abundance of *K*-strategist (Table 11-3).

Percent reductions in abundance for the *r*-strategist were slightly larger under small reductions in adult survival but were slightly smaller under large reductions in adult survival, compared to the *K*-strategist (Table 11-3, Figure 11-4). These results show the sensitivity of the *K*-strategist to > 20% reductions in adult survival. Despite the assumption that all individuals are affected, which tends to make the *r*-strategist more sensitive, small reductions in adult survival resulted in similar population losses, and > 25% reductions in adult survival resulted in larger losses for the *K*-

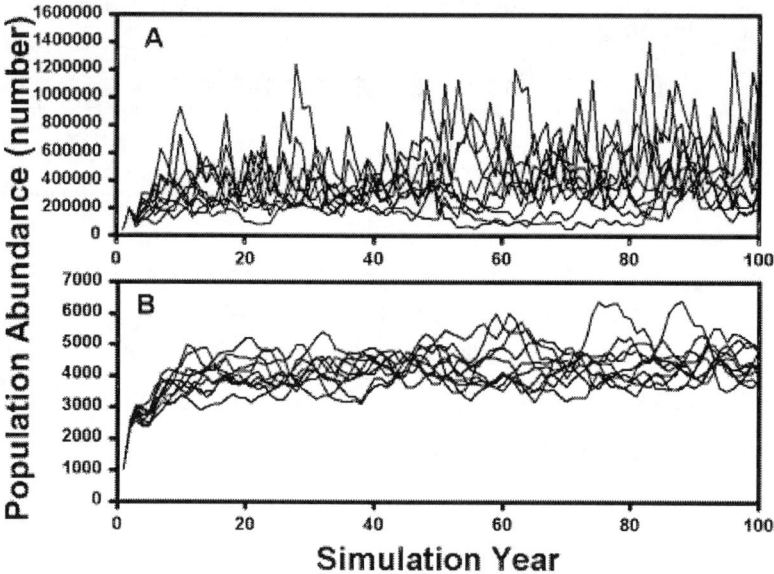

Figure 11-2 Baseline 100-y population trends for hypothetical avian (**a**) *r*- and (**b**) *K*-strategist species

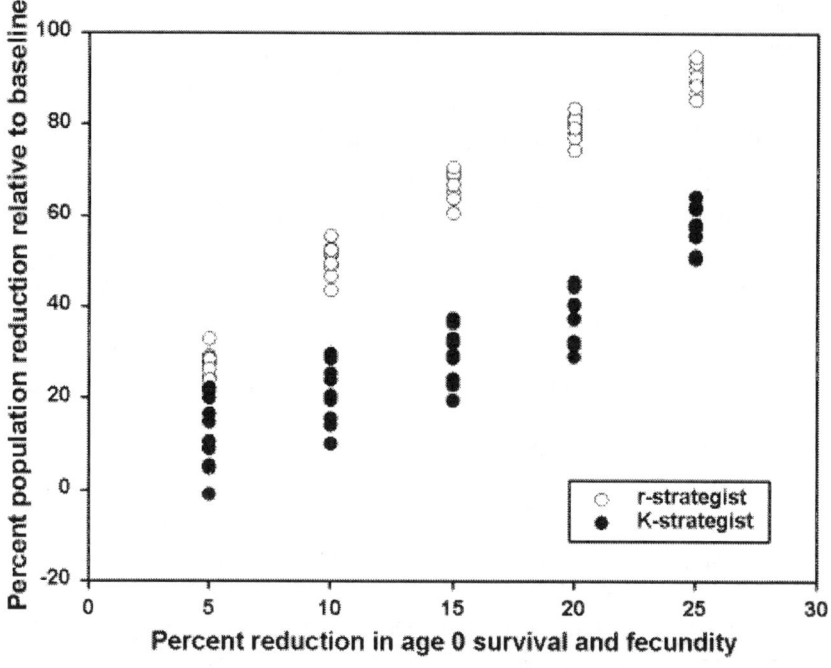

Figure 11-3 Simulation results for reduced age-zero survival and fecundity for hypothetical *r*- and *K*-selected species. Results are expressed as percent reduction of population abundance for each simulation run relative to baseline population abundance

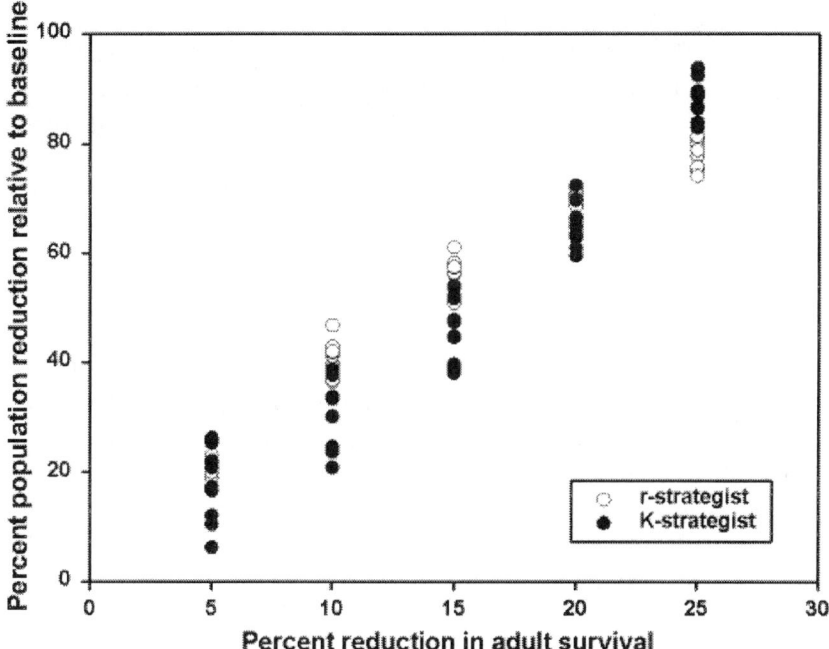

Figure 11-4 Simulation results for reduced adult survival for hypothetical r- and K-selected species. Results are expressed as percent reduction of population abundance for each simulation run relative to baseline population abundance.

strategist. This is in contrast to the reduced fecundity and age-zero survival simulations in which the r-strategist exhibited much larger reductions (typically 2-fold) than did the K-strategist.

Approximately 40% of the r-strategist population would need to be affected in order to yield the same reductions that were predicted for the K-strategist. Under the 5% reduction in age-zero survival and fecundity scenario, a 22% reduction in population abundance was predicted when 40% of the r-strategist was assumed to be affected. This is the same percentage reduction that was predicted for the K-strategist when all individuals were assumed to be affected. Similarly, under the 25% reduction in age-zero survival and fecundity scenario, assuming that 40% of the r-strategists were affected resulted in a 48% reduction in population abundance, which is comparable to the 50% reduction predicted for the K-strategist under the same scenario.

CVs for all scenarios remain roughly constant for each strategist, except for the higher contaminant-effects scenarios, when simulations begin to predict extinction (Table 11-3). In these scenarios, including the zero-population abundances inflated the CVs.

Varying stochasticity in response to contaminant effects affected population responses. A 10% reduction in age-zero survival and fecundity that was associated with stochasticity equivalent to baseline resulted in a 50% population reduction (Figure 11-5). In contrast, the same contaminant effect with reduced stochasticity resulted in an average reduction of 68% (65% to 70% over the 10 replicates), while increased stochasticity resulted in an average reduction of only 17% (30% reduction to 15% increase over the 10 replicate simulations).

Discussion

While highly generalized, our simulations demonstrate a scientifically sound approach for extrapolating individual-level contaminant effects to the population level. Matrix-projection models have a long history of useage in ecology (Caswell 1989), for both theoretical analyses and resource management (e.g., harvesting), and have been recommended for application in ecological risk assessment (Emlen 1989). We explicitly simulated population dynamics with the matrix projection models rather than focusing on the asymptotic properties of the matrices under

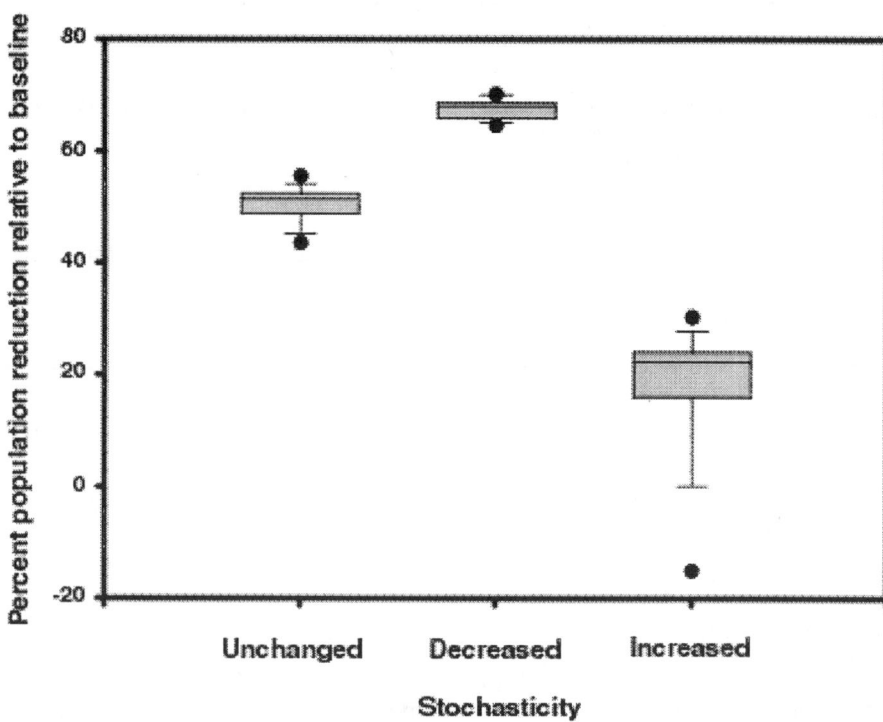

Figure 11-5 Simulation results for 10% reduction in age-zero survival and fecundity for hypothetical r- selected species with stochasticity increased or decreased relative to baseline condition. Results are expressed as percent reduction of population abundance for each simulation run relative to baseline population abundance.

equilibrium conditions. Asymptotic properties that can be derived from matrix-projection models include the population growth rate, the stable age distribution, and the dampening ratio (how quickly the stable age/stage distribution is obtained) (Caswell 1989). The advantage to asymptotic analyses is that results are independent of initial conditions. However, the matrix models used in this paper were both stochastic and density-dependent (nonlinear), which greatly complicates deriving their asymptotic properties (Cushing 1997; Tuljapurkar 1997). We therefore estimated long-term average population abundances by explicitly simulating the populations.

We also envision situations in which individual-based modeling would be appropriate for extrapolating toxic effects to the population level. Individual-based models are the next logical step beyond matrix models in adding more detail to how the population is represented. Age- and stage-based matrix models divide the population into classes, which can differ from each other in their vital rates, but treat all individuals in a class similarly. Individual-based models track the vital rates of individuals in the population. One can think of the individual-based approach as the ultimate definition of classes. Accompanying the increased detail in individual-based models is the demand for additional data and information, which can limit applicability of these models in some situations. With increasing recognition of the importance of inter-individual variation on population and community dynamics, individual-based modeling is gaining popularity (DeAngelis et al. 1994). Individual-based modeling also appears well-suited for scaling contaminant effects to the population level. Toxic effects are often reported as individual-level responses, and contaminant exposures that are spatially localized or temporally episodic are difficult to represent in more aggregated models but can be straightforward to simulate in individual-based models. Individual-based bridge models can be developed that focus on selected life stages and processes. The bridge models can be used to convert sublethal contaminant effects into changes in growth, mortality, and reproduction, which are then used to change the parameters of a matrix-projection model to predict long-term population consequences (Rose et al. 1999).

Our analyses generally conform to the classical interpretation of how r-strategist and K-strategist species respond to stress. K-strategist species are thought to be more sensitive to stress, but r-strategists are especially sensitive to short-term reproductive failures (Barnthouse 1993). For example, we estimated that complete failure of the first age of breeding would result in a 99% population decline for the r-strategist, as opposed to a 33% decline for the K-strategist. We also interpret our results to imply that the K-strategist was relatively sensitive to reductions in survival and fecundity when contaminant exposures are localized, while the r-strategist was sensitive when the contaminant was broadly distributed in the environment. To get the same population-level response for both the r- and K-strategists orders of magnitude, more r-species individuals than K-species individuals must be affected. For example, 40% of r-species individuals (or ~114,000 age-one individuals) would need to be affected under the reduced age-zero survival and fecundity scenario to obtain reductions in abundance that are similar to those obtained by affecting 100% of the K-species individuals (2100 individuals). We argue that 40% of the r-strategist is a large percentage that, given the large number of individuals, likely involves a large geographic area. If the contaminant is ubiquitous in the environment, implying that many individuals are exposed, our results suggest that the r-strategist is relatively sensitive. If the contaminant is localized, we interpret that our results imply that the K-strategist, which is more likely to have a high percentage of individuals affected (because of its small population size), is relatively more sensitive.

While our modeling approach produced effects on mean population abundance, inter-annual variability in population abundance generally was not affected. This was expected because, while we modeled the toxic effects to reduce survival rates or fecundity, inter-annual stochasticity was not altered. As a result, CVs remained constant across contaminant scenarios for each strategist, except at the highest contaminant-effects levels, when predicted population extinctions inflated the CVs.

When we varied stochasticity in response to contaminant exposure, the magnitude of population response increased with decreasing stochasticity. In practical terms, this means that the less variable the environment, the greater the magnitude of the population response to a given level of contaminant effect. The narrower probability distributions for survival rate and fecundity under reduced variability decreased the likelihood of periodic high survival or high fecundity events that are critical for offsetting the chronic effects of the contaminant. With less-frequent high survival or fecundity events, a larger reduction in average population abundance was predicted. Wider probability distributions, which are due to contaminants increasing the variability in survival and fecundity, increased the likelihood of exceptionally good years, resulting in a smaller reduction in population abundance. These results demonstrate that contaminant effects on the variability in survival rates and fecundity can influence the magnitude of population responses.

Our initial modeling results have important ecological risk-management implications by suggesting that current wildlife risk-assessment practices may not be adequately protective of wildlife populations. The current practice is to assume that risk estimated based on an individual-level toxicity-test endpoint that has population-level relevance (e.g., mortality or reproduction) is directly equivalent to population-level risks. Our results clearly invalidate this assumption for situations of chronic (year after year) exposure to contaminants. Under the simple assumption that reduced fecundity and decreased age-zero survival act independently, we can take the product of their reductions to determine their combined effect. For example, under independence, a 20% reduction in fecundity and a 20% decrease in age-zero survival would be expected to result in a 36% population reduction (1− (0.8 × 0.8) = 0.36). Percent population reductions predicted by the model were greater than expected for all scenarios that involved chronic effects for the r-strategist and for all scenarios involving reduced adult survival for the K-strategist (Table 11-3).

Based on our simulations, in order to prevent the occurrence of population reductions of 20%, adult survival should not be reduced more than 5% for either the r- or K-species, and age-zero survival and fecundity should not be reduced more than 5% for the r-species. Clearly, basing population-level risk-management decisions on individual-level risk estimates can underestimate population responses in many situations.

Perhaps the greatest obstacle preventing widespread use of population dynamics models for contaminant effects assessment is the incompatibility between commonly reported toxicological endpoints and population model inputs. For example, incorporation into population models of toxicity-test endpoints that do not directly relate to mortality or reproduction is problematic. Additionally, most wildlife toxicity tests have been designed to test hypotheses of no effect rather than to estimate magnitudes of prescribed effects; therefore, their endpoints are not really directly useful for changing the values of population-dynamics model inputs. Studies with limited dose levels or that report only no-observed-adverse-effect levels (NOAELs) generally are not suitable for incorporation into population models. Ideally, dose-response functions should be fit to the toxicity-test results so that magnitudes of effects associated with exposures can be estimated. In addition to changes in magnitudes, contaminant-related changes in the variability of effects should also be included. As a solution, Rose et al. (1999) and Barnthouse et al. (1990) suggest that a suite of models be used that act as a bridge between bioassay endpoints and the inputs of population-dynamics models. Detailed, individual-based models of selected life stages would be used to translate lethal and sublethal (e.g., behavioral) laboratory endpoints into changes in survival, growth, and reproduction that could then be used by the population-dynamics model.

This paper represents our initial attempt to develop population models for evaluation of contaminant risks to wildlife. Future work will include parameterizing of models using specific avian and mammalian species and chemical-specific toxicity data; expressing results as the likelihood of populations decreasing below critical thresholds; and refining the modeling approach to include stage-based versions where appropriate for the life history of the species of interest. In addition, we hope to improve the models to explicitly deal with spatial heterogeneity, meta-population issues, and multiple species (community) interactions. We believe that the tools for more-realistic wildlife risk assessment are available. What is needed is a coordinated, multidisciplinary effort to combine the various pieces into a coherent methodology. Our project is one small step toward this goal.

Acknowledgements—Support of this research was provided by the U.S. Department of Energy, Environmental Management Science Program to Oak Ridge National Laboratory. We are very grateful to Heather Brooks for the literature searching and data acquisition and Christine Arenal for editorial review.

References

Akcakaya HR, Burgman MA, Ginzburg LR. 1997. Applied Population Ecology. Setauket NY: Applied Biomathematics.

Barnthouse LW. 1993. Population-level effects. In: Suter GW II, editor. Ecological risk assessment. Boca Raton FL: Lewis. p. 247–274.

Barnthouse LW, Suter GW II, Rosen AE. 1990. Risks of toxic contaminants to exploited fish populations: Influence of life history, data uncertainty and exploitation intensity. *Environ Toxicol Chem* 9:297–311.

Barnthouse LW, Suter GW II, Rosen AE, Beauchamp JJ. 1987. Estimating responses of fish populations to toxic contaminants. *Environ Toxicol Chem* 6:811–824.

Botkin DB, Miller RS. 1974. Mortality rates and survival of birds. *Am Nat* 108:181–192.

Bryant DM. 1989. House martin. In: Newton I, editor. Lifetime reproduction in birds. New York, NY: Academic Press. p 89–106.

Caswell H. 1989. Matrix population models: Construction, analysis, and interpretation. Sunderland MA: Sinauer Associates.

Cushing JM. 1997. Nonlinear matrix equations and population dynamics. In: Tuljapurkar S, Caswell H, editors. Structured-population models in marine, terrestrial, and freshwater systems. New York: Chapman and Hall. p 205–243.

DeAngelis DL, Rose KA, Huston MA. 1994. Individual-oriented approaches to modeling populations and communities. In: Levin SA, editor. Frontiers in mathematical biology. Lecture Notes in Biomathematics Vol. 10. New York, New York: Springer-Verlag. p 390–410.

Dobson A, Hudson P. 1994. Assessing the impact of toxic chemicals temporal and spatial variation in avian survival rates. In: Kendall RJ, Lacher TE, editors. Wildlife toxicology and population modeling. Boca Raton FL: Lewis. p 85–98.

Emlen JM. 1989. Terrestrial population models for ecological risk assessment: A state-of-the-art review. *Environ Toxicol Chem* 8:831–842.

Emlen JM, Pikitch EK. 1989. Animal population dynamics: Identification of critical components. *Ecol Model* 44:253–273.

Grue CE, Hunter CC. 1984. Brain cholinesterase activity in fledgling starlings: Implications for monitoring exposure of songbirds to ChE inhibitors. *Bull Environ Contam Toxicol* 32:282–289.

Grue CE, Shipley BK. 1984. Sensitivity of nestling and adult starlings to dicrotophos, an organophosphate pesticide. *Environ Res* 35:454–465.

Newton I. 1989. Lifetime reproduction in birds. San Diego CA: Academic.

O'Connor RJ. 1994. Population patterns and process parameters: Issues in integrating monitoring and models. In: Kendall RJ Lacher TE editors. Wildlife Toxicology and Population Modeling. Boca Raton FL: Lewis. p 283–300.

Pianka ER. 1970. On r- and K-selection. *Am Nat* 104:592–597.

Ricklefs RE. 1973. Fecundity, mortality, and avian demography. In: Farner DS, editor. Avian biology. Washington DC: National Academy of Sciences. p 366–447.

Rose KA, Brewer LW, Barnthouse LW, Fox GA, Gard NW, Mendonca M, Munkittrick KR, Vitt LJ. 1999. Ecological responses of oviparous vertebrates to contaminant effects on reproduction and development. In: Di Giulio RT, Tillitt DE, editors. Reproductive and developmental effects of contaminants in oviparous vertebrates. Pensacola FL: Society of Toxicology and Chemistry (SETAC). p 225–281.

Rose KA, Cowan JH, Houde ED, Coutant CC. 1993. Individual-based modeling of environmental quality effects on early life stages of fish: A case study using striped bass. *Am Fish Soc Symp* 14:125–145.

Sample BE, Aplin MS, Efroymson RE, Suter GW II, Welsh CJE. 1997. Methods and tools for estimation of the exposure of terrestrial wildlife to contaminants. Oak Ridge TN: Oak Ridge National Laboratory. ORNL/TM–13391.

Sample BE, Hinzman RL, Jackson BL, Baron LA. 1996. Preliminary assessment of the ecological risks to wide-ranging wildlife species on the Oak Ridge Reservation: 1996 Update. Oak Ridge TN: Oak Ridge National Laboratory. DOE/OR/01–1407&D2.

Sample BE, Opresko DM , Suter GW II. 1996. Toxicological benchmarks for wildlife: 1996 Revision. Oak Ridge, TN: Oak Ridge National Laboratory. ES/ER/TM–86/R3.

Stearns SC. 1977. The evolution of life history traits: A critique of the theory and a review of the data. *Ann Rev Ecol Syst* 8:145–71.

Suter GW II, Rosen AE, Linder E, Parkhurst BF. 1987. Endpoints for responses of fish to chronic toxic exposures. *Environ Toxicol Chem* 6:793–809.

Tuljapurkar S. 1997. Stochastic matrix models. In: Tuljapurkar S, Caswell H, editors. Structured-population models in marine, terrestrial, and freshwater systems. New York: Chapman and Hall. p 59–67.

[USEPA] U.S. Environmental Protection Agency. 1997. Ecological risk assessment guidance for superfund: Process for designing and conducting ecological risk assessment, interim final. Edison NJ: Environmental Response Team. EPA 540-R-97-006.

Winemiller KO, Rose KA. 1992. Patterns of life-history diversification in North American fishes: Implications for population regulation. *Can J Fish Aquat Sci* 49:2196–2218.

Wooller RD, Bradley JS, Skiran IJ, Serventy DL. 1989. Short-tailed shearwater. In: Newton I, editor. Lifetime reproduction in birds. New York: Academic. p 405–418.

CHAPTER 12

Effects of Environmental Contaminants in Spatially Structured Environments

John F. McLaughlin and Wayne G. Landis

Context

Contaminants are distributed unevenly in space at most scales of observation. Vertebrates and vertebrate populations also are distributed heterogeneously (Horne and Schnieder 1995). This heterogeneity, or spatial structure, affects both contaminant exposure and vertebrate responses to exposure. These effects are ignored in traditional assessments of contaminant impacts on vertebrate populations, which use non-spatial analysis.

Population-level ecological risk assessments (ERAs) usually extrapolate contaminant effects from mortality and reproduction of individuals to effects on populations (See Chapters 7 and 11). This extrapolation often applies models that contain detailed information regarding the age structure and life history of the vertebrate of interest. While this information may be essential to accurate prediction of contaminant impacts, spatial structure in the distributions of contaminants, populations, and habitats can be of equal importance. Unfortunately, most population toxicity models ignore spatial structure.

Environments with different spatial patterns lead to different dynamics in the populations that inhabit them. Environmental structure also mediates contaminant effects on vertebrate populations. Influences of spatial structure on contaminant effects are analogous to species-specific dose–response curves. When dose–response curves vary widely among species, it becomes important to know which curve pertains to the species of concern. Similarly, prediction of contaminant effects on spatially structured populations may require knowledge about which responses

Chapter Preview
Context 245
Kinds of spatial structure 246
Modeling spatial structure 249
Results 255
Discussion 265

pertain to the structure of interest. Hence, this chapter considers several kinds of spatial structures relevant to vertebrates and discusses influences of these structures on processes at the population level. Because the study of the role of spatial structure in mediating contaminant effects has been limited, this review draws heavily from the ecological and conservation literatures.

The chapter is organized into 4 parts. First, we describe the basic kinds of spatial heterogeneity that are relevant to contaminant effects. Next, because models have played a central role in the study of spatial population structure, we summarize various modeling approaches. Third, we review results from theoretical and empirical study of spatially structured populations. Finally, we discuss implications of these results to field systems, including when to consider spatial influences and when to ignore them. We also suggest directions needed for further progress in understanding the roles of heterogeneity in contaminant effects. The chapter focuses on individual species, with some applications to multi-species systems. Although this volume addresses contaminant effects on vertebrates, there has been limited field study of the effects of spatial structure on any taxon. Consequently, many of the empirical examples we cite are from invertebrate populations.

Kinds of Spatial Structure

Although the spatial structure of each system is unique, 5 general categories can be defined for the purpose of understanding contaminant effects on terrestrial vertebrates:
- isolated populations,
- classical metapopulations,
- mainland-island or source-sink metapopulations,
- patchy populations, and
- continuous populations.

The first 4 categories, depicted in Figure 12-1, contain discrete habitat patches surrounded by non-habitat. In this chapter, we define habitat as areas that provide resources necessary for survival and reproduction. We distinguish between "habitat" that supports resident organisms and areas that are used primarily as migration routes, although we recognize that both may be important to the viability of individuals and populations.

Isolated populations

The simplest case of a spatially structured environment is a collection of discrete habitat patches with no inter-patch dispersal. Populations residing in such a system are isolated from each other and undergo independent local population dynamics. Contaminants in 1 isolated patch do not affect populations in other patches. Conversely, contaminated patches cannot be recolonized after local extinctions.

Kinds of Patch Structure

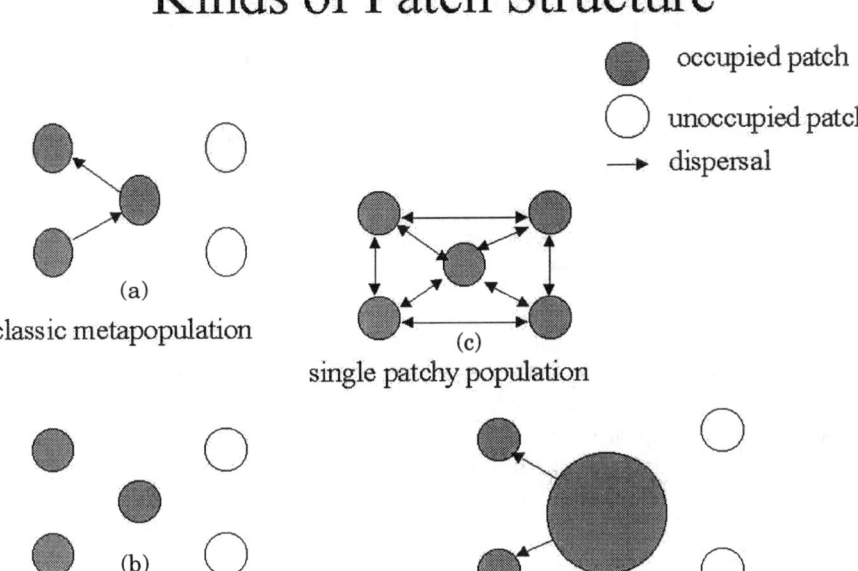

Figure 12-1 Kinds of population structure in patchy environments. Shaded circles represent occupied habitat patches; open circles, unoccupied patches; arrows, dispersal. (a) Metapopulation; (b) isolated populations; (c) single patchy population; d) mainland-island or source-sink metapopulation.

Metapopulations

Regional metapopulations result when populations in discrete patches are linked by low to intermediate rates of migration. Migration affects the dynamics of local populations, including the rate of re-establishment following an extinction (Hanski and Simberloff 1997). As the metapopulation concept increases in popularity, it is used to describe almost any kind of system containing discrete habitat patches (Harrison 1991, 1994; Hanski and Simberloff 1997). Some authors narrowly define metapopulation as a collection of local populations, each vulnerable to extinction, that persists regionally through recolonization (Harrison and Taylor 1997). We use a broader definition: a set of geographically or ecologically distinct populations in which migration has a discernable effect on local demography (Stacey et al. 1997). In metapopulations in which there is relatively rare inter-patch dispersal, the effect of migration is to colonize empty patches. More-frequent inter-patch dispersal can create a "rescue effect," preventing local extinctions (Brown and Kodric-Brown 1977).

Persistence of a metapopulation requires inter-patch migration sufficient to offset local extinction rates. When migration is prevented, isolated patches result; when migration is high, the system becomes a single patchy population. The metapopulation concept commonly has been applied to individual species or to a few interacting species. When multi-species communities are linked through dispersal of 1 or more of their member species, a "metacommunity" results (Karieva and Wennergren 1995; Holt 1997).

Source-sink and mainland-island metapopulations

Source-sink and mainland-island structures result when local populations differ markedly in their relative vulnerability to extinction. The regional dynamics of these structures differ qualitatively from the dynamics of metapopulations. Source-sink and mainland-island systems are distinguished by the causes of local extinction. Size is often the primary difference between mainland and island habitats. Small size alone may predispose island populations to recurrent extinctions. In contrast, sink habitats differ qualitatively from sources in ways that reduce rates of survival and reproduction. Sink habitats generally cannot support populations in the absence of immigrants from sources (Pulliam 1988; Pulliam and Danielson 1991). Contaminants are unlikely to reduce mainland habitats to islands, but they can degrade source habitats into sinks.

Patchy populations

Patchy populations are characterized by high rates of inter-patch migration. In patchy populations, migration exerts a stronger effect on local dynamics than does local productivity. Patchy populations may bear superficial resemblance to metapopulations, but the difference between them can be identified by examining organism life histories. When an average individual resides in more than 1 patch during its lifetime, migration is frequent enough to create a single patchy population (Harrison 1991).

Although the habitat patches represented in Figure 12-1 are discrete, the categories of patch structure are not. A continuum of structures exists, from isolated populations to patchy populations. Between these extremes lie a variety of metapopulations with lesser or greater rates of inter-patch migration. Similarly, variance in patch size or productivity spans a continuum, from metapopulations of essentially equivalent sub-populations to systems whose dynamics are dominated by a few mainland or source populations.

Continuously varying environments

Contaminant effects on vertebrate populations are mediated by the distributions of organisms, resources, and contaminants. Each of these distributions is heterogeneous in most systems, even within a single habitat type (Levin 1992; Bell et al. 1993; Clifford et al. 1995). In many cases, these distributions are continuous at local and/or regional scales. In addition, vertebrate dispersal is a continuous process, and it is often best studied as such. Unlike patchy systems, in which inter-patch dispersal can be summarized as a rate, continuous systems require an explicit description of dispersal behavior. The nature of this behavior determines whether vertebrate distributions reflect, amplify, or blur resource and contaminant heterogeneity.

Continuous variation affects even patchily distributed populations. Although most metapopulation research to date has ignored effects of heterogeneity in the inter-patch "matrix," matrix structure is potentially important (Wiens 1996, 1997). Conditions outside habitat boundaries may influence dispersal rates, which in turn may strongly affect population processes. Hence, some authors call for an integration of metapopulation approaches with landscape ecology (Wiens 1997).

Modeling Spatial Structure

Isolated populations
In the absence of inter-patch migration, processes that occur within habitat patches govern population responses to contaminants. These responses can be studied using conventional non-spatial population models with contaminant effects on individual survival and fecundity. Individual effects are integrated to the population level by using simple models or structured models that contain contaminant effects specific to different ages, sexes, or life stages (See Chapter 7).

Metapopulations
The original metapopulation concept was a theoretical construct, and the development of metapopulation theory is far ahead of laboratory and field study of metapopulations (McCullough 1996). Many models have been developed to study diverse factors that potentially influence metapopulation dynamics, although few address contaminants explicitly. Metapopulation models can be sorted into 4 categories, according to the amount of detail they include about local population dynamics and patch characteristics:

1) Patch-occupancy models assume that patches are equivalent, and these models ignore characteristics of individual patches and local population dynamics.
2) Incidence-function models ignore local population dynamics, but include detail about some patch properties, e.g., patch location or area.
3) Structured metapopulation models describe population processes as a function of characteristics of local populations or patches, e.g., age or size.
4) Simulation models explicitly describe local dynamics, as determined by patch characteristics.

In the rest of this section, we review these 4 kinds of metapopulation models and their relevance to contaminant effects on vertebrate metapopulations.

Patch-occupancy metapopulation models

Patch-occupancy models are the oldest and simplest metapopulation models, dating to Levins' (1969) original formulation. In these models, patches are either occupied or vacant, and different patches are assumed to be equivalent in all other ways. Patch-occupancy models track the total number of occupied patches in a system, as determined by the rates of colonization and extinction that are identical for all patches. This leads to the following model:

$$\frac{dP}{dt} = cP(1-P) - eP \qquad \text{(Equation 12-1)},$$

in which P represents the fraction of occupied patches ($0 \leq P \leq 1$), c is related to the rate at which empty patches are colonized, and e is related to the rate at which occupied patches go extinct. Gotelli (1998) provides a clear derivation of this and related patch-occupancy models.

The model produces a steady-state fraction of occupied patches, $\hat{P}$, in terms of the colonization and extinction parameters:

$$\hat{P} = 1 - \frac{e}{c} \qquad \text{(Equation 12-2)}.$$

In this equilibrium, the status of individual patches alternates between states of occupancy and vacancy, but the overall number of occupied patches does not change.

Because extinction is the only impact in patch-occupancy models, these models are relevant to systems with severe local contamination and cannot evaluate sublethal effects of contaminants. In addition, a system should satisfy the following 4 conditions before patch-occupancy models are considered (Hanski 1997):

- Suitable habitat occurs in discrete patches, and most individual organisms spend their entire lives in their natal patch. If most organisms disperse to other patches, the system functions as a single patchy population.
- All local populations are vulnerable to extinction.
- All habitat patches may be colonized if vacant. Patches too distant to be recolonized should be treated as isolated. If contaminants prevent recolonization for some time, Equation 12-1 may not be appropriate.

- Dynamics of local populations are not all synchronous. Otherwise, the system is a single patchy population, or patches are replicates of each other and do not function as a metapopulation.

Incidence-function metapopulation models

Incidence-function models contain considerably more realism than do simple patch-occupancy models. Incidence-function models predict occupancy of individual patches, as affected by patch characteristics such as size and location. Two approaches have been developed: quantitative incidence-function models (Hanski 1994) and transition incidence-function models (Sjögren-Gulve and Ray 1996).

Quantitative incidence-function models predict patch occupancy resulting from the opposing processes of extinction and colonization, which in turn are determined by patch area and inter-patch distances (Hanski 1994, 1997). Hanski's model is 1 of the few models that have been solved analytically. Because the parameters of Hanski's model can be fitted using relatively simple empirical occupancy data, the model potentially has broad application to field metapopulations. The model predicts that the fraction of occupied patches may obtain multiple equilibria (Hanski et al. 1995). Empirical evidence for this result has been found for the Glanville fritillary (*Melitaea cinxia*) (Hanski, Moilanan, Pakkala et al. 1996). One limitation of the quantitative incidence-function model relative to environmental contaminants is that it assumes the number of occupied patches is at equilibrium. This assumption may be valid for systems in which the extent of contamination has not changed over long periods, but it is problematic for systems affected by habitat degradation (Thomas 1994b), including degradation resulting from contaminants. Hence, the model may have limited application to metapopulations in environments with recent changes in contaminant availability, e.g., systems with newly contaminated sites or sites undergoing remediation.

In contrast, transition incidence-function models do not assume metapopulation equilibrium and can be applied to systems affected by changes in contaminants or other stressors. Unfortunately, this flexibility is obtained with a loss of analytical tractability. Transition incidence-function models apply logistic regression to fit site-specific colonization and extinction probabilities to presence-absence data. These probabilities are determined by habitat characteristics that may include conditions in the surrounding non-habitat "matrix." After fitting the model, colonization and extinction probabilities are combined to predict transitions in patch occupancy. These stochastic simulations can be compared with data on regional dynamics and changes in the occupancy status of individual patches (Sjögren-Gulve and Ray 1996). Because the use of transition incidence-function models considers multiple factors, including stressors that may change in time, this approach holds much promise for the study of contaminant effects on vertebrate metapopulations.

Structured metapopulation models

Structured metapopulation models treat population processes as functions of local population or habitat characteristics (Hastings 1991; Day and Possingham 1995; Gyllenberg et al. 1997). Gyllenberg et al. (1997) provide a general description of the mathematical formalism involved in structured metapopulation models. By considering population-specific or habitat-specific effects on local population dynamics, structured metapopulation models are more realistic than patch occupancy models or incidence-function models, in the same way that age-structured population models are more realistic than models that treat all individuals identically. Structured population models also allow consideration of contaminant effects that are less severe than local extinction, just as demographic models can allow consideration of sublethal effects of contaminants on individuals. The major advantage of structured models is that their greater realism does not necessarily sacrifice the generality of analytical solutions. The primary limitation of structured models is that they usually preclude consideration of explicit spatial arrangement of habitat patches (Gyllenberg et al. 1997). The main practical disadvantage of using structured models is that they contain many parameters whose estimation may require more data than are available for most field systems.

Numerical metapopulation models

The fourth class of metapopulation models describes individual patches and local population dynamics explicitly. Most models in this category also include some detail on the location and spatial arrangement of habitat patches (e.g., Wu et al. 1993). The behavior of these models is determined using numerical simulations; their complexity precludes finding analytical solutions. With simulations, there is no intrinsic limit on the realism that can be included in a model. Nevertheless, realism requires empirical information to specify the functional form and parameter values in a model. Consequently, most simulation models emphasize detail on structural or biological characteristics for which data or managerial control exist, e.g., the number and configuration of habitat patches (e.g., Lindenmayer and Possingham 1996).

Mainland-island and source-sink models

Ordinary metapopulation models are not appropriate when populations in 1 or more habitat patches have a substantially lower risk of extinction than do other populations. This problem was recognized first in cases where a large patch, or "mainland," functioned as a source of colonists that reestablish populations on nearby "island" habitats (Harrison 1991). Mainland-island models assume that persistence of the mainland population is guaranteed. They focus on occupancy of island habitats, as determined by rates of extinction and colonization, distance from the mainland, and other factors. Colonization may result from migrants traveling directly from the mainland, or via stepping-stone dispersal, in which patch occupancy radiates outward from the mainland as colonists from island populations establish populations on other islands (Harrison et al. 1988).

Source-sink models were developed to study cases in which variance in extinction risk is due to differences in population productivity rather than habitat size. Source-sink models originated with Pulliam (1988), who analyzed a single source-sink pair connected by migration. Subsequently, source-sink models have been developed for increasingly complex habitat configurations.

Patchy populations

Models of patchy populations describe inter-patch migration in greater detail than do models for most other patchy environments. In their simplest form, patchy population models assume complete mixing of progeny among habitat patches (Chesson 1981; Roughgarden and Iwasa 1986). In these models, variation in productivity among patches is homogenized across the pool of dispersing juveniles, and effects of patch location are ignored. Models with more-detailed descriptions of dispersal may include advective flow (Roughgarden et al. 1988; Possingham and Roughgarden 1990), other factors that determine where migrants disperse, and factors that dictate where migrants settle (Alexander and Roughgarden 1996). These models allow one to study effects of patch location on local dynamics. Although most patchy population models focus on the movement of organisms, they could be adapted to study effects of contaminant transport on population processes. This approach would be more appropriate for terrestrial vertebrates, which actively disperse but may be affected by passively transported contaminants.

Continuously varying environments

Models of populations in continuous environments require explicit descriptions of migration behavior. Unlike models of patchy environments, which generally summarize migration as the number or probability of animals moving between patches, models of continuous environments describe both distance and direction of animal movement. This information is important because details of dispersal determine whether population densities blur, amplify, or simply reflect environmental heterogeneity (McLaughlin and Roughgarden 1993). Three modeling approaches that represent different tradeoffs between generality and realism have been developed for continuous environments.

Partial differential equation (PDE) models provide the greatest generality of all the models discussed herein, but they include the least detail about any particular system. These models treat space and time as continua, with space being either uniform or heterogeneous. The models include 1, 2, or 3 spatial dimensions in order to represent environments with linear (e.g., riverbank), planar (e.g., grassland), or vertical (e.g., water column) features. Partial differential equation models describe animal movement by using various forms of diffusion processes. Simple diffusion models (Equation 12-3) describe organisms as moving randomly and over time scales that are short enough to ignore reproduction and mortality.

$$\frac{\partial u(x,y,t)}{\partial t} = D\left(\frac{\partial^2 u}{\partial x^2} + \frac{\partial^2 u}{\partial y^2}\right) \qquad \text{(Equation 12-3)}.$$

In Equation 12-3, $u(x,y,t)$ is population density at location (x,y) and time t, and D is the diffusion coefficient or movement rate. Despite its simplicity, this model has described movement well in various animal populations (Okubo 1980; Karieva 1983), particularly in homogeneous environments.

Equation 12-3 can be modified in at least 6 ways. (See Holmes et al. [1994] for a comprehensive review, and Turchin [1997] for a review and a primer on solving diffusion equations.) First, advection terms (e.g., $v_x \frac{\partial u}{\partial x}$, where v_x is the drift velocity in direction x) can be added to study the effects of wind, water currents, and other factors that bias animal movement. Second, individuals often tend to continue moving in a particular direction, rather than bouncing erratically in the Brownian motion described by the diffusion equation. This tendency can be represented as a correlated random walk, leading to the telegraph equation (Holmes 1993). Third, individuals may avoid or aggregate with other members of the population, as modeled by adding terms for biased movement, $\frac{\partial}{\partial x}\left(ku\frac{\partial u}{\partial x}\right)$, where k is the tendency to move toward ($k < 0$) or away from ($k > 0$) conspecifics. Fourth, reproduction and mortality may affect local population density and total population size. Equation 12-3 becomes a reaction-diffusion model when demographic processes are added to it. Originally developed for spatially structured chemical reactions, these models have many applications to animal populations (reviewed in McMurtrie 1978; Okubo 1980; Murray 1989). Fifth, environmental conditions may vary in space in ways that affect animal movement, reproduction, and mortality. Environmental heterogeneity can be modeled using spatially variable diffusion coefficients (Holmes et al. 1994) or demographic parameters (McLaughlin and Roughgarden 1991). Finally, interactions with other species can be included (reviewed in McLaughlin and Roughgarden 1993).

While these factors strongly affect most vertebrate populations, they markedly increase the difficulty of solving PDEs. Indeed, relatively few PDEs can be solved analytically (Hall 1988). Fortunately, approximation methods are useful in many cases (e.g., Kevorkian and Cole 1980), and algorithms for obtaining numerical solutions are well-developed for the remainder of these cases (Smith 1998).

The second approach to modeling populations in continuous environments uses cellular automata models, which depict the environment as a grid of habitat cells. The models simulate individual movements across the artificial landscapes, which may occur over long distances or be restricted to adjacent cells (e.g., Phipps 1991; Molofsky 1994; McCauley et al. 1996). Cellular automata models can include a wide variety of population processes, e.g., density dependence at local or regional scales; stochastic determinants of survival, reproduction, or dispersal; and interspecific interactions. The environment can be uniform or it can contain spatial patterns of diverse shapes and scales. Hence, these models can be used to study effects of diverse contaminant distributions on various population processes. The primary limitation of this approach is that fitting the models requires data on distances and directions of animal migration, which often are not available.

The third approach studies population processes in realistic landscapes, using Geographical Information Systems (GIS) both to map landscape patterns and to track the fates of individual animals. To date most of this type of research has focused on migration, including the effects of landscape pattern on migration success and the relative importance of different habitat areas as dispersal corridors (e.g., Schumacher 1995; Boone and Hunter 1996; Gustafson and Gardner 1996; Keitt et al. 1997; Richards 1998). The landscape approach develops a habitat base map and then simulates the dispersal of individual animals across that map. Dispersal distances and habitat-specific mortality rates are selected to represent the species of interest. If dispersal data are available or can be estimated, GIS-based models could be used to evaluate effects of particular contaminant distributions, and to predict the effectiveness of various remediation strategies. Little work has been done on population dynamics in realistic landscapes (N. Schumaker, personal communication, but see McKelvey et al. 1993).

Results

Persistence of metapopulations

Habitat and contamination thresholds

Contaminants can cause entire metapopulations to go extinct, just as they can drive individual populations to extinction. Metapopulation extinction is likely when the extent of contaminant exposure exceeds a threshold number of habitat patches (Karieva and Wennergren 1995; Maurer and Holt 1996). Beyond this threshold, populations in all patches go extinct, even though some are never exposed. Maurer and Holt (1996) derived a simple rule of thumb for predicting the extinction threshold for classical metapopulations. Using a simple modification of Equation 12-1, they found that metapopulation extinction occurs when the number of contaminated patches exceeds the number of patches that were occupied before contamination. Attempts to apply this result to field systems should incorporate the recognition that it is an equilibrium solution. The rule of thumb is not appropriate for species whose distributions are changing for reasons unrelated to contamination, e.g., invasive species. Extinction thresholds have 2 implications for risk management. First, unoccupied habitats should be included in management plans. Even if a patch of suitable habitat currently does not support a population, it may play an important role in the future persistence of the species. Second, data on species distributions prior to contaminant exposure may be necessary to predict metapopulation responses after exposure. Other metapopulation models produce similar results for various forms of habitat loss, suggesting that extinction thresholds may be robust to model structure.

Similar results led to the concept of a minimum viable metapopulation, which is defined as the smallest number of suitable habitat patches necessary for metapopulation persistence (Hanski, Moilanen, Gyllenberg 1996). As expected, this minimum is lower for organisms with strong dispersal ability than it is for organisms with weak dispersal ability. Gurney and Nisbet (1978) (see also Nisbet and Gurney 1982) estimated the minimal viable population for a stochastic version of Equation 12-1 applied to a system of M patches. They found that the metapopulation would persist at least 100 times as long as any individual population when the following condition is met:

$$\hat{P}\sqrt{M} \geq 3 \qquad \text{(Equation 12-4)},$$

where $\hat{P}$ is the steady-state fraction of occupied patches. Hence, if a species occupies most available habitats, at least 10 to 20 patches are necessary for long-term persistence. This minimum requirement for long-term persistence will be much larger when habitats are far apart, organisms disperse poorly, environmental stochasticity is high, and local dynamics are spatially correlated (Thomas and Hanski 1997).

This result is supported by field data on butterfly metapopulations. Thomas (1994a) found that his study species occupied areas with at least 15 to 20 habitat patches but was virtually absent from areas with fewer than 10 patches. Hanski, Moilanen, Pakkala et al. (1996) also found similar extinction thresholds for *Melitaea cinxia* butterflies. Hanski et al. extended the work of Gurney and Nisbet (1978) by using numerical simulations of a system of diverse patch areas and fitting their model with butterfly field data.

Threshold areas necessary for persistence also have been found with models of continuous environments. When individuals that disperse beyond habitat boundaries perish, populations cannot persist in habitats smaller than a critical area (Skellam 1951; McMurtrie 1978). Similar results have been obtained for models of species that interact in continuous environments. Unless the habitat area exceeds a threshold, weaker competitors are excluded (Pacala and Roughgarden 1982) and predators drive prey to extinction (McMurtrie 1978; Li 1989). Although area thresholds in vertebrate populations are difficult to demonstrate empirically, there is evidence for effects of area on extinction risk. Newmark (1987, 1995) found a strong negative relationship between the sizes of national parks in western North America and the number of mammal extinctions within the parks.

Toxicity affects persistence

Not surprisingly, the risk of metapopulation extinction increases with contaminant toxicity. Dispersal modulates the relationship between toxicity and persistence because metapopulation persistence is a regional process whose dynamics are governed largely by dispersal patterns and rates. When contaminants are toxic enough to cause extinctions in exposed populations, they also will cause metapopulation extinctions when dispersal is high (Maurer and Holt 1996; Spromberg et al. 1998). Contaminants with less severe or sublethal effects may not cause extinctions, but they are likely to reduce the number of occupied habitats. This reduction results from a decrease in the number of dispersing individuals, which is caused by contaminant-induced decreases in survival or reproduction among exposed individuals.

Asynchrony and metapopulation persistence

The risk of metapopulation extinction increases substantially when local population dynamics become synchronous. There is little safety in numbers (of local populations) if all patches are contaminated at the same time. The importance of asynchrony to metapopulation persistence pertains to both single species (Levins 1969; Thomas and Hanski 1997) and interacting species (McLaughlin and Roughgarden 1993; but see Ives and Settle 1997).

Four main processes can cause synchrony in local dynamics. First, high rates of inter-patch migration can produce spatial correlation among populations in different patches. Second, local dynamics can be entrained by changes in land use and habitat structure, which often are correlated across large areas. Third, regional climatic conditions can synchronize population fluctuations over large geographic regions (Pollard 1991; Pollard and Yates 1993), and extreme climatic events can cause simultaneous extinctions of many local populations (Ehrlich et al. 1980; Harrison et al. 1988). Fourth, anthropogenic factors can cause synchrony. Links between population synchrony and human activities can be indirect, through broadly distributed contaminants, or direct, through simultaneous planting of crops or control of pests (Levins 1969; Ives and Settle 1997).

Three approaches can assure asynchrony in population fluctuations, thereby preventing extinction at the metapopulation level. First, maintaining a sufficiently large system of patches may provide differing regional climates or a vanishingly small risk of simultaneous stochastic declines in all populations. The area required may be prohibitively large for some practical management situations. Second, gradients in environmental conditions allow metapopulations to spread the risk across different habitat patches, thereby preventing simultaneous extinctions of all local populations. For this reason, static measures of habitat quality can be misleading, e.g., areas that appear to be poor habitat may provide essential refugia during extreme environmental events (Ehrlich et al. 1975; Weiss et al. 1988; Kindvall 1994). Third, strict protection of some habitat patches from contamination can provide asynchrony when contaminant effects drive population dynamics. The number of protected patches required increases with migration rate; however contaminants in 1 patch can cause synchronous fluctuations in other patches when both toxicity and migration are high (Spromberg et al. 1998).

Dispersal and persistence

Recognition of the importance of migration to persistence and other metapopulation processes predates the metapopulation concept (Huffaker 1958; Pimental et al. 1963). Research in recent decades has both strengthened this recognition and revealed how little is known about the details of migration (Halley et al. 1996; Ims and Yoccoz 1997). Consistently, model analyses have concluded that both metapopulation persistence and the fraction of occupied patches increase with migration rate (Fahrig and Merriam 1985; Gutierrez and Harrison 1996; Stacey et al. 1997). These results have been supported by empirical studies that included work on insects (Schoener and Spiller 1987; Thomas 1994a; Thomas and Hanski 1997; Morrison 1998), amphibians (Gill 1978; Sjögren-Gulve 1994; Hecnar and M'Closkey 1996), and small mammals (Middleton and Merriam 1981; Henderson et al. 1985; Hansson 1991; Smith and Gilpin 1997). Migration is essential to many metapopulations because it either maintains colonization rates high enough to offset local extinctions or prevents extinction in the first place via the "rescue effect" (Brown and Kodrick-Brown 1977). In many terrestrial vertebrates, most dispersal is done by juveniles, and the success rate of dispersing juveniles can be critical to metapopulation persistence (Breitenmoser et al. 1992; McKelvey et al. 1993; Haight et al. 1998).

Contaminants can reduce persistence by decreasing migration rates in 2 ways:
- Individuals suffering direct exposure before, during, or after migration are less likely to reproduce in their new patches.
- Contaminant-induced declines in local population productivity effectively reduce the number of potential migrants.

Metacommunities

There is little consensus about effects of spatial structure on ecological communities, even though these effects have been studied almost as long as have effects on populations. This lack of consensus occurs is in part because communities themselves are not as well understood as populations (Oksanen 1991). Nevertheless, several themes have emerged, as reviewed by Holt (1993, 1997).

A metacommunity can be defined as a simple multispecies analog of a metapopulation, i.e., a set of local communities in different locations, linked by dispersal of 1 or more of their constituent members (Gilpin and Hanski 1991). The composition of metacommunities is likely to be biased toward species that are either habitat specialists or generalists (Holt 1997). Specialists are predicted to be common among common habitat types. Specialists present in rare habitats are expected to be unusually resistant to extinction or very good at dispersal. Generalist species may be represented disproportionately in rare habitats, because of spillover from common habitats, and in all habitats because of their general tolerances.

Interactions among species complicate the relationship between spatial structure and community composition, which is evident in the effect of migration on persistence. For individual non-interacting species, increased migration rates enhance persistence. When species interactions are locally unstable, e.g., when predators eliminate prey or dominant species exclude their weaker competitors, persistence is enhanced by intermediate rates of migration (McLaughlin and Roughgarden 1993). Prey or subdominant competitors that disperse poorly do not colonize vacant patches before being driven extinct by predators or dominant competitors. Predators or dominant competitors that migrate rapidly produce the same result. Rapidly dispersing predators or dominants effectively synchronize population dynamics in different patches, causing nearly simultaneous extinctions if they cannot coexist locally with their prey or competitors. Moderate rates of migration preserve asynchrony and provide favorable conditions in some patches for all species.

Contaminant effects in metacommunities may be difficult to detect in time to prevent large numbers of extinctions. This concern is due to work by Tilman (1994), who analyzed a metacommunity model containing species in a competitive hierarchy. The model assumed that dispersal ability and competitive strength are inversely related. All species persisted in the un-affected system. When habitat destruction was added, however, competitively dominant species experienced increased risk of extinction. Extinctions accelerated with increasing amounts of habitat loss, but only after a time lag. Hence, when contaminants render habitat unsuitable, their effects at a regional level may be both severe and undetectable. Similar results were obtained when a second tropic level was added to the model (Karieva and Wennergren 1995). The generality of this "extinction debt" is unknown, however, for it rests on the untested assumption that there is a negative correlation between dispersal ability and competitive dominance.

Specific predictions about contaminant effects on metacommunities require information about the structure of those communities, including the strength of interspecific interactions. Data on interaction strengths are few and are difficult to obtain, however (Mills et al. 1993). In Chapter 6 of this book, Holmes and Karieva point out that further progress on understanding contaminant effects at a community level may require direct empirical study of the communities of interest.

Source-sink systems

Persistence in source-sink systems

Source-sink systems are maintained by productivity of source populations and therefore are sensitive to any threat to sources, including threats from contaminants. If contamination of source habitats is severe, extinction of all local populations may occur. If contamination is not severe enough to eliminate source populations, sink populations may go extinct because of a reduction in the number of individuals migrating from sources. Predicting the result of contamination in a particular system requires habitat-specific data on rates of survival, reproduction, and migration.

Contamination in sink habitats may have little effect on regional persistence, i.e., it may cause extinctions of sink populations but not affect sources. Hence, contamination of sinks may reduce the distribution of occupied habitats, without affecting the regional persistence of the species. Contamination of sinks can cause regional extinctions, however, if migration rates are potentially greater than the surplus productivity of source habitats (Maurer and Holt 1996). High toxicity combined with high migration reduces population densities in sinks low enough that the sinks effectively drain sources. For this reason, some authors recommend isolating sinks (Spromberg et al. 1998) or sources (Cantrell and Cosner 1993).

Efforts to manage contaminated systems should maintain a substantial safety margin to accommodate stochastic factors. The importance of stochasticity was demonstrated by Gaona et al. (1998) in an analysis of population processes in the Iberian lynx *(Lynx pardinus)*, the most endangered feline in the world. The lynx persists in a set of source and sink habitats, with sources in Doñana National Park, Spain. Gaona et al. (1998) developed a source-sink model with age structure and parameterized it using field data that had been collected over a decade. A deterministic version of the model predicted metapopulation persistence for at least 100 years. Regional extinction occurred, however, when stochasticity was included in demographic (22% of simulations) or environmental (38% of simulations) parameters. These results are consistent with analysis of a general metapopulation model in which demographic and environmental stochasticity substantially increased risk of regional extinction (Mangel and Tier 1993).

Detecting effects of contamination

Effects of habitat degradation resulting from contaminants or other factors may be difficult to detect using field data. This was shown by Doak (1995), who developed a model of source and sink habitats coupled by migration. Population declines occurred after enough source habitat was degraded into sink habitat. These declines occurred sooner in populations with high migration rates than in populations with relatively sedentary animals. In all populations, Doak found a time lag between when population declines began and when they would be detected, even if census data are unrealistically accurate. This lag exceeded 10 years in a version of the model fitted to the Yellowstone population of grizzly bears *(Ursus arctos)*. Hence, traditional monitoring programs using census data may fail to detect contaminant-induced habitat degradation until it is extensive and populations have declined to dangerously low levels. A more effective approach to monitoring populations focuses on demographic rates rather than on census data (Pease and Mattson, 1999).

Patchy populations

Patchy populations exhibit low rates of local extinction, and their dynamics differ little between local and regional time scales. Dispersal patterns determine local population dynamics; i.e., the kinds of dynamics observed at a particular site depend on the location of that site within the regional dispersal pattern (Roughgarden et al. 1988). When dispersal is passive, as for many sessile marine invertebrates with pelagic larvae, knowledge of physical transport processes is essential to understanding local population dynamics and regional patterns of abundance (Alexander and Roughgarden 1996; Caley et al. 1996). For organisms in which dispersal and habitat selection are active processes, population dynamics and distributions are products of both behavior and landscape pattern (Lima and Zollner 1996). Unfortunately, few data on dispersal behavior in such organisms are available (Ims and Yoccoz 1997); most work on patchy populations has been done using plants, insects, and marine invertebrates, so it will receive limited discussion here.

When localized contamination occurs in patchy populations, many individuals may be exposed, but regional extinction is unlikely unless contaminants are regionally distributed. A corollary to this conclusion is that local censuses may be inadequate for detecting contaminant effects in patchy populations, since the number of organisms found at an exposed site may depend on the number produced at unexposed sites and vice versa.

Effects of structure on population dynamics

Spatial structure can influence population dynamics markedly, making the effects of contaminants more difficult to detect. Habitat fragmentation can cause population dynamics to become more volatile (Karieva 1987; Burkey and Stenseth 1994) or more stable (Wolff 1980). Effects of fragmentation on population dynamics often are indirect and are mediated by direct effects on predator populations (Wolff 1980; Karieva 1987). Because specialist predators with delayed numerical responses can increase prey fluctuations (Hanski et al. 1993; Turchin and Hanski 1997), loss of these predators resulting from fragmentation would stabilize prey dynamics. The reverse would follow from the loss of generalist predators. A similar diversity of effects on population dynamics has been found for continuous habitats (Karieva 1990; Karieva and Wennergren 1995). Although continuous spatial structure may either simplify or complicate dynamics, it generally enhances the persistence of competing species (Gopalsamy 1977; Shigesada 1984; Namba 1989) or predators and prey (Comins and Blatt 1974; Sabelis and Diekmann 1988; Li 1989; McLaughlin and Roughgarden 1991; Comins et al. 1992).

Spatial structure also can cause lengthy transient behavior (Hastings and Higgins 1994). In cases in which thousands of years may pass before transients converge to predictable dynamics, analysis of equilibrium behavior becomes meaningless for ecological time scales. Long and complex transient behavior would render the tasks of monitoring and model-fitting extremely difficult (Karieva and Wennergren 1995). It remains to be seen, however, whether spatial structure in general leads to lengthy transients.

Contaminant effects on patch dynamics

In patchy habitats, inter-patch migration can alter local population dynamics qualitatively. Migration causes a shift in population dynamics from isolated populations with independent trajectories to a regional system of interacting local populations. This shift in dynamical scale from local to regional levels is 1 of the main reasons that spatial structure must be considered in order to understand contaminant effects in some systems. The extent of this shift and the importance of considering spatial structure are proportional to migration rates.

These effects were studied using a numerical model (Spromberg et al. 1998). The model contained 3 local populations connected by dispersal in linear (Figure 12-2) or circular arrangements. Local population growth in the model was logistic, with a lower threshold (minimum viable population size) below which local population growth was negative. Migration rate from 1 patch to another was proportional to population density in the first patch and inversely proportional to the distance between the patches. In each simulation, 1 patch was dosed with the contaminant. Contaminant concentration either remained constant throughout the simulation or decayed exponentially. Exposure of individual organisms within contaminated patches was described as a Poisson distribution because neither organisms nor contaminants are distributed uniformly in nature. Contaminant exposure was the only stochastic factor in the model.

Simulation results were consistent with work on other empirical and theoretical metapopulations, discussed previously. Spromberg et al. (1998) summarized their results in 4 main conclusions:

1) When contaminants were applied to 1 patch, population dynamics in all patches were affected (Figure 12-3).
2) Inter-patch distance affected population dynamics. Dynamics in uncontaminated patches were more strongly affected and dynamics of all patches were more tightly coupled when inter-patch distances were short.

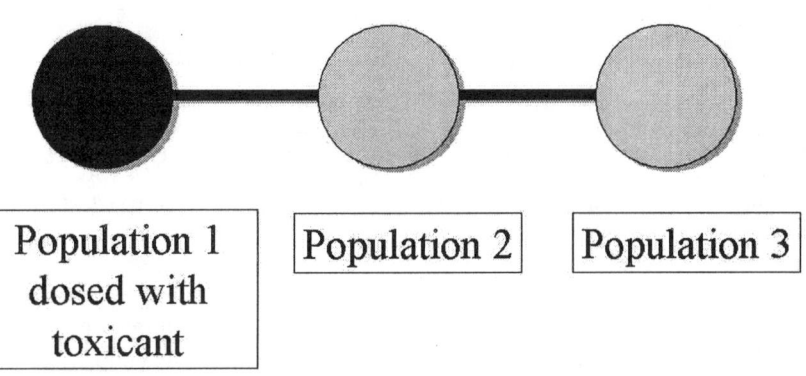

Figure 12-2 Schematic diagram of patch structure in toxicant metapopulation model. Linear arrangement, with contaminated patch at left end.

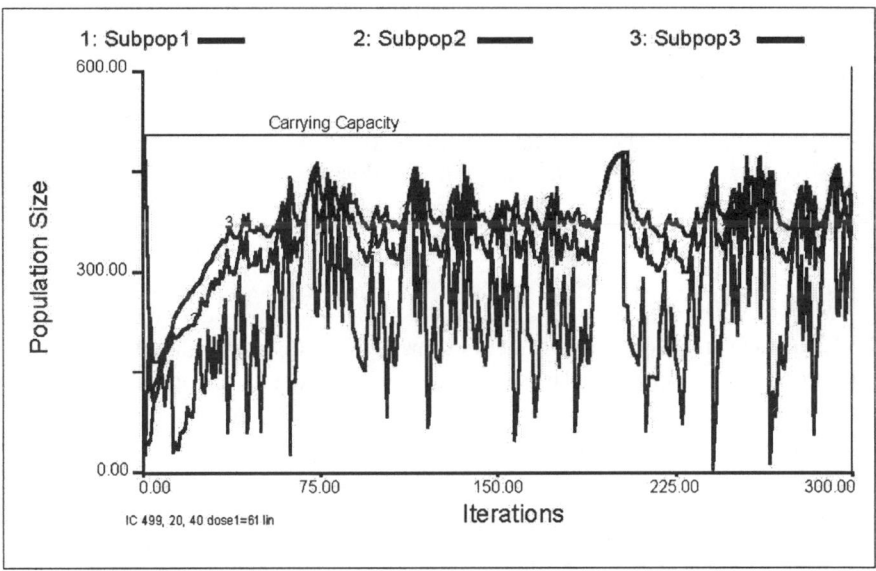

Figure 12-3 Dynamics of the 3-patch linear model analyzed by Spromberg et al. 1998. Numbers of organisms in each of the 3 patches are plotted against time. Initial population sizes were 499, 20, and 40 for patches 1, 2, and 3, respectively. Patch 1 received contaminant dosage at 50% lethality; organism exposure to the contaminant in this patch was the only stochastic factor in the model. The contaminant induced marked population fluctuations in both dosed and non-dosed patches.

Table 12-1 Distribution of outcomes in 3-patch linear toxicant metapopulation model, with toxicant degradation[1]

Fate of populations in patches 1, 2, 3	End patch (1) contaminated % of simulations	Center patch (2) contaminated % of simulations
2 at MVP; 1,3 extinct	50	0
1,2 at MVP; 3 extinct	26	0
1,2,3 at K	14	16
1,2,3 at MVP	10	28
1,3 at MVP; 2 extinct	0	56

Source: Spromberg et al. (1998)

[1] In the absence of contaminants, populations in all patches reached carrying capacity (K). With contaminants, this outcome occurred infrequently, regardless of which patch was contaminated. Outcomes of contaminated and uncontaminated patches alike were affected qualitatively by contaminants in at least 84% of all simulations. The location of the contaminated patch affected the distribution of outcomes. MVP is a viability threshold, below which the population declines toward extinction.

3) Degrading contaminants caused several qualitatively different outcomes. These outcomes occurred with different frequencies, as listed in Table 12-1.
4) Location of the contaminated patch affected both the outcomes and their relative frequencies (Table 12-1).

These results led Spromberg et al. (1998) to develop the "action at a distance" hypothesis, which states that contaminants can affect unexposed patches by altering patterns of migration. This impact occurs without direct exposure to the contaminant of any organism in the unexposed patch. The importance of action at a distance increases with migration rate. A logical corollary to this hypothesis is that true reference sites cannot exist where there is migration. When migration occurs, the dynamics and even the persistence of populations in uncontaminated "reference" sites may be affected by contaminants in other sites. The action at a distance hypothesis was confirmed recently in laboratory experiments using *Tribolium* beetles (Macovsky 1999).

Landscape context

Most empirical and theoretical studies of spatially structured populations ignore structure at a landscape level. The simplest studies assume that habitats are uniform, surrounded by an unsuitable "matrix" over which migration may or may not occur. More-detailed analyses consider that structure may vary within a habitat or that patches of habitat may differ, but even these analyses generally ignore structure in the matrix. Because matrix structure likely influences migration rates and patterns, and because migration consistently emerges as a critical process affecting spatially structured populations, effects of matrix structure should be examined more carefully (Wiens 1996). Preliminary analyses suggest that landscape structure may warrant special attention when the distribution of suitable habitat patches is intermediate between connected and isolated (Wiens 1997).

Unfortunately, little work has been conducted at the interface between population and landscape ecology. Two recently published examples offer conflicting conclusions about the importance of landscape context. Sjögren-Gulve and Ray (1996) analyzed habitat-occupancy data in a pool frog metapopulation along the Baltic coast of Sweden. They found that conditions in matrix habitat were needed to explain extinctions in ponds in the southern portion of the study area. Forestry and associated ditching surrounding southern ponds reduced migration success and led to extinctions in southern ponds. Because conditions in the ponds themselves had little to do with the extinctions, Sjögren-Gulve and Ray described this as a "matrix effect." Similar effects of land-use changes on migrating organisms have been found in other amphibians (Fahrig et al. 1995; Gibbs 1998; Vos and Chardon 1998) and are likely to affect many populations that migrate between 2 or more different kinds of habitats. These results support concerns about the perils of adopting a "reserve mentality," in which suitable habitat is designated and protected, while degradation of surrounding area is ignored (Brussard et al. 1992; Wiens 1996). For populations that depend on migration, such a strategy may lead to "matrix effects" and subsequent extinctions in protected habitats.

The second example offers a more optimistic conclusion. Moilanen and Hanski (1998) used GIS to enhance the quantitative incidence-function model with data on patch quality and landscape structure. Then they predicted patch occupancy in a metapopulation of the Glanville fritillary butterfly (*Melitaea cinxia*) and compared these predictions with those of the original model. The result was surprising in that the additional spatial information did not improve model fit significantly. Their conclusion was hopeful, however, in that patch area and isolation in the basic incidence-function model perhaps are sufficient to describe metapopulation processes and detailed depictions of landscape structure may not be necessary. We advise caution, because several properties of the *M. cinxia* system allow it to be well described by the basic incidence-function model (Hanski et al. 1994). Tests on other systems are needed to assess the generality of the results of Moilanen and Hanski (1998).

Discussion

A consistent theme in the work reviewed in this chapter is that spatial structure markedly influences population processes. These influences complicate efforts to understand and manage effects of contaminants. For example, metapopulation models show that populations collapse when habitat loss exceeds a threshold, that thresholds may be difficult or impossible to determine empirically with precision, and that population declines resulting from habitat loss may not be detected until it is too late to reverse them (Gutierrez and Harrison 1996). Analogous results have been found for populations in environments with continuous habitats and in source-sink structures. In addition, spatial structure alone may increase the complexity of population dynamics, which can be difficult to characterize even without considering space (Ellner and Turchin 1995).

It also is clear that effects of spatial structure need more empirical study, particularly for vertebrates and for contaminated systems. Empirical study alone will not be adequate, however, because of the multiple outcomes possible in a given system and the lengthy datasets necessary to characterize complex dynamics (Ellner and Turchin 1995). Consequently, we suggest that integrating models with field study will be the most effective way to address contaminant effects in spatially structured environments. Furthermore, we suggest that the models used should contain an intermediate level of detail. Most simple patch-occupancy models ignore the environmental features that would inform contaminant management efforts, and fully structured models contain more parameters than can be estimated with available empirical data (Gyllenberg et al. 1997).

Integrating models and empirical study

We suggest that field study should measure most or all of the following 6 features, which also should be represented in companion models. We identified these features based on their consistent importance in diverse field and model systems.

Kind of spatial structure

First, the kind of structure must be identified. Because population responses to contaminants are likely to vary greatly among different structures, appropriate identification is essential. Categories described in this chapter or elsewhere (e.g., Harrison 1991, Harrison and Taylor 1997) should suffice. This step usually requires more than a map of suitable habitat—without information about migration rates, most systems in Figure 12-1 would appear deceptively similar.

Location of contaminant exposure

Contaminant locations must be determined, particularly in relation to source populations, centrally located patches, and important migration routes (Keitt et al. 1997).

Contaminant effects on local population productivity

Contaminant toxicity data are needed to determine effects at both local and regional levels. These data are needed to estimate 3 contaminant properties: 1) dose–response curves at the level of local populations, 2) severity and latency of responses to contaminant exposure, and 3) duration of contaminant persistence. Productivity data are essential in the study of source-sink systems.

Migration rates and patterns

Although existing migration data are adequate for few populations, migration both determines the kind of structure and strongly influences system dynamics. When habitat patches differ qualitatively, e.g., by size or relative location, migration data should be patch-specific.

Pre-exposure dynamics

If effects of contaminants on local population dynamics are of interest, then pre-exposure population data may be necessary. Pre-exposure dynamics may be difficult to determine for many species because at least 15 years of annual data are needed to fully characterize the dynamics of a population (Ellner and Turchin 1995). In some cases, it may suffice to collect enough data to estimate temporal population variability and then use contaminant effects on variability as a surrogate for effects on dynamics. Caution is advised when using this approach because estimates of variability often increase with the amount of data collected (Pimm and Redfearn 1988).

Matrix effects

Because populations within discrete habitats can be affected by conditions in surrounding areas (Sjögren-Gulve and Ray 1996), it may be necessary to consider these conditions. For this reason, contamination of unsuitable habitat may be important. When matrix effects affect migration only, they are included in migration data and need not be measured directly.

Directions for future research

We see 5 primary directions for future research:
1) There is a great need for field tests of contaminant effects using systems that represent a variety of spatial structures.
2) Impacts on population processes that are caused by lingering contaminant presence need to be considered.
3) Spatial models need to be adapted to study effects of multiple stressors.
4) Effects of landscape structure on population processes need further analysis, including impacts of contaminants located outside habitat boundaries.
5) Contaminant effects on multiple interacting species need more theoretical and empirical study. Multispecies systems represent a difficult challenge because they often require consideration of spatial structure and dynamics on several scales.

Field tests

The study of spatially structured populations has progressed much further in theory than it has in the field. While additional theoretical development is needed, especially for contaminant effects on vertebrates, empirical study is needed even more. One immediate practical need is to evaluate how various spatial structures mediate contaminant effects on vertebrate populations.

Of necessity, models make unrealistic assumptions. The utility of a given model depends on the importance of the factors it considers, relative to those it ignores. This judgment must be made empirically. At present, we lack the data required to make this evaluation about spatial structure for most systems. Hence, field study on effects of spatial structure on vertebrate populations should become a research priority. This work should be integrated with model analyses for the reasons discussed above and because many vertebrate population processes occur at spatial and temporal scales that exceed practical constraints of strictly empirical approaches.

Among the diverse kinds of empirical studies needed, 6 stand out as priorities. The greatest priority is to determine the kind of population structure in a given arrangement of habitats. Human activities have isolated or fragmented habitats of most species, but we rarely know which of those species become isolated populations, which species function as source-sink structures, and which species persist through metapopulation dynamics. A second and closely related study need is the measurement of rates and patterns of migration for various organisms and environments. A third need is for the testing of predicted thresholds in habitat area or number of patches necessary for persistence. Fourth, testing of the "action at a distance" hypothesis is needed. Such tests will help determine the importance to local population dynamics of migration relative to within-patch productivity. Fifth, the importance of landscape structure needs further study. Sixth, data are needed to evaluate relationships between competitive dominance and dispersal in metacommunities and to test predictions about potential extinction debts. Results from each of these 6 priorities for empirical research will contribute information that is necessary for effective management of contaminants in spatially structured environments. Management decisions that require this information include designating which habitats should be strictly protected, which habitats should be prioritized for remediation, and which contaminated sites should be fenced off.

Effects of time lags in contaminant degradation

Work on spatially structured populations that explicitly considers contaminants is badly needed. While there is now a small amount of literature on pesticide effects on insects (e.g., Levins 1969; Sherrat and Jepson 1993; Maurer and Holt 1996), work on vertebrates that directly addresses both contaminants and spatial habitat structure is limited. Although the extensive ecological literature provides some insight, contaminant effects differ qualitatively from general stochastic components of mortality, reproduction, and extinction that are considered in ecological research. Most metapopulation work assumes that vacant habitat can be colonized immediately, but a contaminated habitat is likely to remain unsuitable until the contaminants degrade or are removed. Conversely, source and sink designations are static in most ecological models, but contaminated sinks could become source habitats after effective remediation. Hence, there is a need for theoretical and empirical research on effects of persistent contaminants in spatially structured environments. Likely benefits of this work include both improved understanding of contaminant effects and identification of priorities for remediation. Delays between contaminant introduction and removal have the potential to increase the complexity of population processes, just as time lags in density dependence complicate the dynamics of individual populations (Turchin and Taylor 1992).

Multiple stressors in spatially structured environments

Much has been written elsewhere about the need for research on multiple stressors (Gentile et al. 1999). These needs are acute for spatially structured systems because contaminant effects are likely to be confounded with spatial stressors. The large number of parameters needed to describe both multiple stressors and spatial structure suggests that this research need will require integration of theoretical and empirical approaches.

Landscape context

The need for integrating population and landscape approaches is discussed at length by Wiens (1996, 1997). Here we simply emphasize that vertebrate population processes often occur on a landscape scale. Considering spatial structure at this scale is necessary to understanding contaminant effects on vertebrate populations.

Multispecies systems

Understanding contaminant effects within the dual complexities of multiple species and spatial structure is a difficult challenge, and is largely beyond the scope of this review. We concur with Holmes and Karieva (Chapter 6) that this challenge may be addressed most effectively with an empirical approach.

When can spatial structure be ignored?

Given its potentially large influence on population processes, spatial structure should be considered when analyzing contaminant effects. Similar arguments could be made, however, for considering individual differences, age structure, interspecific interactions, and any additional pet factor related to one's research interests. When does spatial structure really matter, and when can it safely be ignored? There is no definitive answer because the appropriate experiments have not been done. In the meantime, we offer the following reasoning.

Questions about when to consider or ignore spatial structure are equivalent to asking "When does understanding contaminant effects at 1 location require information about other locations?" For vertebrate populations, the answer depends on whether local or regional factors dominate population dynamics. When local conditions drive population processes, attention should focus on impacts from contaminants located on-site. When regional process strongly influence local dynamics, one must consider interactions with other populations and impacts from other contaminants at distant sites. Given these answers, it is clear that decisions to consider or ignore spatial structure in a given system should not be determined by the structure itself but rather by the movement behavior of animals within that structure. Hence, we offer the following suggestions:

1) Spatial structure can be ignored when little or no migration occurs between isolated habitats, and large, relatively homogeneous areas are being considered, within which individual movements occur over short distances relative to habitat dimensions.
2) Spatial structure should be considered when moderate rates of migration occur within continuous habitat or between habitat patches.
3) Patterns of migration should be considered explicitly when migration rates are high or dispersal distances are long.

This review could be taken as unwelcome news in that however difficult it is to understand contaminant effects on vertebrate populations, the task becomes more challenging when spatial structure is added. Instead, we offer a more hopeful perspective. Because spatial structure is inherent to most populations, its effects are present whether we consider them or not. If we ignore structure, those effects are counted as experimental error and thus tend to obscure the real effects of contaminants. Considering spatial structure adds complexity to our models, measurements, and experiments, but the effects of contaminants within this complexity will be clearer.

We conclude by recalling Hobbs' (1992) comment that "answers to practical conservation questions can only come from studies of real populations in real landscapes." We suggest that when the spatial structure of those landscapes matters, real populations often will not provide complete answers within practical limits of space and time. Hence, answers to practical questions will require study of real populations in real landscapes, but understanding those answers will require the broader spatial and temporal context provided by models.

Acknowledgements—We thank P. Karieva and L. Macovsky for their insightful discussions, and Pete Albers and 2 anonymous reviewers for comments on the manuscript. K. Short improved the quality of the figures.

References

Alexander SE, Roughgarden J. 1996. Larval transport and population dynamics of intertidal barnacles: A coupled benthic/oceanic model. *Ecol Monogr* 66:259–275.

Bell G, Lechowicz MJ, Appenzeller A, Chandler M, DeBlois E, Jackson L, Mackenzie B, Preziosi R, Schallenberg, M, Tinker N. 1993. The spatial structure of the physical environment. *Oecologia* 96:114–121.

Boone RB, Hunter Jr. ML. 1996. Using diffusion models to simulate the effects of land use on grizzly bear dispersal in the Rocky Mountains. *Landscape Ecol* 11:51–64.

Breitenmoser U, Kaczensy M, Dötterer C, Breitenmoser-Würsten S, Bernhart F, Liberek M. 1992. Spatial organization and recruitment of lynx (*Lynx lynx*) in a re-introduced population in the Swiss Jura Mountains. *J Zoology* 231:449–464.

Brown JH, Kodric-Brown A. 1977. Turnover rates in insular biogeography: Effect of immigration on extinction. *Ecology* 58:445–449.

Brussard PF, Murphy DD, Noss RF. 1992. Strategy and tactics for conserving biodiversity in the United States. *Cons Biol* 6:157–159.

Burkey TV, Stenseth NC. 1994. Population dynamics of territorial species in seasonal and patchy environments. *Oikos* 69:47–53.

Caley MJ, Carr MH, Hixon MA, Hughes TP, Jones GP, Menge BA. 1996. Recruitment and the local dynamics of open marine populations. *Ann Rev Ecol Syst* 27:477–500.

Cantrell RS, Cosner C. 1993. Should a park be an island? *SIAM J Appl Math* 53:219–252.

Chesson PL. 1981. Models for spatially distributed populations: The effect of within-patch variability. *Theor Pop Biol* 19:288–325.

Clifford PA, Barchers DE, Ludwig DF, Sielken RL, Klingensmith JS, Graham RV, Banton MI. 1995. An approach to quantifying spatial components of exposure for ecological risk assessment. *Environ Toxicol Chem* 14:895–906.

Comins HN, Blatt DWE. 1974. Predator-prey interactions in spatially heterogeneous environments. *J Theor Biol* 48:75–83.

Comins HN, Hassell MP, May RM. 1992. The spatial dynamics of host-parasitoid systems. *J Anim Ecol* 61:735–748.

Day JR, Possingham HP. 1995. A stochastic metapopulation model with variability in patch size and position. *Theor Pop Biol* 48:333–360.

Doak DF. 1995. Source-sink models and the problem of habitat degradation: General models and applications to the Yellowstone grizzly. *Cons Biol* 9:1370–1379.

Ehrlich PR, White RR, Singer MC, McKechnie SW, Gilbert LE. 1975. Checkerspot butterflies: A historical perspective. *Science* 188:221–228.

Ehrlich PR, Murphy DD, Singer MC, Sherwood CB, White RR, Brown IL. 1980. Extinction, reduction, stability and increase: The responses of checkerspot butterfly (*Euphydryas*) populations to the California drought. *Oecologia* 46:101–105.

Ellner S, Turchin P. 1995. Chaos in a noisy world: New methods and evidence from time series analysis. *Am Nat* 145:343–375.

Fahrig L, Merriam G. 1985. Habitat patch connectivity and population survival. *Ecology* 66:1762–1768.

Fahrig L, Pedlar JH, Pope SE, Taylor PD, Wegner JF. 1995. Effect of road traffic on amphibian density. *Biol Conserv* 73:177–182.

Gaona P, Ferreras P, Delibes M. 1998. Dynamics and viability of a metapopulation of the endangered Iberian lynx (*Lynx pardinus*). *Ecol Monogr* 68:349–370.

Gibbs JP. 1998. Amphibian movements in response to forest edges, roads, and streambeds in southern New England. *J Wildl Manage* 62:584–589.

Gill DE. 1978. The metapopulation ecology of the red-spotted newt, *Notophthalamus viridescens* (Rafinesque). *Ecol Monogr* 48:145–166.

Gilpin ME, Hanski I, editors. 1991. Metapopulation dynamics: Empirical and theoretical investigations. London: Academic Press.

Gopalsamy K. 1977. Competition, dispersion and coexistence. *Math Biosci* 33:25–33.

Gotelli NJ. 1998. A primer of ecology, 2nd ed. Sunderland MA: Sinauer.

Gurney WSC, Nisbet RM. 1978. Single species population fluctuations in patchy environments. *Am Nat* 112:1075–1090.

Gustafson EJ, Gardner RH. 1996. The effect of landscape heterogeneity on the probability of patch colonization. *Ecology* 77:94–107.

Gutierrez RJ, Harrison S. 1996. Applying metapopulation theory to spotted owl management: A history and critique. In: McCullough DR, editor. Metapopulations and wildlife conservation. Washington DC: Island Press. p 167–185.

Gyllenberg M, Hanski I, Hastings A. 1997. Structured metapopulation models. In: Hanski I, Gilpin ME, editors. Metapopulation biology: Ecology, genetics, and evolution. San Diego CA: Academic Press. p 93–122.

Haight RG, Mladenoff DJ, Wydeven AP. 1998. Modeling disjunct gray wolf populations in semi-wild landscapes. *Cons Biol* 12:879–888.

Hall CAS. 1988. An assessment of several of the historically most influential theoretical models used in ecology and of the data provided in their support. *Ecol Model* 43:5–31.

Halley JM, Oldham RS, Arntzen JW. 1996. Predicting the persistence of amphibian populations with the help of a spatial model. *J Appl Ecol* 33:455–470.

Hanski I. 1994. A practical model of metapopulation dynamics. *J Anim Ecol* 63:151–162.

Hanski I. 1997. Metapopulation dynamics: From concepts and observations to predictive models. In: Hanski I, Gilpin ME, editors. Metapopulation biology: Ecology, genetics, and evolution. San Diego CA: Academic Press. p. 69–91.

Hanski I, Simberloff D. 1997. The metapopulation approach, its history, conceptual domain, and application to conservation. In: Hanski I, Gilpin ME, editors. Metapopulation biology: Ecology, genetics, and evolution. San Diego CA: Academic Press. p 5–26.

Hanski I, Moilanen A, Gyllenberg M. 1996. Minimum viable metapopulation size. *Am Nat* 147:527–541.

Hanski I, Moilanen A, Pakkala T, Kuussaari M. 1996. The quantitative incidence function model and persistence of an endangered butterfly metapopulation. *Cons Biol* 10:578–590.

Hanski I, Pöyry J, Pakkala T, Kuussaari M. 1995. Multiple equilibria in metapopulation dynamics. *Nature* (London) 377:618–621.

Hanski I, Kuussari M, Nieminen M. 1994. Metapopulation structure and migration in the butterfly *Melitaea cinxia*. *Ecology* 75:747-762.

Hanski I, Turchin P, Korpimäki E, Henttonen H. 1993. Population oscillations of boreal rodents: Regulation by mustelid predators leads to chaos. *Nature* 364:232–235.

Hansson L. 1991. Dispersal and connectivity in metapopulations. *Biol J Lin Soc* 42:89–103.

Harrison S. 1991. Local extinction in a metapopulation context: An empirical evaluation. *Biol J Lin Soc* 42:73–88.

Harrison S. 1994. Metapopulations and conservation. In: Edwards PJ, Webb NR, May RM, editors. Large-scale ecology and conservation biology. Oxford: Blackwell. p 111–128.

Harrison S, Taylor AD. 1997. Empirical evidence for metapopulation dynamics. In: Hanski I, Gilpin ME, editors. Metapopulation biology: Ecology, genetics, and evolution. San Diego CA: Academic Press. p 27–42.

Harrison S, Murphy DD, Ehrlich PR. 1988. Distribution of the Bay checkerspot butterfly, *Euphydryas editha bayensis*: Evidence for a metapopulation model. *Am Nat* 132:360–382.

Hastings A. 1991. Structured models of metapopulation dynamics. *Biol J Lin Soc* 42:57–71.

Hastings A, Higgins K. 1994. Persistence of transients in spatially structured models. *Science* 263:1133–1136.

Hecnar SJ, M'Closkey RT. 1996. Regional dynamics and the status of amphibians. *Ecology* 77:2091–2097.

Henderson MT, Merriam G, Wegner J. 1985. Patchy environments and species survival: Chipmunks in an agricultural mosaic. *Biol Conserv* 31:95–105.

Hobbs RJ. 1992. The role of corridors in conservation: Solution or bandwagon? *Tr Ecol Evol* 7:389–392.

Holmes EE. 1993. Are diffusion models too simple? A comparison with telegraph models of invasion. *Am Nat* 142:403–419.

Holmes EE, Lewis MA, Banks JE, Veit RR. 1994. Partial differential equations in ecology: Spatial interactions and population dynamics. *Ecology* 75:17–29.

Holt RD. 1993. Ecology at the mesoscale: The influence of regional processes on local communities. In: Ricklefs R, Schluter D, editors. Species diversity in ecological communities: Historical and geographical perspectives. Chicago IL: University of Chicago Press. p 77–88.

Holt RD. 1997. From metapopulation dynamics to community structure: Some consequences of spatial heterogeneity. In: Hanski I, Gilpin ME, editors. Metapopulation biology: Ecology, genetics, and evolution. San Diego CA: Academic Press. p 149–164.

Horne JK, Schneider DC. 1995. Spatial variance in ecology. *Oikos* 74:18–26.

Huffaker CB. 1958. Experimental studies on predation: Dispersion factors and predator-prey oscillations. *Hilgardia* 27:343–383.

Ims RA, Yoccoz NG. 1997. Studying transfer processes in metapopulations: Emigration, migration, and colonization. In: Hanski I, Gilpin ME, editors. Metapopulation biology: Ecology, genetics, and evolution. San Diego CA: Academic Press. p 1247–265.

Ives AR, Settle WH. 1997. Metapopulation dynamics and pest control in agriculture systems. *Am Nat* 149:220–246.

Karieva P. 1983. Local movement in herbivorous insects: Applying a passive diffusion model to mark-recapture field experiments. *Oecologia* 57:322–327.

Karieva P. 1987. Habitat fragmentation and the stability of predator-prey interactions. *Nature* (London) 326:388–390.

Karieva P. 1990. Population dynamics in spatially complex environments: Theory and data. In: Hassell MP, May RM, editors. Population regulation and dynamics. London: Royal Society. p 53–68.

Karieva P, Wennergren U. 1995. Connecting landscape patterns to ecosystem and population processes. *Nature* (London) 373:299–302.

Keitt TH, Urban DL, Milne BT. 1997. Detecting critical scales in fragmented landscapes. *Conserv Ecol* 1(1):4 (URL:http//www.consecol.org/vol1/iss1/art4).

Kevorkian J, Cole JD. 1980. Perturbation methods in applied mathematics: Applied mathematical sciences No. 34. New York: Springer-Verlag.

Kindvall O. 1994. Habitat heterogeneity and survival in a bush cricket metapopulation. *Ecology* 77:207–214.

Levin SA. 1992. The problem of pattern and scale in ecology. *Ecology* 73:1943–1967.

Levins R. 1969. Some demographic and genetic consequences of environmental heterogeneity for biological control. *Bull Entomol Soc Am* 15:237–240.

Li L. 1989. Global positive coexistence of a nonlinear elliptic biological interacting model. *Math Biosci* 97:1–15.

Lima SL, Zollner PA. 1996. Towards a behavioral ecology of ecological landscapes. *Tr Ecol Evol* 11:131–135.

Lindenmayer DB, Possingham HP. 1996. Modelling the inter-relationships between habitat patchiness, dispersal capability and metapopulation persistence of the endangered species, Leadbeater's possum, in south-eastern Australia. *Landscape Ecol* 11:79–105.

Macovsky LM. 1999. The effects of toxicant related mortality on metapopulation dynamics: A laboratory model. [M.S. thesis]. Bellingham WA: Western Washington University. 74 p.

Mangel M, Tier C. 1993. Dynamics of metapopulations with demographic stochasticity and environmental catastrophies. *Theor Pop Biol* 44:1–31.

Maurer BA, Holt RD. 1996. Effects of chronic pesticide stress on wildlife populations in complex landscapes: Processes at multiple scales. *Environ Toxicol Chem* 15:420–426.

McCauley E, Wilson WG, De Roos AM. 1996. Dynamics of age-structured predator-prey populations in space: Asymmetrical effects of mobility in juvenile and adult predators. *Oikos* 76:485–497.

McCullough DR. 1996. Introduction. In: McCullough DR, editor. Metapopulations and wildlife conservation. Washington DC: Island Press. p 1–10.

McKelvey K, Noon BR, and Lamberson RH. 1993. Conservation planning for species occupying fragmented landscapes: The case of the northern spotted owl. In: Kareiva PM, Kingsolver JG, Huey RB, editors. Biotic interactions and global change. Sunderland MA: Sinauer. p 424–450.

McLaughlin JF, Roughgarden J. 1991. Pattern and stability in predator-prey communities: How diffusion in spatially variable environments affects the Lotka-Volterra model. *Theor Pop Biol* 40:148–172.

McLaughlin JF, Roughgarden J. 1993. Species interactions in space. In: Ricklefs R, Schluter D, editors. Species diversity in ecological communities: Historical and geographical perspectives. Chicago IL: University of Chicago Press. p 89–98.

McMurtrie R. 1978. Persistence and the stability of single-species and prey-predator systems in spatially heterogeneous environments. *Math Biosci* 39:11–51.

Middleton J, Merriam G. 1981. Woodland mice in a farmland mosaic. *J Appl Ecol* 18:703–710.

Mills LS, Soulé ME, Doak DF. 1993. The keystone-species concept in ecology and conservation. *BioScience* 43:219–224.

Moilanen A, Hanski I. 1998. Metapopulation dynamics: Effects of habitat quality and landscape structure. *Ecology* 79:2503–2515.

Molofsky J. 1994. Population dynamics and pattern formation in theoretical populations. *Ecology* 75:30–39.

Morrison LW. 1998. The spatiotemporal dynamics of insular ant metapopulations. *Ecology* 79:1135–1146.

Murray JD. 1989. Mathematical biology. Berlin, Germany: Springer-Verlag.

Namba T. 1989. Competition for space in a heterogeneous environment. *J Math Biol* 27:1–16.

Newmark WD. 1987. A land-bridge island perspective on mammalian extinctions in western North American parks. *Nature* (London) 325:430–432.

Newmark WD. 1995. Extinction of mammal populations in western North American national parks. *Cons Biol* 9:512–526.

Nisbet RM, Gurney WSC. 1982. Modelling fluctuating populations. New York: Wiley.

Oksanen L. 1991. A century of community ecology: How much progress? *Tr Ecol Evol* 6:294–296.

Okubo A. 1980. Diffusion and ecological problems: Mathematical models. Berlin, Germany: Springer-Verlag.

Pacala SW, Roughgarden J. 1982. Spatial heterogeneity and interspecific competition. *Theor Pop Biol* 21:92–113.

Pease CM, Mattson DJ. 1999. Demography of the Yellowstone grizzly bears. *Ecology* 80:957-975.

Phipps MJ. 1991. From local to global: The lesson of cellular automata. In: DeAngelis DL, Gross LS, editors. Individual-based models and approaches in ecology: Populations, communities and ecosystems. New York: Academic Press. p 165–187.

Pimental D, Nigel W, Madden J. 1963. Space-time structure of the environment and the survival of host-parasite systems. *Am Nat* 97:141–166.

Pimm SL, Redfearn A. 1988. The variability of population densities. *Nature* (London) 334:613–614.

Pollard E. 1991. Synchrony of population fluctuations: The dominant influence of widespread factors on local butterfly populations. *Oikos* 60:7–10.

Pollard E, Yates TJ. 1993. Monitoring butterflies for conservation. London: Chapman & Hall.

Possingham HP, Roughgarden J. 1990. Spatial population dynamics of a marine organism with a complex life cycle. *Ecology* 71:973–985.

Pulliam HR. 1988. Sources, sinks, and population regulation. *Am Nat* 132:652–661.

Pulliam HR, Danielson BJ. 1991. Sources, sinks, and habitat selection: A landscape perspective on population dynamics. *Am Nat* 137:S50–S66.

Richards B. 1998. An analysis of connectivity of late-seral forest in western Oregon. MS thesis. Bellingham WA: Western Washington University. 53 p.

Roughgarden J, Iwasa Y. 1986. Dynamics of a metapopulation with space-limited subpopulations. *Theor Pop Biol* 29:235–261.

Roughgarden J, Gaines S, Possingham H. 1988. Recruitment dynamics in complex life cycles. *Science* 241:1460–1466.

Sabelis MW, Diekmann O. 1988. Overall population stability despite local extinction: The stabilizing influence of prey dispersal from predator-invaded patches. *Theor Pop Biol* 34:169–176.

Schoener TW, Spiller DA. 1987. High population persistence in a system with high turnover. *Nature* (London) 330:474–477.

Schumaker NH. 1995. Habitat connectivity and spotted owl population dynamics. [Ph.D. dissertation]. University of Washington. 126 p.

Sherrat TN, Jepson PC. 1993. A metapopulation approach to modelling the long-term impact of pesticides on invertebrates. *J Appl Ecol* 30:696–705.

Shigesada N. 1984. Spatial distribution of rapidly dispersing animals in heterogeneous environments. In: Levin SA, Hallam TG, editors. Mathematical ecology. Lecture notes in biomathematics 54. p 478–491.

Skellam JG. 1951. Random dispersal in theoretical populations. *Biometrika* 38:196–218.

Sjögren-Gulve P. 1994. Distribution and extinction patterns within a northern metapopulation of the pool frog, *Rana lessonae*. *Ecology* 75:1357–1367.

Sjögren-Gulve P, Ray C. 1996. Using logistic regression to model metapopulation dynamics: Large-scale forestry extirpates the pool frog. In: McCullough DR, editor. Metapopulations and wildlife conservation. Washington DC: Island Press. p 111–137.

Smith GD. 1998. Numerical solution of partial differential equations: Finite difference methods. Oxford, UK: Clarendon Press.

Smith AT, Gilpin ME. 1997. Spatially correlated dynamics in a pika metapopulation. In: Hanski I, Gilpin ME, editors. Metapopulation biology: Ecology, genetics, and evolution. San Diego CA: Academic Press. p 407–428.

Spromberg JA, Johns BM, Landis WG. 1998. Metapopulation dynamics: Indirect effects and multiple distinct outcomes in ecological risk assessment. *Environ Toxicol Chem* 17:1640–1649.

Stacey PB, Johnson VA, Taper ML. 1997. Migration within metapopulations: The impact upon local population dynamics. In: Hanski I, Gilpin ME, editors. Metapopulation biology: Ecology, genetics, and evolution. San Diego CA: Academic Press. p 267–291.

Thomas CD. 1994a. Extinction, colonization, and metapopulations: Environmental tracking by rare species. *Cons Biol* 8:373–378.
Thomas CD. 1994b. Difficulties in deducing dynamics from static distributions. *Tr Ecol Evol* 9:300.
Thomas CD, Hanski I. 1997. Butterfly metapopulations. In: Hanski I, Gilpin ME, editors. Metapopulation biology: Ecology, genetics, and evolution. San Diego CA: Academic Press. p 359–386.
Tilman D. 1994. Competition and biodiversity in spatially structured habitats. *Ecology* 75:2–16.
Turchin P. 1997. Quantitative analysis of movement. Sunderland MA: Sinauer.
Turchin P, Hanski I. 1997. An empirically based model for latitudinal gradient in vole population dynamics. *Am Nat* 149:842–872.
Turchin P, Taylor AD. 1992. Complex dynamics in ecological time series. *Ecology* 73:289–305.
Vos CC, Chardon JP. 1998. Effects of habitat fragmentation and road density on the distribution pattern of the moor frog *Rana arvalis*. *J Appl Ecol* 35:44–56.
Gentile JH, Solomon KR, Butcher JB, Harrass M, Landis WG, Power M, Rattner BA, Warren-Hicks WJ, Wenger R. 1999. Linking stressors and ecological responses. In: Foran J, Ferenc S, editors. Multiple stressors in ecological risk and impact assessment. Pensacola FL: Society of Toxicology and Chemistry (SETAC). p 27–50.
Wiens JA. 1996. Wildlife in patchy environments: Metapopulations, mosaics, and management. In: McCullough DR, editor. Metapopulations and wildlife conservation. Washington DC: Island. p 53–84.
Wiens JA. 1997. Metapopulation dynamics and landscape ecology. In: Hanski I, Gilpin ME, editors. Metapopulation biology: Ecology, genetics, and evolution. San Diego CA: Academic. p 43–62.
Weiss SJ, Murphy DD, White RR. 1988. Sun, slope and butterflies: Topographic determinants of habitat quality in *Euphydryas editha*. *Ecology* 69:1486–1496.
Wolff JO. 1980. The role of habitat patchiness in the population dynamics of snowshoe hares. *Ecol Monogr* 50:111–130.
Wu J, Vankat JL, Barlas Y. 1993. Effects of patch connectivity and arrangement of animal metapopulation dynamics: A simulation study. *Ecol Model* 65:221–254.

CHAPTER 13

Disruption of Rodent Assemblages in Disturbed Tallgrass Prairie Ecosystems Contaminated with Petroleum Wastes

Robert L. Lochmiller, Daniel P. Rafferty, Karen McBee, Nick T. Basta, James A. Wilson

Advances toward understanding how contaminants in terrestrial environments affect structural and functional attributes of ecological organization have been nearly non-existent in comparison to our understanding of aquatic ecosystems (Cairns et al. 1995). Research in aquatic systems has shown that making predictions of how terrestrial ecosystems might respond to contaminant inputs will be very difficult. Because of the inherent complexity of ecosystems, a reductionist approach, e.g., focusing on a particular assemblage of species, is most likely to provide an understanding of the effects of contaminants. For terrestrial systems, even focusing on structural and functional attributes of small mammal assemblages poses many difficult problems in detecting or predicting environmental impacts from contamination. In addition to varying spatially, small mammal assemblages vary temporally in that they demonstrate remarkable dynamic change in numbers of species represented and population abundances within seasons and across seasons or years (Rose and Birney 1985). Developing a model that can be robust to such dynamics and complexity has not been attempted for assemblages of small mammals in contaminated terrestrial environments.

Small mammals are thought to be good indicator organisms for biomonitoring (Talmage and Walton 1991) and may also prove to be an appropriate model system for assessing the impacts of contaminants on community-level attributes of ecosystems. Many studies have documented uptake and accumulation of contaminants (e.g., heavy metals, insecticides, and polychlorinated biphenyls [PCBs]) in tissues of small mammals (McBee and Bickham 1990) inhabiting contaminated terrestrial

CHAPTER PREVIEW

Experimental design and methods 278
Results 282
Discussion 293
Conclusions 298

ecosystems. These studies have revealed that exposure risks are variable among species and possibly are associated at least in part with trophic level. These risks also could be in response to habitat specializations and use. Talmage and Walton (1991) suggested that small mammals with insectivorous feeding habits would be at greater risk of exposure to soil contaminants than would those species with omnivorous or herbivorous foraging habits. Provided that accumulations of contaminants in an organism are a reasonable reflection of relative impacts from exposure, results of previous studies would indicate that changes in the structure and function of small mammal assemblages are likely in contaminated environments. Unfortunately, few studies have attempted to explore these predictions in actual contaminated environments that support assemblages of resident small mammals.

In this study we explored the sensitivity of small mammal assembleges to the toxic effects of soils contaminated with petrochemicals derived from the oil-refining process. As with most industrial-waste sites contaminated with petrochemicals, there are several chemical mixtures in the soil, including both organic and inorganic compounds. Crude and refined oils and by-products from the refining process represent highly complex chemical mixtures whose toxicity to mammalian systems remains largely unknown (Coppock et al. 1995). We hypothesized that habitats contaminated from the disposal of petrochemical wastes would demonstrate a decline in sensitive indicator species while demonstrating relative increases in more opportunistic species within resident assemblages of rodents. Consequently, these habitats would demonstrate greater temporal instability in structure compared with assemblages from reference habitats. To explore this hypothesis, we used mark-recapture techniques to count rodent populations within and across seasons in replicated contaminated and reference habitats. Because all petrochemical-contaminated habitats were early successional, disturbed ecosystems, we selected reference study areas based upon their similarity in disturbance vegetation to that present on paired contaminated areas in Oklahoma.

Experimental Design and Methods

Study areas

Location of sites
Our experimental approach was to compare the dynamic changes in structure of assemblages of small mammals inhabiting replicated contaminated and reference habitats. We selected 12 study sites that were known to be contaminated with complex mixtures of petrochemicals in the soil; all the sites consisted of disturbed terrestrial ecosystems (early-seral stage plant species) that supported viable populations of resident small mammals. Each contaminated study site was matched with an ecologically similar reference site in the general vicinity of the site of actual contamination. Reference sites were selected based on a number of criteria, fore-

most being access, because most sites were privately owned. Because all contaminated sites were disturbed, early-successional habitats, we required reference sites to be similarly disturbed and to possess vegetative characteristics, including similar dominant plant species and cover of woody vegetation. These vegetative assessments were made by visual inspection. Finally, reference sites were located in close proximity to the paired contaminated site to control variation resulting from weather variables. Matched reference sites permitted us considerable experimental control over non-pollutant environmental variables (e.g., climate, nutrition, cover), which can frequently confound interpretation of results (Dutilleul 1993). The petrochemical-contaminated study sites that we chose for monitoring were selected from known Superfund Waste Sites and several abandoned oil-refinery sites distributed throughout Oklahoma. Specific study sites were at least 1 ha in size to accommodate a sufficiently large assemblage of rodents for counts within and across seasons. Sites also were selected to represent varying degrees of toxicity (from low to high) based on preliminary soil and groundwater contaminant analyses available from selected unpublished reports and consultations with personnel from the Oklahoma Department of Environmental Quality.

Four contaminated and 4 matched reference study areas were counted across summer and winter seasons of each of 3 years for 1 complete annual cycle. Logistics prevented us from simultaneously conducting a count of all 24 study sites during the same annual cycle. Matched reference study areas were always counted at the same time as their contaminated counterparts. During Year 1 of the study (summer 1995 and winter 1996) 2 contaminated study areas were located on an abandoned oil refinery in Cyril, OK. These areas consisted of a former refinery waste land-treatment (LT) facility (Cyril LT) and earthen levees surrounding a former waste-oil sludge pit (Cyril SP). Two abandoned waste-oil sludge pits were located in Cleveland (Cleveland SP) and Cushing (Cushing SP), OK.

In Year 2 (summer 1996 and winter 1997) we counted a sludge pit on an abandoned re-refining complex, formerly known as Double Eagle Oil Refinery (Eagle SP), and a former petrochemical landfill in Oklahoma City (OKC LF) that was scheduled for cleanup after our study and had been used extensively in the 1950s and 1960s for the disposal of solvents and other aircraft-maintenance wastes. Two contaminated study areas were located 25 miles south of Tulsa, OK, where wastes from refineries were disposed by land treatment on 1 area (Tulsa LT) and where a sludge pit for oil wastes had been filled and capped with contaminated land-treated soil (Cap LT).

In Year 3 (summer 1997 and winter 1998) 3 contaminated study areas were located on a large abandoned oil refinery in Duncan, OK. These sites consisted of a former land-treatment facility (Duncan LT), an asphalt waste pit containing acid-waste sludges (Duncan AP), and a waste-sludge settling pond (Duncan SP). The refinery began operation in the 1920s and shut down in the early 1980s. The fourth site was located in Ponca City, OK, on an active oil refining complex where tank-bottom wastes were land treated (Ponca LT).

Contaminant profiles

Each of the contaminated study areas contained numerous metals and complex mixtures of organic hydrocarbons at varying concentrations in the soil. Soil samples from the top 3 cm of the surface were analyzed for contaminants to characterize the relative degree of toxicity for each site (Schroeder 1998). Heavy metals that potentially posed an immunotoxic risk to the health of resident rodents were Cr and Pb, which were summed to derive an estimate of total metals. Immunotoxicity was chosen as an endpoint for our metal-toxicity index because of the immunotoxic metals Pb and Cr that were present in the soils of our study sites. Our laboratory had previously recorded immune system lesions in *S. hispidus* that had been collected from many of these contaminated sites. Pb and Cr were often the only 2 metals present in the soil at potentially toxic levels. An index of metal toxicity was calculated by dividing total metal content by 17 ppm, which was selected as the average Pb concentration in U.S. reference soils (Bradley et al. 1994). A relative index of toxicity for total petroleum hydrocarbons (TPH) and total carcinogenic polycyclic aromatic hydrocarbon (PAH) levels was calculated by dividing by 100 ppm and 100 ppb, respectively, which are the recommended cleanup levels in soil (Bradley et al. 1994). An overall index of toxicity was calculated as the sum of the above 3 indices for each study site. Details about analytical procedures and sampling of soils can be found in Schroeder (1998).

Vegetation characteristics

All study sites were originally tallgrass prairie ecosystems that were subjected to severe disturbance of the soil by mechanical methods and by deposition of contaminants. As a result of these disturbances, vegetation on nearly all of these study sites was dominated by early seral-plant species. The most prominent plant on these heavily disturbed prairies was johnsongrass (*Sorghum halapense* L.), which occurred in solid stands on 20% to 80% of each trapping grid and reached heights > 2 m. The other major monocots in these disturbed communities included little bluestem (*Schizachyrium scoparium* Nash), big bluestem (*Andropogon gerardii* Vitman), brome (*Bromus* spp.), and Bermuda grass (*Cynodon dactylon* L.). Forbs also were a prominent component of each habitat and included western ragweed (*Ambrosia psilostachya* DC.), common sunflower (*Helianthus annuus* L.), white sage (*Artemisia ludoviciana* Nutt.), aster (*Aster* spp.), marestail (*Conyza canadensis* L.), and partridge pea (*Cassia fasciculata* L.). Woody plants occurred on a few of the study areas but were < 10% canopy coverage and were comprised primarily of sumac (*Rhus* spp.), elm (*Ulmus* spp.), and other miscellaneous species.

Count methodology

The 24 assemblages of rodents (12 contaminated sites and 12 matched reference sites) that we monitored were counted during 1 complete annual cycle during 1 of 3 years. Permanent trapping grids were established on each study site, which consisted of 64 stations spaced at 10-m intervals with 1 Sherman live-catch trap (HB Sherman Traps Inc., Tallahassee, FL) at each station. For each contaminated study site, the grid was square or rectangular in configuration to conform to its often peculiar shape, and the shape of the grid on its corresponding matched reference study site was identical to its matched contaminated site. Traps were baited with whole oats, and cotton bedding was provided in winter. Traps were closed during the day, opened in late afternoon, and checked the following morning for 4 consecutive days during each trapping session. Grids were counted every 3 weeks for 4 sessions in both summer (July through September) and winter (December through February). For a given annual cycle, all 8 grids (4 contaminated and 4 reference) were counted concurrently during each session.

Upon capture, rodents were toe-clipped with a unique identification number, and the trap station, sex, reproductive status, body mass, and species were recorded before their release. Body mass was recorded with a spring scale to the nearest gram. Because of the morphological similarity between *Peromyscus leucopus* and *P. maniculatus*, we recorded them simply as *Peromyscus* spp. Two species of insectivore were captured infrequently and were not included in the final analyses because our count techniques were inappropriate for capturing these animals. Hence, we focus our attention only on the rodent assemblages inhabiting each of the 24 study areas.

Population and community parameters

Size of populations (abundance) was estimated as the minimum number known alive (MNKA) on each sampling grid during a 4-day trapping session. Mean abundance for a season was calculated as the mean MNKA across all 4 trapping sessions within that season. Seasonal abundance of rodent populations in assemblages was calculated as the total number of unique individuals captured over the 4 trapping sessions in summer or winter. Total abundance of the rodent assemblage was calculated as the sum of seasonal abundance for each species in that particular assemblage of rodents. The dynamic changes in population size within seasons ($n = 4$ occasions) or across one annual cycle ($n = 8$ occasions) for each population was expressed as an s-index, which is a relative measure of temporal variability of abundance across trapping occasions and is calculated as the standard deviation of log (abundance) as described by Lewontin (1966). Species richness was determined as the number of unique species captured on a sampling grid. Rodent assemblages were compared using similarity indices (Horn 1966) to generate a matrix of coefficients that was examined for groupings by using the clustering procedure known as the unweighted pair-group method using arithmetic averages (UPGMA); the results were expressed using a dendrogram of relatedness (Krebs 1989).

Data analyses

Demographic (i.e., population abundance estimators, s-index) and community (i.e., assemblage abundance estimators, species richness, s-index) parameters were tested for differences between contaminated and matched reference areas using single classification analysis of variance with a block design (PROC GLM; SAS Institute Inc. 1990). The model consisted of a treatment main effect (contaminated, reference) and block effect (12 paired sites). To increase statistical power, we a priori set significance for these analyses at $P < 0.10$. All mean values presented in the text are accompanied by the standard error.

Logistic regression, using the maximum-likelihood method (PROC LOGISTIC; SAS Institute Inc. 1990), was used to determine if demographic parameters were significant correlates of site contamination. A logistic-regression model was developed for each season for predicting site contamination using stepwise forward selection of variables; variables were permitted to enter the model in a hierarchical fashion when the $\log_e$ likelihood was deemed appropriate ($P < 0.20$). Three subsets of variables were used in developing predictive models for each season; only s-indices, mean abundance estimators, and seasonal abundance estimators were analyzed separately.

Stepwise multiple-linear regression analysis (PROC REG; SAS Institute Inc. 1990) was used to develop best-fit models relating contaminant levels in soil to intrinsic attributes of rodent assemblages. Dependent variables consisting of log-transformed concentrations of TPH, PAH, and total metals and the index of toxicity were regressed against a variety of independent variables consisting of seasonal estimates of abundance and stability (s-index), and species-richness indices, for populations and rodent assemblages. The best single-variable model and 3-variable model were determined for each dependent variable.

Results

Contaminant profiles

Average concentrations of toxic metals (Cr, Pb), TPH, and PAH on each site showed reference areas to be uncontaminated in comparison to their matched contaminated study area (Table 13-1). Concentrations of heavy metals averaged 30.0 ± 3.3 ppm on reference areas, compared with 677 ± 292 ppm on contaminated areas. Total PAH and TPH levels were low on reference areas but averaged 3884 ± 1559 ppb and 1827 ± 843 ppm on contaminated areas, respectively. The overall toxicity index averaged 2.1 ± 0.2 for the 12 reference areas and 97.0 ± 34.1 for the 12 contaminated areas. The Duncan LT, Duncan AP, Duncan SP, and Ponca LT areas were judged to be considerably more toxic than the other study areas (Figure 13-1).

Table 13-1 Average concentration of heavy metals (Cr and Pb), total PAH, and TPH on 12 study sites that were comprised of a suspected petrochemical-contaminated area and a paired reference area in disturbed, tallgrass prairie habitats in Oklahoma

Location	Contaminated grid			Reference grid		
	Metals (ppm)	TPH (ppm)	PAH (ppb)	Metals (ppm)	TPH (ppm)	PAH (ppb)
Cyril LT	294.3	274.50	159.5	26.5	30.5	5.0
Cyril SP	1191.7	645.30	556.5	28.5	72.0	14.5
Cleveland SP	37.4	66.00	733.2	22.0	31.5	78.5
Cushing SP	196.5	295.00	1022.5	49.0	27.0	12.5
OKC LF	184.7	17.80	127.3	40.0	8.5	0.0
Eagle SP	351.0	121.80	136.1	9.5	1.5	5.0
Tulsa LT	134.1	609.80	616.8	25.0	6.0	0.0
Cap LT	73.8	769.60	550.8	49.5	0.0	65.5
Duncan LT	1532.1	3242.60	4996.6	27.0	38.5	0.0
Duncan AP	251.8	774.010	2127.8	19.5	0.0	0.0
Duncan SP	339.1	9443.60	12,941.8	30.0	0.0	0.0
Ponca LT	3542.3	5665.1	12,648.6	34.5	9.0	0.0

Composition of rodent assemblages

A total of 49,152 trap-nights of effort yielded 5696 unique individuals that were captured on 17,730 occasions over the 3-year study. Cumulatively, 10 different species of rodents were captured on the 24 trapping grids over the course of the study. Most species were captured infrequently, but *Baiomys taylori*, *Oryzomys palustris*, and *Neotoma floridana* were extremely rare and were pooled under the category "other species" (Table 13-2). All 24 study areas supported populations of *Sigmodon hispidus* (3958 total unique individuals), *Peromyscus* spp. (659 total unique individuals), and *Reithrodontomys fulvescens* (536 total unique individuals), which collectively comprised 90% of the individuals within rodent assemblages. Rodent assemblages of disturbed prairie habitats in the southern Great Plains are dominated by *S. hispidus*, as was observed on both contaminated and matched reference areas in this study (*S. hispidus* comprised 70% of all individuals captured). Total abundance of individuals within a small mammal assemblage was variable across sites and between summer and winter (Table 13-3). As *S. hispidus* abundance declined from summer to winter, other species, e.g., *R. fulvescens* and *Microtus* spp. increased in abundance (Figure 13-2). We also caught *Mus musculus* and *Chaetodipus hispidus*, 2 species that occur infrequently in disturbed prairie habitats with extensive vegetation cover but often were caught in large numbers on our trapping grids.

Alterations in composition of assemblages

Mean total abundance of individuals within rodent assemblages across 4 trapping occasions was similar between contaminated and reference study areas in both

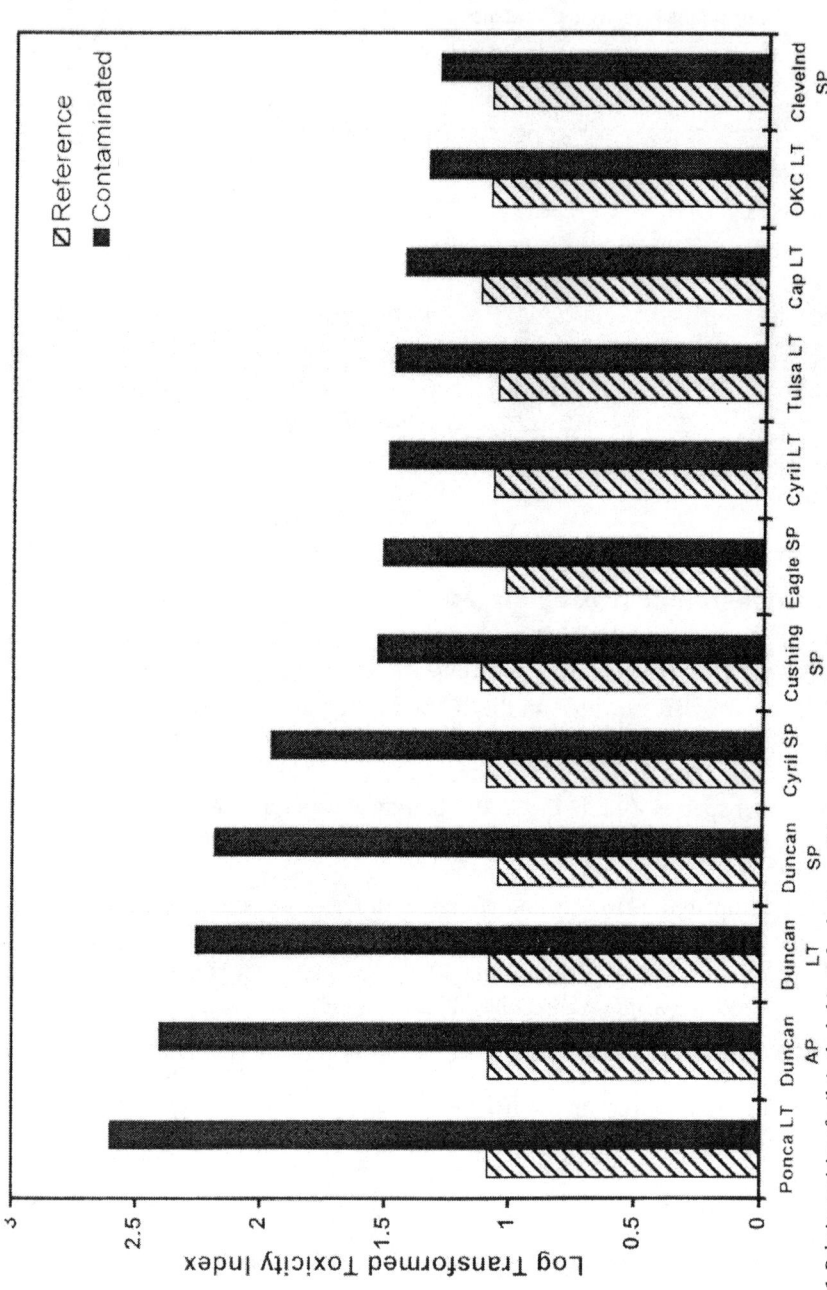

Figure 13-1 Relative toxicity of soils in the habitat of rodent assemblages on 12 study sites that were comprised of a suspected petrochemical-contaminated area and a paired reference area in disturbed, tallgrass prairie in Oklahoma. Average concentration of heavy metals (Cr and Pb), total PAHs, and TPHs on each site were divided by acceptable clean-up levels to derive a relative toxicity index: Index = (metals / 17 ppm) + (PAH / 100 ppb) + (TPH / 100 ppm).

Table 13-2 Seasonal (summer = sum and winter = win) abundance (minimum number known alive) for resident small mammal species occurring on 12 suspected petrochemical-contaminated (C) and 12 matched reference (R) sites in Oklahoma

Location[1]	Site[2]	S. hispidus		Peromyscus spp.		R. fulvescens		M. musculus		M. ochrogaster		M. pinetorum		C. hispidus		Other spp.	
		Sum	Win	Sum	Win	Sum	Win	Sum	Win	Sum	Win	Sum	Win	Sum	Win	Sum	Win
Cyril LT	C	60	35	1	4	2	34	3	24	1	13	0	0	0	0	0	0
	R	106	70	24	41	9	29	1	2	1	0	0	0	0	2	0	0
Cyril SP	C	47	20	12	72	0	16	14	51	0	1	0	0	3	0	0	0
	R	83	28	20	35	0	28	0	2	3	17	6	1	1	0	0	0
Cleveland SP	C	147	135	5	16	2	32	2	5	0	0	0	0	0	0	0	0
	R	99	38	8	8	6	7	1	0	0	0	12	55	0	0	0	0
Cushing SP	C	95	19	24	42	0	14	2	6	1	0	0	0	4	0	1	0
	R	135	45	1	1	5	26	0	1	1	1	0	8	0	0	0	0
OKC LF	C	111	133	0	0	0	1	0	1	0	0	0	0	0	0	0	0
	R	59	35	2	8	0	1	0	22	0	0	0	0	0	0	0	0
Eagle SP	C	96	25	8	14	0	20	10	18	0	0	0	0	0	0	0	0
	R	21	1	11	21	0	25	3	16	1	1	0	0	0	0	0	0
Tulsa LT	C	140	71	2	0	5	10	0	0	0	2	0	3	0	0	0	2
	R	117	57	0	2	3	14	4	0	1	1	0	0	0	0	0	0
Cap LT	C	156	102	0	2	3	11	0	0	0	0	0	0	0	0	0	2
	R	103	31	1	0	0	17	0	3	3	0	0	0	0	0	0	0
Duncan LT	C	77	16	27	62	7	37	12	2	0	0	0	0	11	2	0	1
	R	244	99	2	3	1	40	0	0	1	1	1	0	5	1	1	1
Duncan AP	C	56	31	44	35	3	35	4	1	0	0	0	0	5	0	0	0
	R	260	95	8	12	0	24	0	0	4	5	0	1	1	0	1	0
Duncan SP	C	65	14	31	43	0	53	7	28	0	0	1	0	9	0	0	0
	R	251	151	20	25	0	15	1	0	5	1	1	1	0	0	0	0
Ponca LT	C	193	191	4	0	0	1	37	34	0	2	0	0	0	0	0	0
	R	62	128	12	7	2	7	0	0	0	1	3	1	0	0	3	0

[1]Abbreviations for study sites: LT = land treatment, SP = waste oil sludge pit, AP = waste asphalt pit, LF = land fill for oily wastes.
[2]C = Contaminated, R = Reference

Table 13-3 Total number of small mammals captured and total abundance (minimum number of individuals known alive) in the small mammal assemblage following 1024 trap-nights of census monitoring on each of 12 study sites during summer and again in winter

Location grid	Site[1]	Summer		Winter	
		Total captures	Abundance	Total captures	Abundance
Cyril LT	C	157	67	279	110
	R	458	141	563	142
Cyril SP	C	141	76	422	162
	R	389	113	422	111
Cleveland SP	C	440	156	517	188
	R	430	126	301	108
Cushing SP	C	347	126	318	81
	R	416	142	260	82
OKC LF	C	424	111	291	135
	R	163	61	184	66
Eagle SP	C	274	114	244	77
	R	45	35	102	63
Tulsa LT	C	431	148	189	82
	R	363	124	327	80
Cap LT	C	469	162	348	116
	R	247	104	153	53
Duncan LT	C	269	134	351	119
	R	659	255	537	145
Duncan AP	C	273	112	292	102
	R	755	274	481	137
Duncan SP	C	308	113	308	138
	R	748	278	510	192
Ponca LT	C	664	234	626	228
	R	159	82	677	144

[1]Study sites were comprised of a suspected petrochemical-contaminated area (C) and a paired reference area (R) that were counted over 1 annual cycle during the period 1995 to 1997 in disturbed, tallgrass prairie habitats in Oklahoma.

summer (overall mean = 54.9 ± 3.0) and winter (overall mean = 52.7 ± 2.5; $P > 0.10$; Figure 13-2). Three species showed particular sensitivity to the disturbances of habitat resulting from deposition of petrochemical wastes in the soil; *M. musculus*, *C. hispidus*, and *Microtus*. Populations of these 3 species were usually low in abundance but periodically would demonstrate an increase in winter. The most notable alteration was the prominence of *M. musculus* populations within assemblages of contaminated study areas (mean abundance = 2.60 ± 0.65 in summer, 4.44 ± 0.85 in winter) compared with reference areas (0.27 ± 0.11 in summer, 1.10 ± 0.38 in winter) in summer ($P < 0.048$) and winter ($P < 0.062$). Only 2 of the 12 contaminated study areas were void of *M. musculus*, and on some contaminated areas, this species was more common than *S. hispidus*. Populations of *C. hispidus* demonstrated a trend in abundance that was similar to that of *M. musculus* on contaminated study areas, although numbers of individuals remained comparatively low on all areas and collections. Mean abundance of *C. hispidus* was greater on contaminated areas (0.85 ± 0.21) compared with reference areas (0.14 ± 0.07) in summer ($P < 0.046$) but not in

13: Disruption of rodent assemblages in disturbed tallgrass prairie ecosystems 287

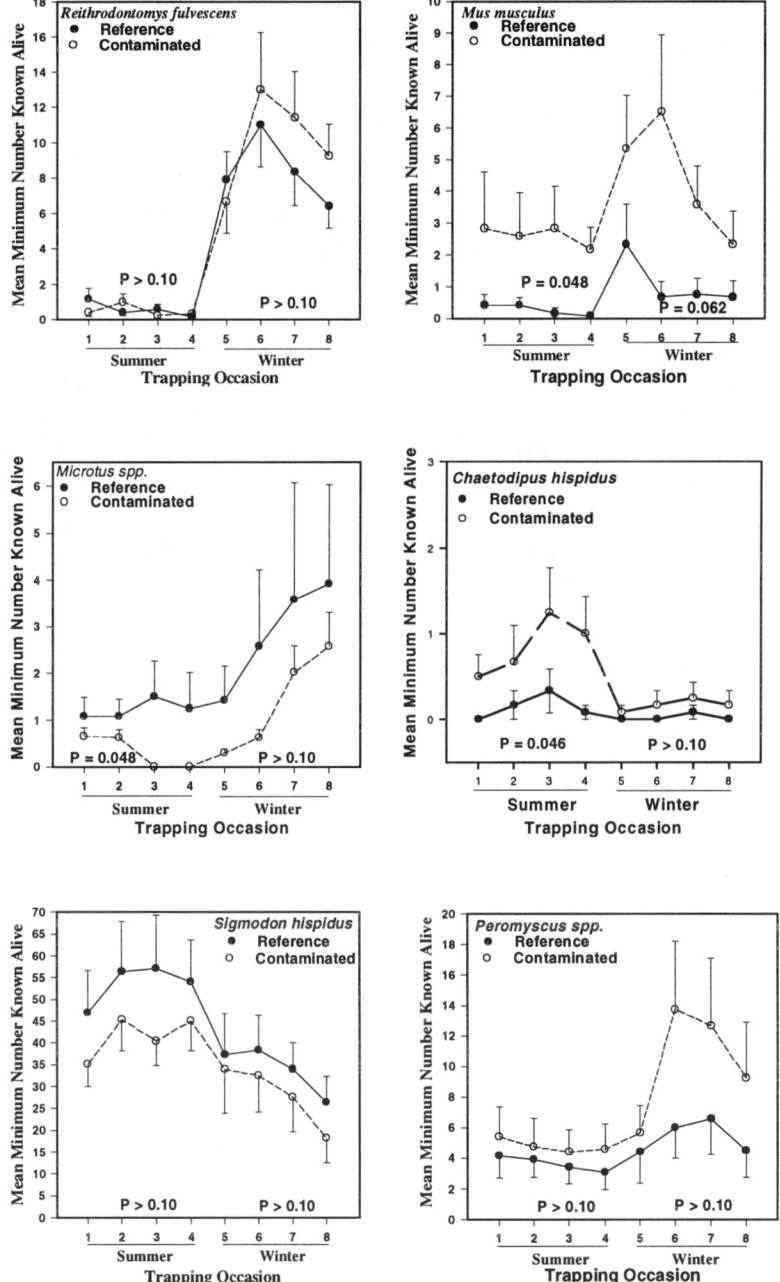

Figure 13-2 Temporal fluctuations in abundance (MNKA) across 8 trapping occasions during 1 complete annual cycle for the common rodent populations inhabiting suspected petrochemical-contaminated areas ($n = 12$) and matched reference areas ($n = 12$) in disturbed, tallgrass prairie in Oklahoma. Values represent mean ± SE; results of statistical comparisons between reference and contaminated areas within a season are depicted as P-values.

winter ($P > 0.10$), while *Microtus* populations were less abundant on contaminated areas (mean abundance = 0.14 ± 0.07) compared with reference areas (1.23 ± 0.30) in summer ($P < 0.048$) but not in winter ($P > 0.10$).

Composition of the rodent assemblage varied across study areas as indicated above, but species richness remained relatively stable and unaffected by contamination of habitats. Mean species richness across the 4 trapping occasions was similar between reference (overall mean = 3.09 ± 0.11 species) and contaminated (3.18 ± 0.12 species) areas in both summer and winter ($P > 0.50$). Similarity indices were used to qualitatively explore compositional characteristics among the 24 study areas (Figures 13-3 and 13-4). Cluster analysis of summer assemblages of rodents demonstrated an interesting cluster (Figure 13-3) comprised of 4 of the most toxic areas (Cyril SP, the 3 Duncan contaminated sites) and a reference area (Eagle SPR). The Eagle SPR area was very unusual in that very few *S. hispidus* were collected and that the total abundance of the assemblage was consistently low. Thus, this reference area had a rodent community similar to 4 of the 5 most toxic contaminated areas as indicated by the overall toxicity index (Figure 13-1); these 5 areas were only 74% similar to all other assemblages. In winter, a cluster of 10 study areas (Figure 13-4) was apparent, which consisted of 7 reference areas and 3 contaminated areas that

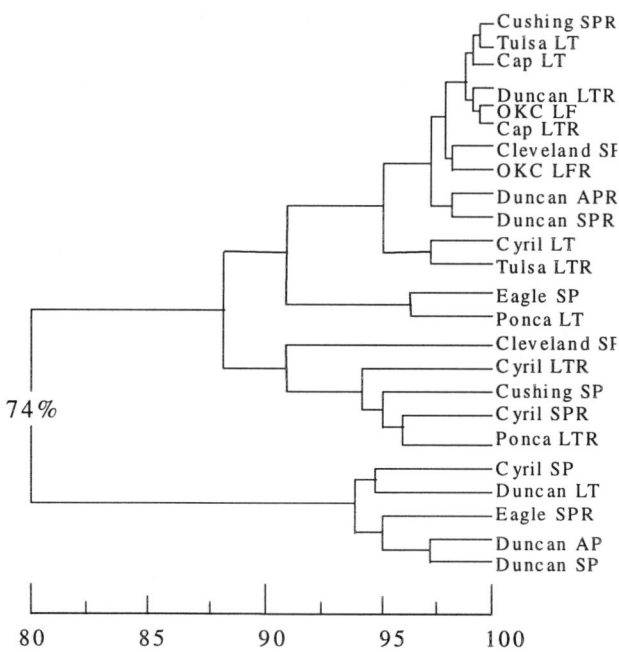

Figure 13-3 Dendrogram resulting from average-linkage cluster analysis of a matrix of similarity coefficients (Horn 1966) derived from the total abundances of species within rodent assemblages in summer on 12 study sites that were comprised of a suspected petrochemical-contaminated area and a paired reference area (designated by the letter "R" at the end of the site name) in disturbed, tallgrass prairie habitat in Oklahoma.

had a low toxicity index (Table 13-1). No other discernable clustering patterns were apparent for either season after using this technique of comparing structural characteristics of rodent assemblages.

Dynamic attributes of rodent assemblages

The relative stability of rodent assemblages across 4 trapping occasions within summer and winter was compared using s-indices (Figure 13-5). All rodent assemblages were relatively stable in summer, showing similar s-indices between contaminated and reference study areas for all species and for the total abundance of individuals within rodent assemblages ($P > 0.10$). In winter, differences in s-indices were observed between the study areas for *M. musculus* and the total abundance of individuals in rodent assemblages. Populations of *M. musculus* were less stable on contaminated (0.065 ± 0.015) compared with reference (0.029 ± 0.013) study areas in winter ($P < 0.05$), although abundances were mostly low on reference areas. The s-index for total abundance of individuals in rodent assemblages was also greater on contaminated (0.135 ± 0.016) compared with reference (0.094 ± 0.013) study areas in winter ($P < 0.047$).

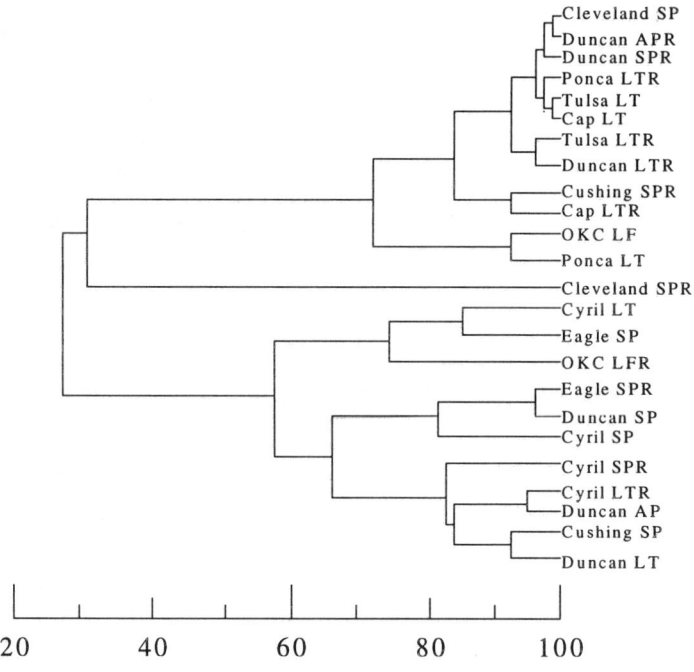

Figure 13-4 Dendrogram resulting from average-linkage cluster analysis of a matrix of similarity coefficients (Horn 1966) derived from the total abundances of species within rodent assemblages in winter on 12 study sites that were comprised of a suspected petrochemical-contaminated area and a paired reference area (designated by the letter "R" at the end of the site name) in disturbed, tallgrass prairie habitat in Oklahoma.

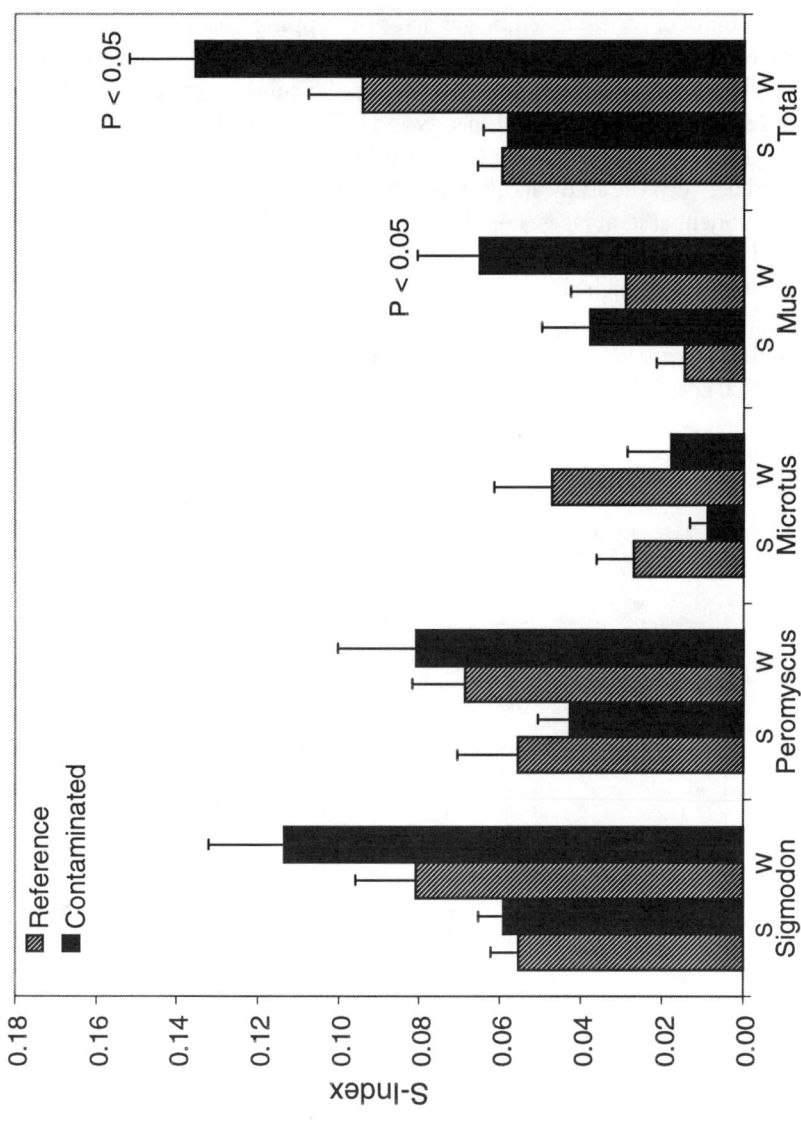

Figure 13-5 Mean (SE bar) measures of population stability across 4 trapping occasions in summer (S) and winter (W) for *S. hispidus*, *Peromyscus* spp., *Microtus* spp., and *M. musculus*. Stability was expressed as an *s*-index calculated as the standard deviation of the log-transformed estimate of abundance as described by Lewontin (1966). Results of statistical comparisons between reference and contaminated areas within a season are depicted as $P < 0.05$.

Predictive relationships

Stepwise logistic-regression analysis was used to ascertain the predictive value of subsets of demographic variables in determining whether a study area was a contaminated or reference site, without considering actual contaminant levels in the soil. Subsets of variables included for model selection were s-indices, mean abundance across 4 trapping occasions, and seasonal abundance (Table 13-4). The overall logistic-regression model selected the s-index for winter populations of *S. hispidus* and *M. musculus* as the best predictor of site contamination in winter when only s-indices of all populations were considered in the analysis ($c^2 = 5.340$, $d.f. = 2$, $P = 0.02$). The s-index was related positively to the probability of site contamination, with less-stable populations (high s-index) being more likely to have come from a contaminated area. Overall concordance for the s-index model for winter was 86.1%.

Table 13-4 Logistic-regression models, analysis of maximum-likelihood estimates, and concordance outcomes for predicting contamination of habitats (overall index of toxicity) for rodent assemblages on 24 study areas that were comprised of suspected petrochemical-contaminated areas and reference areas in disturbed, tallgrass prairie habitats in Oklahoma

Summer							
Equation	Parameter	Estimate	SE	χ^2	P-value	Concordance (%)	
Mean abundance	Overall model			7.724	0.005	69.4	
	Intercept	−4.496	1.908	5.340	0.020		
	M. musculus	3.529	1.872	3.534	0.050		
Seasonal abundance	Overall model			8.678	0.003	70.1	
	Intercept	−2.841	1.323	4.605	0.032		
	M. musculus	1.679	0.809	4.301	0.038		
Winter							
Equation	Parameter	Estimate	SE	χ^2	P-value	Concordance (%)	
s-index	Overall model			10.609	0.005	86.1	
	Intercept	−4.411	1.908	5.340	0.020		
	S. hispidus	28.769	13.503	4.539	0.033		
	M. musculus	35.849	16.138	4.934	0.026		
Mean abundance	Overall model			4.662	0.031	68.1	
	Intercept	−1.879	1.044	3.235	0.072		
	M. musculus	1.139	0.610	3.478	0.062		
Seasonal abundance	Overall model			6.511	0.038	77.1	
	Intercept	−0.011	1.123	0.001	0.992		
	M. musculus	0.465	0.287	2.606	0.106		
	M. ochrogaster	−0.656	0.509	1.660	0.197		

When only mean abundance values for each species within a season were entered into the variable list, *M. musculus* was selected as the best predictor of site contamination for summer ($c^2 = 3.534$, $d.f. = 1$, $P = 0.05$) and winter ($c^2 = 3.478$, $d.f. = 1$, $P = 0.062$). In both models, relationships were positive, with those rodent assemblages that had a greater mean abundance of *M. musculus* being more likely to have been from a contaminated area. Overall concordance for these models was 69.4% in summer and 68.1% in winter. When using only seasonal abundance estimators in the variable list, the stepwise procedure selected *M. musculus* as the best predictor in summer ($c^2 = 4.678$, $d.f. = 1$, $P = 0.038$) and *M. musculus* and *M. ochrogaster* as the best predictors in winter ($c^2 = 0.001$, $d.f. = 2$, $P = 0.992$). The relationship was positive for *M. musculus*, while an inverse relationship existed for seasonal abundance of *M. ochrogaster* (concordance for models in summer and winter was 70.1% and 77.1%, respectively).

Stepwise multiple-linear regression analysis was used to explore the relationship of various demographic and community parameters to actual concentrations of contaminants in the soil on the 24 study areas. These areas varied greatly in concentrations of organics and metals in the soil. Seasonal abundance of summer *M. musculus* populations was selected by the stepwise procedure as the independent variable demonstrating the best-fit, single-variable model relating intrinsic attributes of rodent assemblages to soil concentrations of contaminants (Figure 13-6). The model relating seasonal *M. musculus* abundance to the total concentration of heavy metal in the soil demonstrated the largest coefficient of determination ($R^2 = 0.63$, $P < 0.0001$; Table 13-5). The poorest relationship was derived when attempts were made to relate seasonal *M. musculus* abundance to total PAH concentrations in the soil ($R^2 = 0.33$, $P < 0.0032$).

Stepwise regression revealed that a 3-variable model provided a good fit for estimating actual soil concentrations of contaminants on the 24 study areas (Table 13-6). An estimated 78% of the variation in levels of heavy metals in the soil could be explained by 3 variables: seasonal abundance (summer) and *s*-index (summer) of *M. musculus* and winter *s*-index of *S. hispidus* ($P < 0.0001$). A strong relationship ($R^2 = 0.78$, $P < 0.0001$) also was evident for estimating the overall toxicity index of each site based on the independent variables mean abundance of *C. hispidus* (summer), seasonal abundance of *M. musculus* (summer), and *s*-index of *S. hispidus* (winter). The relationship of intrinsic attributes of rodent assemblages to total PAH concentrations in the soil was significant but weak ($R^2 = 0.52$, $P < 0.0017$). The stepwise regression procedure ultimately developed a 7-variable model to provide the best-fit relationship for predicting the toxicity index of each study area, with a significant coefficient of determination ($R^2 = 0.94$, $P < 0.0001$; data not shown).

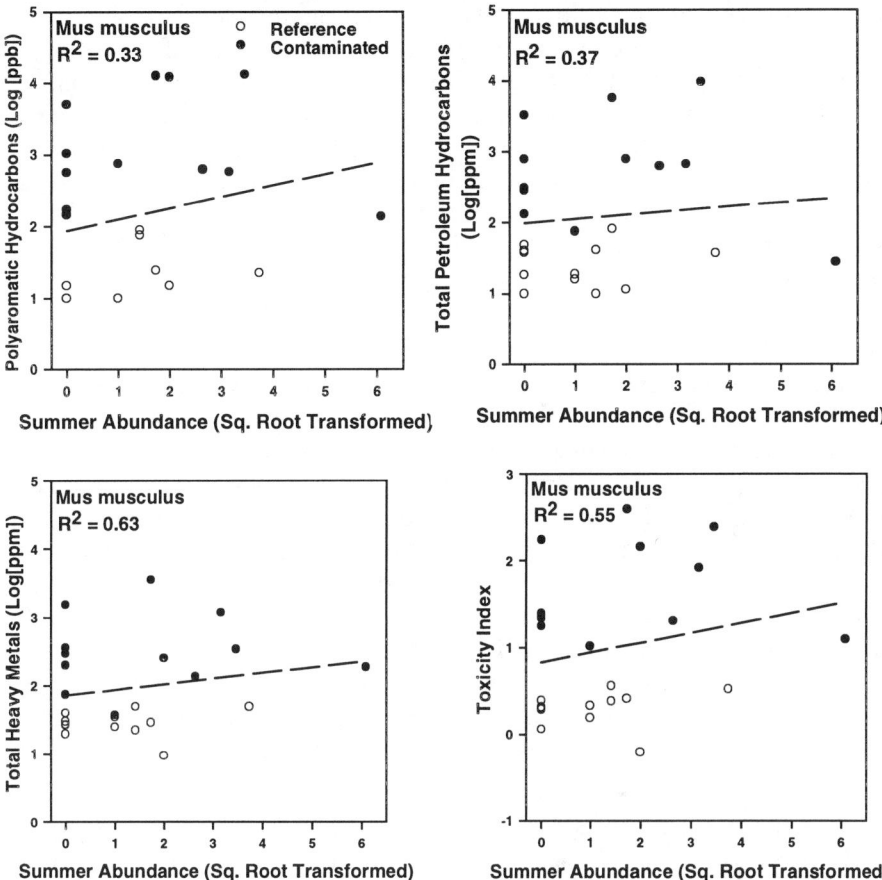

Figure 13-6 Scatter plot of observations depicting the relationship between seasonal abundance (square-root transformed) of *M. musculus* populations in summer and average concentrations ($\log_{10}$ transformed) of heavy metals (Cr and Pb), total PAHs, and TPHs, and a relative toxicity index, for 24 study areas in disturbed, tallgrass prairie habitat in Oklahoma. A regression line derived from equations in Table 13-5 has been plotted.

Discussion

General

The 12 reference areas were similar in soil concentrations of organics and metals, whereas the 12 contaminated areas varied greatly in the relative levels of TPH, PAH, and total heavy metals. Limited historical information for these contaminated areas indicated that a wide variety of refinery wastes was stored in impoundments or incorporated into the soil. Consequently, it is accurate to assume that no 2 areas had the same contaminant mixtures present in soil and that all areas differed remarkably in concentrations of each contaminant. This variation, coupled with differences in soils, degrees of phytotoxicity, and vegetation among study sites, accounted for much of the variability observed in the rodent communities. Some of this variation was controlled experimentally through the use of paired reference locations, but the assessment procedure itself must be robust toward unavoidable differences in the type and concentration of contaminants. Utilizing paired reference and contaminated sites allowed changes in the population that were due to extrinsic factors to be identified and reduced the number of variables that could be responsible for the effects observed. Only changes between the paired sites were assumed to be a result of site contamination.

Alterations in rodent assemblages

The most common alteration that we observed on contaminated study areas was the low abundance of microtines and elevated abundance of *M. musculus* and *C. hispidus* relative to reference areas. These differences became more apparent in the winter when *S. hispidus* populations declined in all study areas. Populations of *M. musculus* are not normally abundant on disturbed tallgrass prairie habitat whose densities are normally less than 1 individual per ha (Grant and Birney 1979; Kaufman and Kaufman 1990). Regression analyses suggested that structural (abundance estimators for *M. musculus* and *C. hispidus*) and dynamic (*s*-indices for *M. musculus* and *S. hispidus*) attributes of rodent assemblages may provide some insight into how petrochemical wastes in the environment can alter vertebrate communities in terrestrial systems. Logistic regression models and goodness-of-fit tests suggested that selected combinations of intrinsic demographic variables were associated with

Table 13-5 Best-fit single-variable models relating contaminant concentrations in soil to intrinsic attributes of rodent assemblages inhabiting 24 study areas in disturbed tallgrass prairie habitats in Oklahoma using stepwise multiple-linear regression analysis. TPH (ppm), total PAH (ppb), THM (ppm), and a relative toxicity index (calculated from levels of the other 3 contaminant groups) were regressed against a variety of independent variables consisting of seasonal estimates of total assemblage and population abundance and stability (*s*-index), and species-richness indices. Coefficients of determination (R^2) and significance of the regression relationship are indicated for each model.

Dependent/independent variables[1]	Estimate	SE	d.f.	P-value	R^2
Toxicity index			1,23	0.0001	0.55
Intercept	0.270	0.186		0.1612	
Seasonal abundance	0.424	0.081		0.0001	
TPH			1,23	0.0015	0.37
Intercept	0.867	0.334		0.0167	
Seasonal abundance	0.532	0.146		0.0015	
THM			1,23	0.0001	0.63
Intercept	1.178	0.154		0.0001	
Seasonal abundance	0.417	0.067		0.0001	
PAH			1,23	0.0032	0.33
Intercept	0.589	0.450		0.2044	
Seasonal abundance	0.653	0.197		0.0032	

[1]Dependent variables were log-transformed. Seasonal abundance (minimum number known alive over 4 trapping occasions) of summer populations of *M. musculus* was selected as the best-fit for all models and values were square-root transformed.

assemblages of rodents on contaminated areas. The significant relationships between these intrinsic variables and actual soil concentrations of contaminants (especially heavy metals) suggested a link between contamination and observed alterations in rodent assemblages.

There are a few previous studies examining the sensitivity of small mammal assemblages to the presence of petrochemical contaminants in the environment. Rowley et al. (1983) noted that *Microtus pennsylvanicus*, a resident species, appeared sensitive to the contaminants (i.e., insecticides and aromatic hydrocarbons) present in the soil at Love Canal and that this sensitivity resulted in greater species richness on a reference site but greater diversity on the contaminated site. Investigators working with our laboratory have documented changes in the demographics of *S. hispidus* populations on the Oklahoma Refining Company (McMurry 1993) and Royal Hardage (Elangbam et al. 1989) Superfund waste sites in Oklahoma. In both of these studies, peak densities were consistently greater on reference study sites compared with those sites contaminated with complex mixtures of petrochemicals. McMurry (1993) reported the absence of certain uncommon species of small mammals on contaminated study sites (i.e., *M. ochrogaster*, *M. pinetorum*, and

Table 13-6 Best-fit 3-variable models relating contaminant concentrations in soil to intrinsic attributes of rodent assemblages inhabiting 24 study areas in disturbed tallgrass prairie habitats in Oklahoma using stepwise multiple-linear regression analysis. TPH (ppm), total PAH (ppb), THM (ppm), and a relative toxicity index (calculated from levels of the other 3 contaminant groups) were regressed against a variety of independent variables consisting of seasonal measures of total assemblage, population abundance, stability (s-index), and species-richness indices.

Dependent/independent variables[1]	Estimate	SE	d.f.	P-value	R^2
Toxicity index			3,23	0.0001	0.78
Intercept	−1.189	0.342		0.0024	
Mean abundance CH (sum)	1.061	0.242		0.0003	
Season-abundance MM (sum)	0.351	0.061		0.0001	
s-index SH (win)	3.560	1.317		0.0137	
TPH			3,23	0.0001	0.65
Intercept	−1.557	0.662		0.0291	
Mean abundance CH (sum)	1.731	0.470		0.0015	
Season-abundance MM (sum)	0.414	0.120		0.0025	
s-index SH (win)	6.253	2.551		0.0236	
THM			3,23	0.0001	0.78
Intercept	0.697	0.195		0.0019	
Season-abundance MM (sum)	0.753	0.115		0.0001	
s-index SH (win)	2.033	1.183		0.1013	
s-index MM (sum)	−13.740	4.194		0.0038	
PAH			3,23	0.0017	0.52
Intercept	−2.020	1.011		0.0595	
Mean abundance CH (sum)	1.868	0.718		0.0171	
Season-abundance MM (sum)	0.526	0.183		0.0094	
s-index SH (win)	6.685	3.895		0.1016	

[1]Dependent variables were $\log_{10}$-transformed. CH = *Chaetodipus hispidus*; MM = *Mus musculus*; SH = *Sigmodon hispidus*

shrews). However, another normally rare species (i.e., *M. musculus*) on reference sites was observed to periodically reach remarkably high densities on certain contaminated study sites. The only other relevant study of petrochemical-contaminated environments of which we are aware was conducted in Houston, Texas, by Flickinger and Nichols (1990), who failed to find any adverse effects of soil contaminants on densities of resident *S. hispidus* populations.

Similar patterns of interspecific sensitivities to contaminants have been noted in other studies related to pesticides and metal-polluted sites. Kataev et al. (1994) noted that the density of the most common microtine (*Clethrionomys rufocanus*) was lowest at sites close to a smelter and increased with distance from the source of contamination. They also noted a complete lack of uncommon microtines (i.e., *C. glareolus*, *C. rutilus*, and *Lemmus lemmus*) at sites close to the smelter. Microtines also have been observed to be more sensitive to carbamate pesticides than is *M. musculus*, which increased in abundance in treated enclosures (Barrett 1968, 1988). In both of these studies it appeared that reproductive function may have been adversely affected in microtines but not in *M. musculus*. Laboratory studies have demonstrated interspecific differences in sensitivity to organophosphate insecticides between microtines and *M. musculus* (Meyers and Wolff 1994).

Tendency towards instability

The overall degree of stability across a season, as measured by the s-index, for the rodent assemblages in this study were generally characteristic of what is found in tallgrass prairie habitats in the midwest. As noted by Ostfeld (1988), very low s-indices are indicative of constancy in numbers but are not necessarily a measure of long-term equilibria. With that caveat in mind, fluctuations in populations and the total abundance of individuals in rodent assemblages supported our initial hypothesis that those species inhabiting petrochemical-contaminated habitats would be prone to demonstrating erratic fluctuations in abundance as sensitive species decline and as opportunistic species increase in response to direct or indirect effects. These tendencies toward instability were more prominent in winter than in summer and were evident in *M. musculus* populations and in the total rodent assemblage. Even fluctuations in abundance for the dominant member of these assemblages, *S. hispidus*, showed a significant correlation with soil contamination. These relationships suggest that population and interspecific relationships are more dynamic and fluid in response to soil contamination and could provide a useful endpoint for assessing the degree of ecotoxicity.

Nearly all ecotoxicological efforts have neglected to incorporate the inherent dynamics of small mammal assemblages across a temporal scale into their assessments of how environmental contaminants affect such systems. A notable exception has been the work of Kataev et al. (1994), who observed that microtine populations fluctuated cyclically, as predicted, on less-polluted areas, while those populations nearest a smelter showed no clear regularity in their fluctuations, partly because of low densities.

Direct mechanisms of influence

It is not known whether observed differences in the structure and function of rodent assemblages were a result of direct toxic effects on selected species or if they were mediated through a more complex set of indirect effects on competition, predation, habitat quality, and other processes. There is evidence that the complex mixtures of contaminants present on many of these study sites could have had direct toxic effects on resident small mammals. Elevated hepatic cytochrome P-450 isoenzyme induction was observed in resident *S. hispidus* collected from the Cyril, OK, oil-refinery study site, and differences were most evident in summer collections (Lochmiller et al. 1998). McMurry et al. (1998) reported significant alterations in selected measures of immunotoxicity in resident *S. hispidus* from these same study sites. These alterations included differences in organ mass and cellularity, hematology, in vivo T-cell hypersensitivity, macrophage metabolic activity, natural killer cell tumorcidal activity, and T- and B-lymphocyte lymphoproliferative responsiveness between contaminated and reference populations. Concurrent studies using mesocosms established on specific sites of contamination at Cyril, OK, where laboratory-bred *S. hispidus* were housed for up to 6 weeks of exposure, did not show dramatic alterations in immune system function (Propst et al. 1995, 1998). Rafferty et al. (1998) observed immune system lesions in *S. hispidus* that were collected from land-treatment study sites where petrochemical wastes were tilled into the soil for biodegradation.

Other evidence for direct exposure effects on resident small mammals has been gathered from postmortem examinations. Remarkable tooth lesions that are consistent with dental fluorosis have been observed on several of our study sites in nearly all rodent species sampled, but those residing in close association to the soil (e.g., *S. hispidus*) and those consuming insects (e.g., shrews) tended to have the highest levels of fluoride in bone tissue and should possess more severe forms of dental lesions (Paranjpe et al. 1994; Rafferty 1998; Schroeder et al. 1998). In addition, *S. hispidus* collected from many of our study areas were found to contain elevated levels of heavy metals and fluoride in bone and liver tissue (Schroeder 1998). Similarly, Rattner et al. (1993) observed alterations in liver mass of *S. hispidus* residing on petrochemical-contaminated industrial sites in Texas.

Indirect mechanisms

The likelihood that indirect effects of petrochemical wastes influence rodent assemblages appears to be high on our study areas. Competitive interactions and other interspecific encounters are assumed to be pervasive functional attributes of all rodent communities (Rose and Birney 1985), and these interspecific interactions within rodent assemblages can become extremely complex owing to strongly connected multispecies assemblages (Hanski and Henttonen 1996). A decline in abundance of 1 species undoubtedly has the potential to influence other species through altered competitive relationships or resources across space and time. The winter increase in microtines and *R. fulvescens* on study areas was probably a response to the reductions in abundance of *S. hispidus*, which tends to be more dominant in these interspecific interactions (Cameron 1977; Glass and Slade 1980). The significant increases in abundance of *M. musculus* populations on contaminated areas in summer, when *S. hispidus* populations were most abundant and similar on both areas, were not consistent with the competition hypothesis. It is possible that food resources differed across treatments, although we attempted to match reference areas and contaminated areas with respect to vegetation composition and structure. It is also possible that changes in the abundance of other, less dominant members (e.g., microtines) of the rodent assemblage reduced competition for key resources and permitted increases in *M. musculus* populations. However, little is known about competitive interactions between *M. musculus* and other rodents. Documenting such indirect effects would be extremely difficult because the dynamics and interrelationships among species of most rodent assemblages have been poorly studied relative to population ecology of individual species, especially those involving *M. musculus* (Rose and Birney 1985).

Conclusions

Although a number of environmental factors can initiate fluctuations in rodent populations in both density-independent and density-dependent fashions, most of these factors should have been controlled by our experimental design, which incorporated the use of local reference study sites that were similar in vegetation and pattern of the count grid. This is important because habitat types are known to have a profound influence on distribution and demographic patterns of *S. hispidus* and other rodents (McMurry et al. 1994, Goertz 1964). Consequently, we propose that petrochemical contamination of habitats was the primary factor responsible for the observed structural and functional attributes that differed from reference areas and that these differences were associated with soil-contaminant levels. It was apparent that rodent assemblages were inherently variable across spatial and temporal scales, which complicates efforts directed toward identifying the ultimate consequences of petrochemical contaminants on their structure and function. These efforts are made difficult because of the historical neglect of basic community ecology of rodents in the literature. Despite these difficulties, several interesting structural characteristics were unique to reference or contaminated study areas, although none of the characteristics were universally diagnostic of prior contamination of habitats. The dominance and stability (s-index) of *M. musculus* populations in rodent assemblages were the most consistent alterations observed on contaminated areas. However, these results also indicate that populations of *C. hispidus* and microtines may be important indicator species of petrochemical contamination. The strong relationships between various demographic attributes of rodent assemblages and actual concentrations of contaminants in soil provide promising insights into how models may be used to assess site contamination in disturbed, tallgrass prairies. It is important to stress that proper validation of models with other unique data sets should be done to evaluate the predictive powers of these models. We do not imply that there is a direct causal relationship with these models but merely that alterations in rodent assemblages do occur in response to some disturbance mechanism associated with the disposal of petrochemical wastes in such habitats.

Acknowledgments—The authors greatly appreciate the assistance of Lee Jones in the laboratory and Dr. C. L. Goad, Department of Statistics, Oklahoma State University, with data analysis. Financial support for this research was primarily provided through the U.S. Air Force Office of Scientific Research. Additional support was obtained through the National Science Foundation (IBN-9318066) and the Department of Zoology, Oklahoma State University. This research was approved by the Oklahoma State University Institutional Animal Care and Use Committee as Protocol 141.

References

Barrett GW. 1968. The effects of an acute insecticide stress on a semi-enclosed grassland ecosystem. *Ecology* 49:1019–1035.

Barrett GW. 1988. Effects of Sevin on small-mammal populations in agriculture and old-field ecosystems. *J Mammal* 69:731–739.

Bradley LJN, Magee BH, Allen SL. 1994. Background levels of polycyclic aromatic hydrocarbons (PAH) and selected metals in New England urban soils. *J Soil Contam* 3:1–13.

Cairns J Jr, Niederlehner BR, Smith EP. 1995. Ecosystem effects: Functional end points. In: Rand GM, editor. Fundamentals of aquatic toxicology: Effects, environmental fate, and risk assessment. Second ed. Bristol PA: Taylor and Francis. p 589–607.

Cameron GN. 1977. Experimental species removal: Demographic responses by *Sigmodon hispidus* and *Reithrodontomys fulvescens*. *J Mammal* 58:488–506.

Coppock RW, Mostrom MS, Khan AA, Semalulu SS. 1995. Toxicology of oil field pollutants in cattle: A review. *Vet Human Toxicol* 37:569–576.

Dutilleul P. 1993. Spatial heterogeneity and the design of ecological field experiments. *Ecology* 74:1646–1658.

Elangbam CS, Qualls CW, Lochmiller RL, Novak J. 1989. Development of the cotton rat (*Sigmodon hispidus*) as a biomonitor of environmental contamination with emphasis on hepatic cytochrome P-450 induction and population characteristics. *Bull Environ Contam Toxicol* 42:482–488.

Flickinger EL, Nichols JD. 1990. Small mammal populations at hazardous waste disposal sites near Houston, Texas, USA. *Environ Pollut* 65:169–180.

Glass GE, Slade NA. 1980. Population structure as a predictor of spatial association between *Sigmodon hispidus* and *Microtus ochrogaster*. *J Mammal* 61:473–485.

Goertz JW. 1964. The influence of habitat quality upon density of cotton rat populations. *Ecol Monogr* 34:359–381.

Grant WE, Birney EC. 1979. Small mammal community structure in North American grasslands. *J Mammal* 60:23–36.

Hanski I, Henttonen H. 1996. Predation on competing rodent species: A simple explanation of complex patterns. *J Anim Ecol* 65:220–232.

Horn HS. 1966. Measurement of overlap in comparative ecological studies. *Am Nat* 100:419–424.

Kaufman DW, Kaufman GA. 1990. House mice (*Mus musculus*) in natural and disturbed habitats in Kansas. *J Mammal* 71:428–432.

Kataev GD, Suomela J, Palokangas P. 1994. Densities of microtine rodents along a pollution gradient from a copper-nickel smelter. *Oecologia* 97:491–498.

Krebs CJ. 1989. Ecological methodology. New York NY: Harper Collins.

Lewontin RC. 1966. On the movement of relative variability. *Systematic Zoology* 15:141–142.

Lochmiller RL, McMurry ST, McBee K, Qualls CW, Rafferty DP. 1998. Hepatic cytochrome P-450 isoenzyme induction as an indicator of exposure in cotton rats (*Sigmodon hispidus*) inhabiting petrochemical waste sites. *Environ Pollut* (in press).

McBee K, Bickham JW. 1990. Mammals as bioindicators of environmental toxicity. In: Genoways HH, editor. Volume 2, Current mammalogy. New York NY: Plenum Press. p 37-88.

McMurry ST. 1993. Development of an in situ mammalian biomonitor to assess the effect of environmental contaminants on population and community health [dissertation]. Stillwater OK: Oklahoma State University. Available from: Ann Arbor MI: University Microfilms.

McMurry ST, Lochmiller RL, Boggs J, Leslie DM, Engle DM. 1994. Demographic profiles of populations of cotton rats in a continuum of habitat types. *J Mammal* 75:50–59.

McMurry ST, Lochmiller RL, McBee K, Qualls CW. 1999. Indicators of immunotoxicity in populations of cotton rats (*Sigmodon hispidus*) inhabiting petrochemical-contaminated environments. *Ecotoxicol Environ Saf* 42:223-235.

Meyers SM, Wolff JO. 1994. Comparative toxicity of azinophos-methyl to house mice, laboratory mice, deer mice, and gray-tailed voles. *Arch Environ Contam Toxicol* 26:478–482.

Ostfeld RS. 1988. Fluctuations and constancy in populations of small rodents. *Am Nat* 131:445–452.

Paranjpe MG, Chandra AM, Qualls CW, McMurry ST, Rohrer MD, Whaley MM, Lochmiller RL, McBee K. 1994. Fluorosis in a wild cotton rat (*Sigmodon hispidus*) population inhabiting a petrochemical waste site. *Toxicol Pathol* 22:569–578.

Propst TL, Lochmiller RL, Qualls CW. 1995. Using mesocosms to assess immunotoxicity risks in terrestrial environments contaminated with petrochemicals. In: Frink L, Ball-Weir K, Smith A, editors. The effects of oil on wildlife: Response, research and contingency planning. Newark DE: Tri-State Bird Rescue & Research, Inc. p 160–166.

Propst TL, Lochmiller RL, Qualls CW, McBee K. 1999. In situ (mesocosm) assessment of immunotoxicity risks to small mammals inhabiting petrochemical waste sites. *Chemosphere* 38:1049-1067.

Rafferty DP. 1998. Immunotoxicity risks to small mammals from land treatment of petrochemical wastes [master science thesis]. Stillwater OK: Oklahoma State University. Available from: Ann Arbor MI: University Microfilms.

Rafferty DP, Lochmiller RL, McBee K, Qualls CW, Basta NT. 2000. Immunotoxicity risks associated with land treatment of petrochemical wastes revealed using an in situ rodent model. *Environ Pollut* (in press).

Rattner BA, Flickinger EL, Hoffman DJ. 1993. Morphological, biochemical, and histopathological indices and contaminant burdens of cotton rats (*Sigmodon hispidus*) at three hazardous waste sites near Houston, Texas, USA. *Environ Pollut* 79:85–93.

Rose RK, Birney EC. 1985. Community ecology. In: Tamarin RH, editor. Biology of New World *Microtus*. American Society of Mammalogists: Special Publ. No. 8. p 310–339.

Rowley MH, Christian JJ, Basu DK, Pawlikowski MA, Paigen B. 1983. Use of small mammals (voles) to assess a hazardous waste site at Love Canal, Niagara Falls, New York. *Arch Environ Contam Toxicol* 12:383–397.

Statistical Analysis System Institute Inc [SAS]. 1990. SAS user's guide: Statistics. Cary NC: SAS.

Schroeder JL. 1998. Bioaccumulation and exposure pathways of soil contaminants to cotton rats on petrochemical sites [master science thesis]. Stillwater OK: Oklahoma State University. Available from: Ann Arbor MI: University Microfilms.

Schroeder JL, Basta NT, Rafferty DP, Qualls CW, Lochmiller RL, McBee K. 1999. Soil and vegetation fluoride exposure pathways to cotton rats on a petrochemical-contaminated landfarm. *Environ Toxicol Chem* 18:2028-2033.

Talmage SS, Walton BT. 1991. Small mammals as monitors of environmental contaminants. *Rev Environ Contam Toxicol* 119:47–145.

Index

A

Abundance
    calculation, rodent assemblages, 281
    factors which affect, 22, 169
    predictive ratios of taxa, 222
    relative, effect of mortality rate of one species, 157–159, 160
    rodent assemblages in Oklahoma prairie, 285, 286, 287, 296, 297–298
    from survival and fecundity changes, 227, 235–241
Accident assessment, 102–104, 205
Accipiter cooperii, 33
Accipiter nisus, 31, 49
Accipiter striatus, 33
AChE. *See* Cholinesterase inhibition
Acid deposition, 44–45
    community-level effects, 78
    ecosystem-level effects, 79
    effects on amphibians, 141
    effects on boreal forest, 10
    mercury from, 38–39
Adaptation, genetic, 209
Adaptive management, 15
Additive hunting, 5
African penguin. *See Spheniscus demersus*
Agency for Toxic Substances and Disease Registry (ATSDR), 83
Aggregation, definition, 111
Agricultural land use. *See also* Pesticides
    crop rotation, 34
    effects on bird population dynamics, 27
    hectares, 1960-2000, 11
    organic, 34–35
    three main ecological effects of pesticides, 34
Agricultural runoff. *See* Selenium
Aimophila aestivalis, 27

Air pollutants
    effects on amphibians/reptiles, literature review, 136
    effects on mammals, literature review, 128
    emissions from sulfide ore smelters, 36–37
ALAD. *See* Aminolevulinic acid dehydratase
Alaska Department of Fish and Game, wolf study, 5–6
Alauda arvensis, 34–35
Albinism, 44
Alces alces, 3, 5, 10
Aldrin, 115
Algae. *See* Primary producers
Alien species, 4, 12
Alkyl-mercury, 31
Alligator mississippiensis, 71, 72, 74
Allometric scaling, 195–196
Aluminum, 72
Ambrosia psilostachya, 280
American avocet. *See Recurvirostra americana*
American coot. *See Fulica americana*
American crocodile. *See Crocodylus acutus*
American kestrel. *See Falco sparverius*
American robin. *See Turdus migratorius*
American toad. *See Bufo americanus*
Aminocarb, 29
Aminolevulinic acid dehydratase (ALAD), 46, 67, 84, 210, 219
Amphibian(s)
    deformities, 43, 71
    effects of acid deposition, 141
    as indicator species, 137, 138, 139
    population-level effects, literature review, 123–134
    reproductive success, 74
Anas platyrhynchos, 62, 76, 116, 195
Anas rubripes, 116
Anatomical endpoints, 67
Andropogon gerardii, 280
Animal rights, 63
Anolis carolinensis, 139
Anser spp., 31

Anthropogenic factors. *See also* Agricultural land use
    on peregrine falcon populations, 115
    role in ecosystem management, 9
    synchrony among populations and, 257–258
    water use on the Everglades, 10
    winter feeding, 8–9
Antilocapra americana, 3
Apollo Sea oil spill, 40
Aquatic studies, population-level effects, 177–187, 208, 222
Arsenic, 36
Artemisia ludoviciana, 280
Asia spp., 155
"Assessment endpoints," 62
Aster spp., 280
Asynchrony and metapopulation persistence, 257–258, 259
Athene cunicularia, 15
Aythya collaris, 37, 44
Aythya valisineria, 38

B

Bachman's sparrow. *See Aimophila aestivalis*
Bacteria
    decomposition of contaminants, 194
    heavy metal exposure and infections, 13
Baiomys taylori, 283
Balance. *See* Equilibrium
Bald eagle. *See Haliaeetus leucocephalis*
Baseline data. *See* Reference norm
Bats, 128
Beaver. *See Castor canadensis*
Behavior, 159
    aggression, 159
    effects of heavy metals, 13
    endpoints, 241
    predator avoidance, 33
    reproductive, 7, 33
Belted kingfisher. *See Ceryle alcyon*
Bermuda grass. *See Cynodon dactylon*
Big bluestem. *See Andropogon gerardii*
Bighorn sheep. *See Ovis canadensis*

Bioaccumulation, 22, 127
Bioassays, 62, 241
Biochemical indicators, 62–63, 67
Bioconcentration
    heavy metals, 13
    of pesticides in falcons, 115
Biodiversity
    as an endpoint, 48
    role in ecosystem management, 9
Bioeffects monitors, 62–63
Bioindicator responses, 62
Biomarkers
    biochemical, 46, 67, 69, 192
    definition, 62
    as endpoints for assessment, 186, 219
    functions, 46, 49
    further needs, 84
    genetic, 67
Bird(s)
    difficulties in field measurements, 31
    direct effects of pesticides, 23, 28–34, 41–43, 62. *See also* Eggshell thinning
    effects of heavy metals, 36, 37–39
    effects of oil spills, 39–41, 71
    effects of PCBs, 42–43, 70–71
    effects of radioactive fallout, 43–44
    indirect effects of pesticides, 34–36, 115
    interpopulation extrapolation, 212
    interspecies extrapolation, 195
    population-level effects, 114–123, 207
    population modeling, 26–27
    in reef food-web, 167
    surveys, 212
    teratogenesis, 70–71
Birth rate, 21, 75
Bison. *See Bison bison*
Bison bison, 3
Black-crowned night heron. *See Nycticorax nycticorax*
Black duck. *See Anas rubripes*
Black-footed ferret. *See Mustela nigripes*
Black-legged kittiwake. *See Rissa tridactyla*
Black-necked stilts. *See Himantopus mexicanus*

Black oystercatcher. *See Haematopus bachmani*
Black rail. *See Laterallus jamaicensis*
Black-tailed prairie dog. *See Cynomys ludovicianus*
Blocking, experimental, 99–100
Bobwhite quail. *See Colinus virginianus*
Bone disorders, 70, 297
Boreal forest, 9–10, 29
Brachyramphus marmoratus, 7–8
Branta canadensis, 31
Brevoortia patronus, 183, 184
Brine fly, 165
Brine shrimp, 165
British Trust for Ornithology, 116
Brome. *See Bromus* spp.
Bromus spp., 280
Brooding success, 73. *See also* Reproductive success
Brown pelican. *See Pelecanus occidentalis*
Brown thrasher. *See Toxostoma rufum*
Brown trout. *See Salmo trutta*
Bubo virginianus, 76
Budworm. *See Choristoneura* spp.
Bufo americanus, 71
Bullfrog. *See Rana catesbeiana*
Burrowing owl. *See Athene cunicularia*
Buteo regalis, 15
Buteo swainsoni, 32, 69

C

Calcarius lapponicus, 31
Calcium
    bioavailability and acid rain, 44–45
    translocation in eggshell formation, 72
California condor. *See Gymnogyps californianus*
Calmodulin, 72
Canada geese. *See Branta canadensis*
Canis latrans, 3
Canis lupus, 2–4
Canvasbacks. *See Aythya valisineria*
Captive-breeding programs, black-footed ferret, 15

Carbamates. *See also* Cholinesterase inhibition
    AChE biomarker for, 68
    clinical signs of poisoning, 134
    effects on amphibians/reptiles, literature review, 136
    effects on birds, 31
    effects on mammals, literature review, 77, 124, 128, 133–134
    effects on rodent assemblages, 296
    genotoxicity study, 77
Carbofuran, 31–32
Cardinalis cardinalis, 33–34
Caribou. *See Rangifer tarandus*
Carp. *See Cyprinus carpio*
Carrying capacity
    effects of habitat degradation and loss, 21
    on organic *vs.* conventional farmland, 35
    use of hunting to maintain, 5, 6
Caspian tern. *See Sterna caspia*
Cassia fasciculata, 280
Castor canadensis, 12
Cathartes aura, 38
Cattle, effects of grazing, 14, 15
Cellular automata models, 254–255
Centrocercus urophasianus, 7, 8, 14, 32, 76
Cepphus columba, 40–41
Cervus elaphus, 3, 8–9
Ceryle alcyon, 81
Chaetodipus hispidus
    effects of petroleum wastes, 286, 288, 294, 295
    seasonal abundance in Oklahoma prairie, 283, 285
Chelydra serpentina, 43
Chesapeake Bay striped bass. *See Morone saxatilis*
Chloredecone, 72
Cholinesterase inhibition, 68–69
    effects in birds, 30, 32, 211
    effects on small rodents and insectivores, 133–134
    in extrapolation to population, 211
    as indicator of mortality, 210
Choristoneura spp., 29

Chromium
    concentrations in Oklahoma prairie sites, 283
    effects on rodent assemblages, 280, 292–293
    in toxicity index of Oklahoma prairie sites, 284
Chronomids, 207, 209
Cinclus cinclus, 44
Circus spp., 155
Circus cyanus, 33
Clark Fork River, Mont., 12–13, 213
Clean Water Act, 132
Clethrionomys glareolus, 296
Clethrionomys rufocanus, 296
Clethrionomys rutilus, 296
Cluster analysis, 288, 289
Clutch size, and life history strategy, 28
Coking facilities, 43
Colinus virginianus, 76, 194
Colonization. *See* Dispersal
Colormarking, 49
Columba oenas, 31
Columbia National Fisheries Research Laboratory, 179
Common loon. *See Gavia immer*
Common sunflower. *See Helianthus annuus*
Community-level effects, 78, 81–82
    assessment by individual-level endpoints. *See under* Extrapolation
    endpoints, 190, 221–222
    intercommunity extrapolation, 213–214
    Lotka-Volterra model, 153
    management issues, 2–4
    mortality rate as indicator, 152–159, 160–170
    quantification of shift from equilibrium, 157
Community structure, evolution of components toward current, 6
Community structure analysis, rodent assemblages in Oklahoma prairie, 288
Compensatory hunting, 5

Competition
    "contest," 21
    dispersal and, 259
    interplot, 98
    intraspecific, 153, 169
    metapopulation persistence and, 259
    predator, 169
    in rodent assemblages, effects of petroleum wastes, 297–298
Comprehensive Environmental Response, Compensation and Liability Act (CERCLA), 48, 132
Conservation *vs.* biomonitoring, 65
Contaminant Hazard Review Series, 83
Contaminants. *See also* Metals; Pesticides; Petroleum products
    "action at a distance," 264
    additive or synergistic effects, 75
    characteristics, 22
    community-level effects. *See* Community-level effects
    continuously varying environments, 248–249, 253–255
    direct *vs.* indirect effects, 22, 68, 204
    ecological aspects. *See* Ecological risk assessment; Ecosystem-level effects
    factors contributing to vulnerability of a species, 46
    individual-level effects. *See* Individual-level effects
    landscape-level aspects. *See* Landscape-level effects
    lessons learned, 45–47
    life history strategies and impact of, 24–25, 26, 206, 225–241
    mixtures and multiple stressors, 198–199, 269
    population-level effects. *See* Models, population; Population-level effects
    quantification of exposure, 22–23
    regulatory standards, 79–83
    research needs, 47–49
    resistance to, 76–77

spatial distribution and habitat. *See* Spatial structure
vs. natural population-level stressors, 23–24, 123, 141
water pollution, 7, 8
"Contest competition," 21
Continuously varying environments, 248–249, 253–255
Control measures, 99
Convention on Biodiversity, 48
Conyza canadensis, 280
Cooper's hawk. *See Accipiter cooperii*
Corn bunting. *See Miliaria calandra*
Corvus frugilegus, 31
Cost-benefit analysis, 177
Cotton rat. *See Sigmodon hispidus*
Coyote. *See Canis latrans*
Criteria Maximum Concentrations (CMC), 81–82
Crocodilians, 136
Crocodylus acutus, 72
Cutthroat trout. *See Oncorhynchus clarki bouvieri*
Cyclodiene, 31
Cygnus olor, 37
Cynodon dactylon, 280
Cynomys spp., 213
Cynomys ludovicianus, 15
Cyprinus carpio, 12
Cytochrome P450 induction, 67, 84, 297
Cytotoxicity, 70

D

Dampening ratio, 239
Data
missing, "analog-selection" of, 198
need for cross-seasonal collection, 7–8
toxicology, problems and solutions, 178–182
DDE. *See* Dichlorodiphenyldichloroethylene
DDT. *See* Dichlorodiphenyltrichloroethane
Decision-making, 177, 187
Deer mice, 159
Deformities, 43, 71, 210–211
Degradation of contaminants, 194, 268–269
Deme, definition, 111

Density-dependence, in population modeling, 230–231
Dental fluorosis, 297
Design
experimental. *See* Study design
of toxicity tests, 181–182, 184–185
Diazinon, 178
Dichlorodiphenyldichloroethylene (DDE)
effects on alligators, 71
effects on birds. *See* Eggshell thinning
mechanism of action, 72
Dichlorodiphenyltrichloroethane (DDT)
effects on birds, 23, 32–34, 62, 207
modeling predictors of effects, 179
use in England, 115–116
use in New Brunswick forests, 29–30
Dicofol, 74
Dieldrin, 31, 32, 72, 115
Die-offs
fish kills as grounds for regulation, 178
from pesticides, 69, 122
Dimethoate, 32
Dioxins. *See also* 2,3,7,8-Tetrachlorodibenzo-*p*-dioxin
effects on birds, 70–71, 120–121
"mechanistic similarity" to, 199
Dipper. *See Cinclus cinclus*
Disease, 5, 13. *See also* Immunosuppression
Dispersal
competitive strength and, 259
in continuously varying environments, 248–249, 253–255
corridors, 24
incidence of patch occupancy, 250–251, 265
inter-patch, 21, 247, 262
to island populations, 252–253
persistence and, 257–258
between populations. *See* Population, source and sink
in study design, 100
DNA analysis, 67, 209. *See also* Genotoxicity
Dose-response relationship, 182, 186, 191, 206, 209, 245–246, 266

Double-crested cormorant. *See Phalacrocorax auritus*
Dreissena polymorpha, 213
Ducks, 37, 39, 119

E

Eared grebes. *See Podiceps nigricollis*
Earthworm, 213
Eastern cottontail. *See Sylvilagus floridanus*
Ecological risk assessment
    development of methods, 178
    hypothesis testing, 205–206, 241
    multi-level, 187
    population-level, 186
    precautionary principle, 186, 206
    predictive *vs.* retrospective, 109–110
Ecosystem-level effects, 78–79. *See also* Landscape-level effects
    assessment by individual-level endpoints. *See under* Extrapolation
    endpoints, 190
    Great Lakes water quality, 81–82
    interecosystem extrapolation, 213–214
    management, 9–16
    trophic interactions, 131, 207, 213–214
Ecotoxicology
    definition and use of term, 19, 83
    landscape, 19, 109
    single-species measurements and, 149, 151–152
Education of risk assessors, 185, 186
Eggs. *See also* Eggshell thinning
    clutch size, 28
    "egg transfer" technique, 74
    "sample egg method," 47, 73–74
Eggshell thinning, 32–33, 38
    birds in the Great Lakes ecosystem, 41–42, 81–82, 120–122
    grey partridge, 35–36, 49, 117–119
    lessons learned from, 117
    mechanism of action to produce, 72
    peregrine falcon, 23, 33, 62, 114–117
    pre-DDT usage thickness, 116
    as predictor of reproductive failure, 211
    Thickness Index, 72
    usefulness as an endpoint, 71–72
"Egg transfer" technique, 74
Elk. *See Cervus elaphus*
Ellman colorimetric assay, 68
Elm. *See Ulmus* spp.
Emberiza citrinella, 35
Embryotoxicity, of DDT/DDE on birds, 32, 42–43
Emissions. *See* Air pollutants
Emmigration. *See* Dispersal; Migration
Endangered species
    black-footed ferret, 15
    role in ecosystem management, 9
    Yuma clapper rail, 6
Endangered Species Act, 79
Endocrine disruptors
    endpoints, 67, 74
    oil, 40
    in population modeling, 28, 47
Endpoints, 61–85, 218–219
    behavioral or sublethal, 241
    biodiversity, 48
    biomarkers, 186, 219
    biotic, 62–66
    community- and ecosystem-level, 190, 221–222
    evaluative criteria, 63
    habitat, 219
    molecular, 66
    most commonly used, literature review, 126
    population-level, 111–114, 190–191
    relationship to level of biological organization, 64–65, 193
    selection, 95, 189–191
    sustainability, 10, 48
    of toxicity tests, limitations for population modeling, 241

Endrin, 76
Energy allocation to reproduction, 24
Enhydra lutris
    effects of petroleum products, 43, 132–133
    genotoxic effects, 43
    literature review, 124, 126, 128, 132–133
Environmental fate and transport, 22, 138, 194, 205
Environmental Impact Statement, wolf reintroduction, 2
Epidemiological studies, 209
Equilibrium
    of an ecosystem, 13–14, 153–154
    density-dependent mechanisms, 21
    quantification of shift from, 157
Equilibrium strategist, 228
Error, in study design, 96, 99–100, 106
Estimated environmental concentration (EEC), 80
European Community Directive on the Conservation of Wild Birds, 48
European starling. *See Sturnus vulgaris*
Everglades, 10
Everglades snail kite. *See Rostrhamus sociabilis plumbeus*
Exotic species, 4, 12
Experimental blocking, 99–100
Experimental design. *See* Study design
Exposure
    chronic, lack of data for, 178
    inhalation, 83
    location of, in spatial structure, 266
    measures of. *See* Endpoints
    "significant" or "meaningful," 83, 142
    timing, and teratogenicity, 70
Exposure assessment, 47
    extrapolation in, 110
    predictive, 48, 109, 291
    ranking of contaminants, 22–23
    retrospective, 48, 109
    role in ecological risk assessment, 62
Exposure-response model, 178. *See also* Dose-response relationship

Extinction
    local, 247, 255
    of metapopulations, 255–258, 259
    population dynamics of, 27
    regional, 260
    variability in vulnerability to. *See* Population, source and sink
Extrapolation, 194–199, 210, 221–223
    of AChE inhibition to survival, 69, 76
    among life stages, 179
    among major taxa, 196
    categories and credibility, literature review, 123–141
    in exposure assessment, 110
    from individuals to communities/ecosystems, 120–122, 125, 140, 212–213
    from individuals to populations, 125, 130, 196–197, 211–212
        birds, 114, 119–120, 123
        in ecological risk assessment, 62
        effects of selenium, 119–120
        influence of life history strategies, 225–241
        regulatory aspects, 85
        reptiles, 140
    intercommunity, 213–214
    interecosystem, 213–214
    interpopulation, 125, 212
    interspecies, 82, 125, 139, 179, 195–196, 211
    intra-individual, 194–195, 211
    intraspecies, 195
    from laboratory to field, 130, 140, 196, 197–199
    from laboratory to population models, 182, 184–185
    landscape-level effects, 214
    multiple, 179
    scaling factors, 66
    from single-species to community, 149–174
    of sublethal effects, 134
    suborganism to organism, 210–211
    of unreported responses from available data, 179

Exxon Valdez oil spill, 40–41, 43, 78, 79, 132–133

F

Falco peregrinus, 23, 33, 62, 114–117
Falco sparverius, 62, 116
Falco tinnunculus, 31
Fathead minnow, 179
Fecundity, 152, 159
    combined effect with decreased survival, 240–241
    estimates in study design, 101
    impact of changes in, 162–168, 169, 171
    population-level effects, 226, 228–241
    sensitivity, and usefulness in modeling, 182, 184
Federal Insecticide, Fungicide, and Rodenticide Act, 79, 80, 83
Fenitrothion, 29, 30
Ferruginous hawk. *See Buteo regalis*
Ficedula hypoleuca, 37
Field experiments, 48–49
    integration with models, 266–267
    need for spatial structure analysis, 267–268
    study design, 191–192, 203
Fire, 11, 15
Fish
    effects of acid deposition, 10
    effects of halogenated hydrocarbons in the Great Lakes, 122
    exotic species, 12
    extrapolation among life stages, 179, 182
    modeling population-level effects, 182–183
    taxonomic regression for, 180
Fish and Wildlife Coordination Act, 79
Fishing
    cutthroat trout, 4
    gill net, 8
    harvest aspects, 185–186, 190, 206
Fitness, 113
Fluorosis, 297

Food and Drug Act, 83
Food Quality Protection Act, 83
Food sources
    changes during life history, 25
    effects of contaminants *vs.* stochastic events, 23–24
    effects of heavy metals, 12–13
    effects of pesticides on birds, 34, 35, 118
    loss from water pollution, 7, 8
    winter-feeding by humans, 8–9
Food-web. *See also* Bioconcentration
    California salt marsh, 155, 159, 160, 166, 173
    French Frigate Shoals, 156, 157, 173–174
    Mono Lake, Calif., 155, 159, 160, 165, 172
    reef, 167
    sensitivity to perturbations, 162–168
    six-species model, 172
    stability, 150–151, 153–157
    theory and experiments, 149–152
    three-species model, 171
    two archetypes, 154
Forest Inventory Assessment, 192
Forestry
    boreal forest, 10
    old growth forest, 8
    use of pesticides, 29, 36
Forests
    boreal, 9–10, 29
    hardwood-hemlock, 45
    north-temperate, 29
    old growth, 8
Forster's tern. *See Sterna forsteri*
French Frigate Shoals, food-web, 156, 157, 173–174
Frogs, 135, 136, 265
Fulica americana, 39, 119
Furans, 199

G

Game Conservancy, 49

Gavia immer, 37, 39, 221
Geese, 178
Genetic variation, 76–77
Genotoxicity, 43–44
    as a biomarker, 67
    of oil, 43
    in risk assessment, limitations, 209
    of smelting emissions, 43
Geographic Information System (GIS), 255, 265
Glanville fritillary. *See Melitaea cinxia*
Global warming, 10
Glyphosate, 36
Gray wolf. *See Canis lupus*
Grazing, 14, 15
Great horned owl. *See Bubo virginianus*
Great Lakes Embryo Mortality, Edema, Growth Retardation, and Deformities Syndrome (GLEMEDS), 70
Great Lakes water quality criteria, 81–82
Great Lakes Water Quality Protection Act, 79
Great tit. *See Parus major*
Green frog. *See Rana clamitans*
Grey geese. *See Anser* spp.
Grey partridge. *See Perdix perdix*
Grizzly bear. *See Ursus arctos*
Growth rate, 25, 75
Gulf menhaden. *See Brevoortia patronus*
Gull, 165
Gulo gulo, 10
Gymnogyps californianus, 38
Gypsy moth. *See Lymantria dispar*

H

Habitat
    dispersal corridors, 24
    effects of grazing, 14, 15
    endpoints, 219
    life cycle phases and changes in, 6–8, 14
    migratory/seasonal changes in, 7, 14
    patchiness. *See* Spatial structure
    population-effects of, 6–9
    specialists *vs.* generalists, 259, 261–262

Habitat degradation and loss
    detection, 260–261
    effects of petroleum wastes, 298–299
    effects on population dynamics, 21, 49
    prairie, 15
    through agricultural development. *See* Agricultural land use
    through pesticide use, 34–35
Habitat Evaluation Procedures, 222
Habitat fragmentation, 261, 265
Habitat Suitability Index, 222
Haematopus bachmani, 41
Haliaeetus leucocephalis
    effects of DDT/DDE, 32, 33, 41, 62, 121–122
    effects of oil, 41
    population modeling, 27
Harrier. *See Circus cyanus*
Hatching success, 73, 119–120. *See also* Reproductive success
Helianthus annuus, 280
Hemolytic anemia, 40
Hepatotoxicity of petroleum products, 40, 297
Heptachlor, 31, 32, 115
Herring gull. *See Larus argentatus*
Hexachlorobenzene (HCB), 42
High-production-volume chemicals (HPV), 83
Himantopus mexicanus, 39, 119
Hirundo rustica, 43–44
Home range, 4, 222
Hormone levels, as endpoints, 67
House mouse. *See Mus musculus*
House sparrow. *See Passer domesticus*
Human impact. *See* Anthropogenic factors
Human sperm count, decline, 159
Hunting
    additive *vs.* compensatory, 5
    elk, 9
    grey partridge, 117–118
    level of harvest, 185–186, 190
    population-level effects, 5–6
    prairie, 15
    shot used for. *See* Lead
Hydrolysis, 194

I

Iberian lynx. *See Lynx pardinus*
Immigration. *See* Dispersal; Migration
Immunosuppression, 159, 162
    by organochlorines, 42, 43
    by petroleum products, 40, 297
    study design and, 220
Immunotoxicity, as an endpoint, 67, 280
Incidence-function metapopulation models, 251, 265
Index of Biotic Integrity, 222
Indicator microcosms, 170
Indicator species, 65
    amphibians/reptiles as, 137, 138, 139
    mortality rate, correlation with community impact, 152–159, 160–170
    rodents as, 299
    selection, 218
Individual-level effects, 28
    biotic endpoints, 62–66, 67, 191
    bridge models, 239
    future research, 49
    intra-individual extrapolation, 194–195
    limitations in research, 46–47
Infections, 13
Interspecies extrapolation. *See* Extrapolation, interspecies
Intraspecies extrapolation, 195
Intraspecific competition, 153, 169
Introduced species, 4, 12
Invertebrates, 10, 166, 167, 199, 213

J

Johnsongrass. *See Sorghum halapense*
Joint Nature Conservation Committee, 34

K

Kesterson National Wildlife Refuge. *See* Selenium
Kestrel. *See Falco tinnunculus*
Keystone species, 15, 65, 170, 213, 221
K-strategists
    attributes, 25, 26, 226
    California condor, 38
    effects of oil spill, 40
    in modeling, 206, 227–241

L

Lake trout. *See Salvelinus namaycush*
Landscape ecotoxicology
    definition, 19
    efforts in development, 109
Landscape-level effects, 15
    effects of contaminants on, 24–25
    extrapolation and, 214
    Geographic Information System for modeling, 255, 265
    in population modeling, 27
    spatial structure, 264–265, 269
Lapland longspur. *See Calcarius lapponicus*
Larus argentatus, 42, 43, 122
Laterallus jamaicensis, 7
Lead, 37–38
    biochemical marker for. *See* Aminolevulinic acid dehydratase
    concentration in U.S. reference soils, 280
    concentrations in Oklahoma prairie sites, 283
    diagnostic signs, 71
    effects on rodent assemblages, 280, 292–293
    mechanism of action, 37
    in toxicity index of Oklahoma prairie sites, 284
Lemmus lemmus, 296
Lepus americanus, 10
Level of biological organization, relationship to endpoints, 64–65
Lewis woodpecker. *See Melanerpes lewis*

Life history
    changes in habitat, 6–8, 14
    extrapolation among life stages, 179, 182, 195
    impact of contaminants and, 24–25, 26, 231–241
    influence on individual-population extrapolation, 225–241
    r-K continuum, 25, 26, 206, 226, 227–241
    role in toxicity tests, 184–185, 196
    triangular continuum, 227–228
Life span, changes in, 170, 171
Life-table model, 27, 196
Lindane, 31, 72
Little bluestem. *See Schizachyrium scoparium*
Livestock, effects of grazing, 14, 15
Lizards, 135, 136, 139
Logging. *See* Forestry
Loon. *See Gavia immer*
Love Canal, 76, 295
Lutra canadensis, 4, 12, 13, 81, 128, 212, 213–214
Lymantria dispar, 29
Lynx pardinus, 260

M

Magnesium, bioavailability and acid rain, 44
Malformations. *See* Teratogenesis
Mallard. *See Anas platyrhynchos*
Mammals, literature review of population-level effects, 123–134
Management
    adaptive, 15
    ecosystem, 9–16
    extinction thresholds and habitat patchiness, 255–258, 265
    landscape level. *See* Landscape-level effects
    recommendations for population-based, 186–187
    spatial structure issues, 268
Marbled murrelet. *See Brachyramphus marmoratus*

Marestail. *See Conyza canadensis*
Mark-recapture techniques, 76, 278
Marsh, 7
"Matrix effect," 265, 267
Matrix-projection models, 196–197, 226, 228–241
Mechanistic toxicology, 180–181, 210
Melanerpes lewis, 11
Melitaea cinxia, 251, 256, 265
Mercury
    effects on amphibians/reptiles, 136
    effects on birds, 31, 33, 38–39, 72
    effects on mammals, 124, 128
    water quality criteria, 82
Merlin, 33
Metabolism of contaminants, 194
Metacommunity, 248, 258–259
Metals, 12–13, 36–39. *See also specific metals*
    bioavailability and acid rain, 44
    effects on amphibians/reptiles, 136
    effects on mammals, 124, 126
Metapopulations
    definition, 111, 192, 247–248
    modeling, 27–28, 197, 249–252, 262–264
Methyl parathion, 76
Microtus spp.
    effects of pesticides, 134
    effects of petroleum wastes, 286, 288, 290
    reproductive endpoints, 74
    in salt marsh food-web, 155
    seasonal abundance in Oklahoma prairie, 283
Microtus ochrogaster
    effects of petroleum wastes, 291, 292, 295
    seasonal abundance in Oklahoma prairie, 285
Microtus pennsylvanicus, 76, 295

Microtus pinetorum, 76–77, 285, 295
Migration
    changes in habitat, 7
    from contaminated to uncontaminated site, 27
    dispersal corridors, 24
    loss of fat reserves through, 207
    rates and patterns, 264, 265, 266–267, 268, 270
Migratory Birds Convention Act, 48
Migratory Bird Treaty Act, 79
Miliaria calandra, 35
Mimus polyglottos, 34
Minimum number known alive (MNKA), 281, 287
Mining, heavy metals from, 12–13, 33, 39
Mink. *See Mustela vison*
Models
    cellular automata, 254–255
    "demographic," 182, 282
    environmental fate and transport, 194
    exposure-response, 178
    to extrapolate from individual to population, 119
    field validation for, 193
    food-web, 150–151
    future needs, 48
    incidence-function, 251, 265
    individual-based, 239
    integration with empirical study, 266–267
    Leslie matrix, 101
    life-table, 27, 196
    Lotka-Volterra, 153
    metapopulation, 27–28, 197, 249–252, 262–264
    Monte Carlo, 27, 101, 179
    partial differential equation, 253–254
    patch-occupancy, 250–251
    physiologically based pharmacokinetic, 195, 196, 199, 210
    population, 25–28, 185–187, 204–205, 206–207, 228–241
    problems and solutions, 182–183
    projection matrices, 196–197, 226, 228–241
    six-species food web, 172
    spatial structure of populations, 249–255
    stochasticity, 230, 233, 238, 240
    study design, 193
    sublethal effects, 162–168
    survival, 75
    Sussex, 118
    three-species food web, 171
Monachus spp., 167
Mongoose, 12
Monitoring
    study design, 191–192
    Utility Index *vs.* Vulnerability Index, 65
    vs. conservation, 65
Monk seal. *See Monachus* spp.
Mono Lake food-web, 155, 159, 160, 165, 172
Monte Carlo model, 27, 101, 179
Moose. *See Alces alces*
Morone saxatilis, 183
Morphological aberrations, 69–71
Mortality
    additive *vs.* compensatory, 21
    sublethal effects *vs.* lethal effects, 159, 162, 170–171
Mortality-incident monitoring, 77
Mortality rate, 21, 75
    acceptable level, managed harvest *vs.* contaminant effects, 185–186
    as indicator of community-level effects, 152–159, 160–170
    life history strategy and, 226
    modeling impact of changes in, 162–168
Mountain goat. *See Oreamnos americanus*
Mourning dove. *See Zenaida macroura*
Mouse. *See Mus* spp.
Mule deer. *See Odocoileus hemionus*
Multiple stressors, 198–199, 269
Murre. *See Uria* spp.

Mus spp., 134, 155
Muskrat. *See Ondatra zibethica*
Mus musculus
    effects of pesticides, 77
    effects of petroleum wastes, 78, 286, 289–293, 294–296, 298
    seasonal abundance in Oklahoma prairie, 283, 285
Mustela nigripes, 15
Mustela vison
    effects of heavy metals, 12, 13
    as indicator species, 81, 218
    in interpopulation extrapolation, 212
    literature review, 124, 126, 128, 129
    reproductive endpoints, 74
Mute swans. *See Cygnus olor*
Mycocaster coypus, 12
M=ycteria americana, 10

N

National Academy of Science, 83
National Agricultural Statistical Survey, 192
National Wetlands Inventory, 192
Natural population-level stressors
    in study design, 126
    vs. contaminants, 23–24, 123, 141
Natural Resource Damage Assessment, 67, 79, 80–81, 132
Natural variation, 97–98
Nature Conservancy, 115
Neotoma floridana, 283
Nest-box technique, 73
No-adverse-effect level, 203
No-observed-adverse-effect level (NOAEL), 241
No-observed-effect concentration (NOEC), 183
No-observed-effect level (NOEL), 82, 183
Northern bobwhite. *See Colinus virginianus*
Northern cardinal. *See Cardinalis cardinalis*
Northern leopard frog. *See Rana pipiens*
Northern mockingbird. *See Mimus polyglottos*
Norway rat. *See Rattus norvegicus*
Null hypothesis, 98, 103
Numerical metapopulation models, 252
Nutria. *See Mycocaster coypus*
Nycticorax nycticorax, 218

O

Oak Ridge National Laboratory, 178, 209, 227
Odocoileus hemionus, 3
Odocoileus virginianus, 3
Oil. *See* Petroleum products
Oklahoma, effects of petroleum wastes on rodents, 277–299
Oklahoma Department of Environmental Quality, 279
Old growth forests, 8
Omnivory, and stability of the food-web, 150–151
Oncorhynchus clarki bouvieri, 4
Ondatra zibethica, 128, 131
Opportunistic strategist, 228
Oreamnos americanus, 3
Organically-managed fields, 34–35
Organism-level effects. *See* Individual-level effects
Organochlorines, 41–43. *See also* Eggshell thinning
    effects on amphibians/reptiles, literature review, 136
    effects on mammals, literature review, 124, 128
    Great Lakes water quality, 81–82
Organophosphates. *See also* Cholinesterase inhibition
    effects on amphibians/reptiles, literature review, 136
    effects on birds, 31, 32
    effects on mammals, literature review, 124, 128
    effects on rodent assemblages, 296
Oryzomys palustris, 283
Osprey. *See Pandion haliaetus*
Ovis canadensis, 3
Oxidative stress, measures of, 67
Ozone depletion, 10

P

Pandion haliaetus, 4, 33, 62, 122
Parasites, 13, 24
Partial differential equation models (PDE), 253–254

Partridge pea. *See Cassia fasciculata*
Partridge Survival Project, 118
Parus major, 45
Passer domesticus, 6
Patchiness. *See* Spatial structure
Patch-occupancy metapopulation models, 250–251
Pelecanus erythrorhynchos, 4
Pelecanus occidentalis, 33, 40, 62, 65, 122
Perdix perdix, 35–36, 49, 117–119, 213
Peregrine falcon. *See Falco peregrinus*
Periodic strategist, 228
Peromyscus spp.
    effects of pesticides, 74
    effects of petroleum wastes, 290
    in Oklahoma prairie, 281, 283, 285
Peromyscus leucopus, 281
Peromyscus maniculatus, 281
Persistence of metapopulations, 257–258, 260
Persistent bioaccumulative contaminants, 22, 23. *See also* Dioxins; Organochlorines; Polychlorinated biphenyls; *specific pesticides*
Perturbations. *See* Food-web, stability; Stochastic events
Pesticides
    direct effects on birds, 28–34, 41–43. *See also* Eggshell thinning
    effects on rodent assemblages, 296
    USEPA registration protocols, 222
    field observations of population effects, 29–38
    Great Lakes water quality, 81–82
    indirect effects, 34–36, 118
    sublethal effects, 32, 33, 76
    three main ecological effects, 34
Petroleum products
    accident assessment of spills, 102–104
    community-level effects, 78
    direct effects, 43, 297
    effects on amphibians/reptiles, literature review, 136
    effects on birds, 39–41, 71
    effects on mammals
        literature review, 124, 128, 132–133
        rodent assemblages, 277–299
        sea otters, 43, 132–133
    genotoxicity, 43
    indirect effects, 297–298
    mammal rehabilitation from exposure, 133
    regulatory criteria for cleanup, 80–81
Phalacrocorax auritus, 33, 42–43, 62, 122
Pharmacokinetics, 195, 196, 199, 210
Phenotype, and reproductive success, 25
Philohela minor, 29–30
Phosphamidon, 29
Photolysis, 194
Physiologically based pharmacokinetic models, 195, 196, 199, 210
Pied flycatchers, 37
Pigeon guillemot. *See Cepphus columba*
Pine vole. *See Microtus pinetorum*
Plants. *See* Primary producers
Plegadis chihi, 33
Plethodon cinereous, 45
Plover, 165
Podiceps nigricollis, 39, 119
Political aspects, 220. *See also* Regulatory aspects
Polychlorinated biphenyls (PCB)
    effects on amphibians/reptiles, literature review, 136
    effects on birds, 42–43, 72, 120–121
    effects on mammals, literature review, 124, 128
    mechanistic similarity to TCDD, 199
    water quality criteria, 82
Polycyclic aromatic hydrocarbons (PAH). *See also* Petroleum products
    cleanup levels in soil, 280
    concentrations in Oklahoma prairie sites, 283
    effects on rodent assemblages, 292–293, 294, 295
    in toxicity index of Oklahoma prairie sites, 284

Population(s)
    definitions and characteristics, 20, 110–112, 192, 207–208
    as "evolutionary significant unit," 46
    island, 247, 248, 252–253
    isolated, 246–247, 249, 268
    local, definition, 111
    minimum viable, 256
    modeling. See Models, population
    patchy, 261
    source and sink, 21–22, 41, 98, 212, 248, 252–253, 260–261, 268
    status, determination of, 191
    study, 218
    types of spatial structure, 246–249
Population estimation
    bird surveys, 212
    methodologies, 75
Population growth, to interpret toxicity, 196, 227
Population-level effects
    advantages of analysis, 177
    amphibians, overview, 123–134
    aquatic studies, 177–187, 208, 222
    assessment by individual-level endpoints. See under Extrapolation
    assessment on spatially-relevant scale, 46
    below no-observed-effect concentration, 183
    birds, overview, 114–123
    definitions and criteria, 48
    endpoints, 67–77, 190–191
    USEPA regulatory authority, 113
    factors which induce, 20–22
    integration with toxicology data, 182–183. See also Models, population
    mammals, literature review, 123–134
    management issues, 4–9
    of pesticides, field observations, 29–38
    reptiles, overview, 123–134
    response variables, 100–101
    stability of natural systems, 4

Population Limitation in Birds, 20
Population viability analysis (PVA), 101
Porzana carolina, 7
Power. *See* Statistical power
Power analysis, 48, 105, 183, 209
Power plants
    selenium in fly ash metals, 39
    thermal pollution by, 177
Prairie dog. *See Cynomys* spp.
Prairie ecosystem
    effects of pesticides on birds, 36
    effects of petroleum wastes on rodents, 277–299
    keystone species, 15
    Oklahoma vegetation, 280
Prairie vole. *See Microtus ochrogaster*
Precautionary principle, 186, 206
Predator/prey relationships
    impact of changes in predation rates, 162–168
    impact on community abundances, 169
    metapopulation persistence and, 259, 261–262
    prey resistance, 169
    wolf reintroduction program, 2–4
Prey-capture rate, 152
Primary producers, 155, 156, 165, 166, 167, 207
Principle components analysis (PCA), 179
Procyon lotor, 12, 13
Projection matrices, 196–197, 226, 228–241
Pronghorn. *See Antilocapra americana*
Prospective power analysis, 48
Prostaglandin, 72
Public perception, 208
Puffinus tenuirostris, 230
Pyridine-2-aldoxime methochloride, 69

R

Rabbit, 128, 131
Raccoon. *See Procyon lotor*

Radionuclides
    effects on amphibians/reptiles, literature review, 136
    effects on birds, 43–44
    effects on mammals, literature review, 128
    effects on turtles, 43
Radiotelemetry, 49
    for bird monitoring, 32, 75–76
    effects on behavior, 99
    study design and, 99, 100
Rails, 166. *See also* Laterallus; Rallus spp.
Rainbow trout, 179
Rallus limicola, 7
Rallus longirostris yumanensis, 6–7
Rana catesbeiana, 71
Rana clamitans, 71
Rana pipiens, 71
Randomization, in study design, 96, 98–99, 102
Range, 22
Rangifer tarandus, 5, 10
Rat, 166. *See also Rattus norvegicus*
Rattus norvegicus, 12
Recreational aspects. *See also* Fishing; Hunting
    prairie, 15
    water skiing on rivers, 7
Recurvirostra americana, 39, 119
Red-backed salamander. *See Plethodon cinerous*
Reef food-web, 167
Reef shark, 167
Reference norm, 191
    acetylcholinesterase activity, 69
    pre-exposure population dynamics, 267
    teratogenesis, 69
Reference site, 27, 205, 220, 264, 278–279
Refugia, 257–258
Regional aspects, 10, 130, 260
Regression analysis, 179–181, 182, 282, 291–292, 294
Regulatory aspects, 79–83, 186–187, 204, 208, 222
Regulus calendula, 30
Reithrodontomys spp., 155
Reithrodontomys fulvescens, 283, 285, 298

Remediation, 226, 268
Replication, in study design, 96, 97–98, 100, 102, 220
Reproductive behavior, 7, 33
Reproductive-potential index, 183
Reproductive success
    as an endpoint, 73–75
    eggshell thinning as predictor, 211
    phenotype *vs.* genotype and, 25
Reptiles
    deformities, 71
    as indicator species, 137, 138, 139
    population-level effects, literature review, 134–141
    reproductive success, 74
"Rescue effect," 247, 258
Research
    future needs, 48–49, 84, 142, 199–200, 267–269
    study design recommendations, 130–131, 134, 141
Resistance to contaminants, 76–77
Resistance to predators, 169
Retrospective power analysis, 48
Rhus spp., 280
Ring-necked ducks. *See Aythya collaris*
Risk assessment. *See* Ecological risk assessment
Risk characterization, 62
Risk managers, training, 185, 186
Rissa tridactyla, 41
River contamination, 12–13
River otter. *See Lutra canadensis*
Rivers, 7
Rodents
    effects of petroleum products, 277–299
    trapping methodology, 280–281
Rook. *See Corvus frugilegus*
Rostrhamus sociabilis plumbeus, 10
R-strategists
    attributes, 25, 26, 226
    in modeling, 206, 227–241
Ruby-crowned kinglet. *See Regulus calendula*

S

Safety factors. *See* Uncertainty factors
Sage grouse. *See Centrocercus urophasianus*
Salamanders, 135, 136
Salmo trutta, 12
Salt marsh food-web, 155, 159, 160, 166, 173
Salvelinus namaycush, 4
"Sample egg method," 47, 73–74
Sample size calculations, 104–105, 106
Scale, in population dynamics, 112, 113
Scaling factors, 66, 185, 195
Scatter plot, 293
Schizachyrium scoparium, 280
Seals, 151, 167
Sea otter. *See Enhydra lutris*
Sea turtle, 139
Selectivity, of contaminants, 23
Selenium, 39
    bioaccumulation in eggs, 33
    ecosystem-level effects, 79
    population-level effects, 119–120
    teratogenicity, 70–71
Sentinel species. *See* Indicator species
Sharp-shinned hawk. *See Accipiter striatus*
Sherman live-catch trap, 281
Short-tailed shearwater. *See Puffinus tenuirostris*
Shrew, 134, 166
Sigmodon hispidus
    abundance in Oklahoma prairie, 78, 283, 285
    effects of metals, 280
    effects of petroleum wastes, 286, 288, 290, 291, 292, 294–298
s-index, 282, 289–290, 291, 294, 295, 296
Sink populations. *See* Population, source and sink
Skylark. *See Alauda arvensis*
Slider turtle. *See Trachemys scripta*
Sludge-pits, 279
Smelting operations
    effects on rodent assemblages, 296, 297
    genotoxicity of emissions, 43
    heavy metals from, 12–13, 36–37, 39, 79

Snakes, 135, 136, 138
Snapping turtle. *See Chelydra serpentina*
Snowshoe hare. *See Lepus americanus*
Social value of wildlife, 3
Somateria fischeri, 38
Sora. *See Porzana carolina*
Sorghum halapense, 280
Source populations. *See* Population, source and sink
Space limitation, 21
Sparrowhawk. *See Accipiter nisus*
Spatial aspects
    of population decline, 112–113, 130
    of study design, 126, 127
Spatial structure
    effects on population dynamics, 261–262
    identification of type, 266
    incidence of occupancy, 250–251, 265
    landscape-level effects, 24, 27
    for minimum viable population, 256
    modeling, 249–265
    need for field tests, 267–268
    of populations, types, 246–249
    similarity to dose-response relationships, 245–246
    source and sink habitats, 21–22
    threshold of contaminants for extinction, 255–258, 265
    when to ignore, 269–270
Species-level effects. *See also* Indicator species; Keystone species
    extrapolation to community, 149–174
    intraspecies extrapolation, 195
    management issues, 1
    "most-sensitive species," 65
    selectivity of contaminants, 23
Species richness
    effects of acid rain, 44, 78
    effects of petroleum, 41, 288, 294
    and food-web stability, 150–151
Spectacled eider, 38
Spheniscus demersus, 40
Spills. *See* Accident assessment; Petroleum products

Stakeholders
    decision-making by, 177
    determination of study population, 217, 218
    short-term perspective of, 11
Starvation, 5
Statistical power, 48, 105, 183, 209, 282
Statistical significance, *vs.* biological relevance, 209
Statistics. *See* Cluster analysis; Community structure analysis; Regression analysis; Study design
Sterna caspia, 42, 43
Sterna forsteri, 42
Stochastic events
    as density-independent factor, 21
    emergency response, 47
    in the literature, 126
    in population modeling, 230, 233, 238, 240
        safety margin in management for, 260
        synchrony among populations and, 257–258
        vs. contaminants, 23–24
Stock dove. *See Columba oenas*
Striped bass, 184
Structure-activity relationships (SAR), 198, 200
Structured metapopulation models, 252
Study design, 95–106, 219–220
    duration, 220
    experimental studies, 193, 204
    field experiments, 101–102, 191–192
    modeling studies, 193
    principles of, 96–101, 203–204
    projection matrix simulation, 232–235
    recommendations for improved, 130–131, 134, 141
    sample selection, 192, 217
    study performance, 104–105
    survey and monitoring studies, 191–192
    testable hypotheses *vs.* speculation, 140, 141
    three types of studies, 191

Study performance, 104–105
Study sites, selection criteria, 98
Sturnus vulgaris, 12, 76
Sublethal effects, 24, 28
    extrapolation to survival, 134
    modeling, 162–168
    of pesticides on birds, 32, 33, 76, 123
    shortage of toxicological data, 178
    vs. lethal effects, 159, 162, 170–171
Subpopulation, definition, 111
Succession, 11
Sulfur dioxide, 36
Sulfuric acid, 74
Sumac. *See Rhus* spp.
Superfund sites, 225, 279, 295
Surveys
    bird, 212
    national-scale, 192
    study design, 191–192
Survival
    combined effect with decreased fecundity, 240–241
    estimates in study design, 100–101
    extrapolation of AChE inhibition to, 69
    population-level effects and, 75–77, 226, 229–241
Sussex Model, 118
Sustainability
    definition of endpoints and goals, 10, 48
    equilibrium state of ecosystem, 13–14
Sustainable yield, prairie, 15
Swainson's hawk. *See Buteo swainsoni*
Swift fox. *See Vulpes velox*
Sylvilagus floridanus, 76

T

Tachycineta bicolor, 44
Temporal aspects
    of contaminant degradation, 268–269
    of population decline, 112–113
    of study design, 126, 127, 130–131

Teratogenesis, 69–71
    measurement of frequency, 71
    mechanisms of action, 69–70
    organochlorines, 42
    selenium, 39
Territory, 30
2,3,7,8-Tetrachlorodibenzo-p-dioxin (TCDD)
    effects on birds, 42
    "mechanistic similarity" to, 199
    toxic equivalency factors, 42, 43, 199
    water quality criteria, 82
Threshold concentrations, 83, 84, 265
Tiger shark, 167
Timber harvest. *See* Forestry
Total polycyclic aromatic hydrocarbons (TPH)
    concentrations in Oklahoma prairie sites, 283
    effects on rodent assemblages, 294, 295
    in toxicity index of Oklahoma prairie sites, 284, 292
Tourism. *See* Fishing; Hunting; Recreational aspects
Toxic equivalency factors (TEF), 42, 43, 199
Toxicity tests, 62–63, 66
    of chronic exposure, 178, 182, 185
    design recommendations, 184–185
    effects of design, 181–182
    endpoints, limitations for population modeling, 241
    population growth to interpret, 196
    population-level, lack of, 225
    standardized protocols, 185
Toxicological data, problems and solutions, 178–182
Toxic Substances Control Act, 79, 83, 130
Toxostoma rufum, 34
Trachemys scripta, 43
Transgenerational effects, 220
Tree swallow. *See Tachycineta bicolor*
Tribolium, 264
Trichlorphon, 29
Trophic casades, 169
"Trophic groups," 131

Trophic level, as predictor of community impact, 159, 161, 207, 213–214
Tumors, 210–211
Tuna, 167
Turdus migratorius, 112, 207
Turkey, 9
Turkey vulture. *See Cathartes aura*
Turtles
    as indicator species, 138, 139
    literature survey, 135, 136
    in reef food-web, 167
Type II error, 46, 48

U

Ulmus spp., 280
Ultraviolet radiation, 10, 207
Uncertainty factors
    to extrapolate Criteria Maximum Concentrations, 82
    extrapolation to functionally-similar contaminants, 142
    in fish models, 184
    with interspecies extrapolation, 195–196
    mechanisms of toxicity, 198
    when using multiple extrapolations, 110, 179, 182–183
Urban/suburban environment, 9
Uria spp., 40
Ursus arctos, 3, 261
U.S. Department of Commerce, 80
U.S. Department of Energy, 227
U.S. Department of the Interior, 79, 80, 83
U.S. Environmental Protection Agency (USEPA)
    diazinon restriction, 178
    Ecological Incident Information System, 77
    Environmental Monitoring and Assessment Program, 192
    Great Lakes water quality criteria, 81–82
    Index of Biotic Integrity, 222

National Center for Environmental
    Research and Quality Assurance,
    211
  regulatory authority, 80, 113
  Superfund site policy, 225
U.S. Fish and Wildlife Service
  endangered species listing, 15
  Great Lakes water quality criteria, 81
  Habitat Evaluation Procedures, 222
  Habitat Suitability Index, 222
  "migratory nongame bird of
    management concern" listing, 33
  "threatened species" listing, 38
  wolf reintroduction program, 2–4

V

Validation
  field, for models, 193, 205
  for interspecies extrapolation, 196
Variance, 97–98, 99, 104–105
Viral infections, 13
Virginia rail. See *Rallus limicola*
Vole. See *Microtus* spp.
Vulpes velox, 15

W

Water quality criteria, Great Lakes, 81–82
Weasel, 128, 131
Weight-of-evidence approach, 46, 205, 219
Western ragweed. See *Ambrosia psilostachya*
White-crowned sparrows. See *Zonotrichia albicollis*
White-faced ibis. See *Plegadis chihi*
White pelican. See *Pelecanus erythrorhynchos*
White sage. See *Artemisia ludoviciana*
White-tailed deer. See *Odocoileus virginianus*
Wolf
  gray. See *Canis lupus*
  as keystone species, 221
  predator/prey relationships, Alaska,
    5–6
Wolverine. See *Gulo gulo*
Woodcock. See *Philohela minor*

Wood stork. See *Mycteria americana*

Y

Yellowhammer. See *Emberiza citrinella*
Yellowstone National Park, 2–4
Yuma clapper rail. See *Rallus longirostris yumanensis*

Z

Zebra mussel. See *Dreissena polymorpha*
Zenaida macroura, 76
Zinc, 36, 79
Zonotrichia albicollis, 30
Zooplankton, 167

SETAC

A Professional Society for Environmental Scientists and Engineers and Related Disciplines Concerned with Environmental Quality

The Society of Environmental Toxicology and Chemistry (SETAC), with offices in North America and Europe, is a nonprofit, professional society that provides a forum for individuals and institutions engaged in the study of environmental problems, management and regulation of natural resources, education, research and development, and manufacturing and distribution.

Specific goals of the society are:

- Promote research, education, and training in the environmental sciences.
- Promote systematic application of all relevant scientific disciplines to the evaluation of chemical hazards.
- Participate in scientific interpretation of issues concerned with hazard assessment and risk analysis.
- Support development of ecologically acceptable practices and principles.
- Provide a forum for communication among professionals in government, business, academia, and other segments of society involved in the use, protection, and management of our environment.

These goals are pursued through the conduct of numerous activities which include:

- Hold annual meetings with study and workshop sessions, platform and poster papers, and achievement and merit awards.
- Sponsor a monthly scientific journal, Environmental Toxicology and Chemistry, SETAC newsletter, and special technical publications.
- Provide funds for education and training through the SETAC Scholarship/Fellowship Program.
- Organize and sponsor chapter forums for the presentation of scientific data and for the interchange and study of information about local concerns.
- Provide advice and counsel to technical and nontechnical persons through a number of standing and ad hoc committees.

SETAC membership is currently composed of nearly 6,000 individuals from government, academia, business, and public-interest groups with technical backgrounds in chemistry, toxicology, biology, ecology, atmospheric sciences, health sciences, earth sciences, and engineering.

If you have training in these or related disciplines and are engaged in the study, use, or management of environmental resources, SETAC can fulfill your professional affiliation needs.

All member receive a newsletter highlighting environmental topics and SETAC activities, and reduced fees for the Annual Meeting and SETAC special publications.

All members except Students and Senior Active Members receive monthly issue of Environmental Toxicology and Chemistry (ET&C), a peer-reviewed journal of the Society. Student and Senior active Members may subscribe to the journal. Members may hold office and, with the Emeritus Members, constitute the voting membership.

If you desire more information, contact the appropriate SETAC Office.

1010 North 12th Avenue	Avenue de la Toison d'Or 67
Pensacola, Florida, USA 32501-3367	B-1060 Brussels Belgium
T 850 469 1500 F 850 469 9778,	T 32 2 772 72 81 F 32 2 770 53 83
E setac@setac.org	E setac@setaceu.org

http://www.setac.org

Environmental Quality Through Science®